C0-ALK-985

# Industrial Environmental Chemistry

## Waste Minimization in Industrial Processes and Remediation of Hazardous Waste

# INDUSTRY-UNIVERSITY COOPERATIVE CHEMISTRY PROGRAM SYMPOSIA

---

*Published by Texas A&M University Press*

ORGANOMETALLIC COMPOUNDS
Edited by Bernard L. Shapiro

HETEROGENEOUS CATALYSIS
Edited by Bernard L. Shapiro

NEW DIRECTIONS IN CHEMICAL ANALYSIS
Edited by Bernard L. Shapiro

---

APPLICATIONS OF ENZYME BIOTECHNOLOGY
Edited by Jeffery W. Kelly and Thomas O. Baldwin

CHEMICAL ASPECTS OF ENZYME BIOTECHNOLOGY: Fundamentals
Edited by Thomas O. Baldwin, Frank M. Raushel, and A. Ian Scott

DESIGN OF NEW MATERIALS
Edited by D. L. Cocke and A. Clearfield

FUNCTIONAL POLYMERS
Edited by David E. Bergbreiter and Charles R. Martin

INDUSTRIAL ENVIRONMENTAL CHEMISTRY: Waste Minimization in Industrial Processes and Remediation of Hazardous Waste
Edited by Donald T. Sawyer and Arthur E. Martell

METAL-METAL BONDS AND CLUSTERS IN CHEMISTRY AND CATALYSIS
Edited by John P. Fackler, Jr.

OXYGEN COMPLEXES AND OXYGEN ACTIVATION BY TRANSITION METALS
Edited by Arthur E. Martell and Donald T. Sawyer

# Industrial Environmental Chemistry

## Waste Minimization in Industrial Processes and Remediation of Hazardous Waste

Edited by
**Donald T. Sawyer** and
**Arthur E. Martell**

*Texas A&M University*
*College Station, Texas*

**Plenum Press • New York and London**

Library of Congress Cataloging-in-Publication Data

Industrial environmental chemistry : waste minimization in industrial processes and remediation of hazardous waste / edited by Donald T. Sawyer and Arthur E. Martell.
p. cm. -- (Industry-university cooperative chemistry program symposia)
"Proceedings of the Texas A&M University, IUCCP tenth annual symposium on Industrial Environmental Chemistry: Waste Minimization in Industrial Processes and Remediation of Hazardous Waste, held March 24-26, 1992, at Texas A&M University, College Station, Texas"--T.p. verso.
Includes bibliographical references and index.
ISBN 0-306-44303-1
1. Factory and trade waste--Congresses. 2. Waste minimization--Congresses. 3. Hazardous wastes--Purification--Congresses. I. Sawyer, Donald T. II. Martell, Arthur Earl, 1916- III. Texas A & M University. Industry-University Cooperative Chemistry Program. IV. Symposium on Industrial Environmental Chemistry: Waste Minimization in Industrial Processes and Remediation of Hazardous Waste (1992 : Texas A & M University) V. Series.
TD896.I614
628.5--dc20 92-28759
CIP



Proceedings of the Texas A&M University, IUCCP Tenth Annual Symposium on Industrial Environmental Chemistry: Waste Minimization in Industrial Processes and Remediation of Hazardous Waste, held March 24–26, 1992, at Texas A&M University, College Station, Texas

ISBN 0-306-44303-1


A Division of Plenum Publishing Corporation
233 Spring Street, New York, N.Y. 10013



Printed in the United States of America

# SPONSORS

BASF Corporation

BF Goodrich

Dow Chemical Company

Hoechst-Celanese Corporation

Monsanto Chemical Company

Institute of Biosciences & Technology,
Texas A&M University System

Texas A&M University

## PREFACE

This monograph consists of manuscripts submitted by invited speakers who participated in the symposium "Industrial Environmental Chemistry: Waste Minimization in Industrial Processes and Remediation of Hazardous Waste," held March 24-26, 1992, at Texas A&M University. This meeting was the tenth annual international symposium sponsored by the Texas A&M Industry-University Cooperative Chemistry Program (IUCCP). The program was developed by an academic-industrial steering committee consisting of the co-chairmen, Professors Donald T. Sawyer and Arthur E. Martell of the Texas A&M University Chemistry Department, and members appointed by the sponsoring companies: Bernie A. Allen, Jr., Dow Chemical USA; Kirk W. Brown, Texas A&M University; Abraham Clearfield, Texas A&M University; Greg Leyes, Monsanto Company; Jay Warner, Hoechst-Celanese Corporation; Paul M. Zakriski, BF Goodrich Company; and Emile A. Schweikert, Texas A&M University (IUCCP Coordinator).

The subject of this conference reflects the interest that has developed in academic institutions and industry for technological solutions to environmental contamination by industrial wastes. Progress is most likely with strategies that minimize waste production from industrial processes. Clearly the key to the protection and preservation of the environment will be through R&D that optimizes chemical processes to minimize or eliminate waste streams. Eleven of the papers are directed to waste minimization. An additional ten papers discuss chemical and biological remediation strategies for hazardous wastes that contaminate soils, sludges, and water.

Acknowledgement is made to others whose participation in the Conference is not reflected in this book. Thanks and appreciation are due to Dr. Michael B. Hall, Head, Chemistry Department, and to Dr. Herbert H. Richardson, Chancellor, Texas A&M University System, who gave official welcoming remarks. There were three keynote lectures, that were very well received: Laurie King, Chief of the Oklahoma-Texas Permits Section, EPA, on "Upcoming Changes in Regulations that May Have an Impact on Remediation of Hazardous Waste", Jerry B. Martin, Director, Environmental Affairs, Dow Chemical, U.S.A. on "No More Room for Dinosaurs" and Jim Conner, Manager, Environmental Projects, Hoechst-Celanese Corporation on "WARR - Implementing a Waste Reduction Strategy". All participants at the Banquet enjoyed the theme lecture by James A. Fava, Roy F. Weston, Inc. on "Should Improvement to the Environmental Quality of Products Address Product Life-Cycle Concepts". Also very much appreciated are the services of Chairs of the Sessions: Sa V. Ho (Monsanto), Paul M. Zakriski (BF Goodrich), Jay Warner (Hoechst-Celanese), Jams R. Ryffel (BASF Corporation) and Aydin Akgerman (Texas A&M).

We thank Mary Martell for assistance with logistics of the symposium and for her help and that of Debbie Shepard with the organization and preparation of the final manuscript.

Donald T. Sawyer
Arthur E. Martell

College Station, Texas
May, 1992

## CONTENTS

# INTRODUCTION: STRATEGIES FOR THE PROTECTION OF THE ENVIRONMENT

Donald T. Sawyer and Arthur E. Martell

Department of Chemistry
Texas A&M University
College Station, Texas 77843

Although the contamination of our environment by industrial wastes has been recognized as a major problem since the 1960s, progress towards its resolution is very slow. Until recently hazardous wastes have been stored or dumped at land-fills and waste-ponds, and into rivers, lakes, and oceans. There is increasing evidence that storage and disposal in land-fills is not an effective strategy, and that incineration has the potential to disperse a liquid/solid waste into the atmosphere. The Environmental Protection Agency has catalogued thousands of sites in the United States that are severely contaminated with hazardous materials (PCBs, heavy metals, pesticides, chlorinated solvents, and wastes from the petroleum, chemical, pharmaceutical, electronics, and manufacturing industries), and has convinced Congress to appropriate in excess of $100 billion for the remediation of these Superfund sites. To date these funds have been used mainly to investigate, to evaluate, and to litigate Superfund sites, rather than to remediate.

This Symposium was organized to bring together representatives from Universities, Industry, and Government that are concerned with strategies (a) to minimize waste production and (b) to remediate existent hazardous wastes with effective and economic technologies. The "Theme" talk was given by Jerry B. Martin (Director, Environmental Affairs, Dow Chemical USA), *No More Room for Dinosaurs,* which provided compelling evidence that antiquated waste-producing processes are bad for business economics and will soon make an industry non-competitive. This, coupled with the presentations from major industries that have adopted a zero-effluent philosophy, represents a major advance and one that is driven by economics rather than regulations. We believe that this strategy for environmental protection is in the economic interest of all manufacturing and production industries. The leadership that is exemplified by the papers in this volume is impressive.

*Industrial Environmental Chemistry*, Edited by D.T. Sawyer and A.E. Martell, Plenum Press, New York, 1992

An example that illustrates the merit of waste minimization strategies is the replacement of PVC (polyvinyl chloride) with polybutylene for the manufacture of plastic pipe. The production of vinyl-chloride monomer has an inevitable waste stream that includes hexachorobenzene ($C_6Cl_6$, a human carcinogen), polychloro-dibenzofurans, and related polychloro-aromatics. Currently these wastes are processed by high-temperature, oxygen-rich chemical incinerators. Because $C_6Cl_6$ does not support combustion, one must have concern as to its fate from pyrolysis in an oxygen-rich atmosphere. Representatives of the EPA and operators of such an incinerator believe that $C_6Cl_6$ is transformed to $CO_2$ and HCl (Federal Courthouse, Little Rock, Arkansas, June 7, 1991); more chemically reasonable products include phosgene [$Cl_2C(O)$], chlorine oxide ($Cl_2O$), chlorine ($Cl_2$), and hypochlorous acid (HOCl). By development of a pure-hydrocarbon polymer (polybutylene from polymerization of butene-1) as a substitute for PVC, the polymer industry has eliminated a major hazardous-waste problem through a strategy of waste minimization. In so doing a superior product is on the market, and one that has better profitability and zero chlorinated wastes.

Matters are much less encouraging with respect to the remediation of existent hazardous wastes. Although it is human to want to forget the past, many hazardous waste sites are persistent and are being dispersed into the ground water and the atmosphere. Their effective remediation is limited by an adequate match of technology and economics. If the entire contaminated site (e.g., 30,000 tons of soil that contains 500 ppm PCBs) is to be remediated by incineration, the costs and the concerns with the toxicity of (a) the exhaust gases and particulates and (b) the ash (non-volatile residue) has prompted most to seek restraint in the courts. The dispersed nature of hazardous materials in soils, sludges, and waste ponds dictates, in the interests of economics and energy conservation, that separation and concentration technologies be applied prior to their remediation. Thermal desorption, distillation, solvent extraction, ion exchange, and supercritical-fluid extraction (see Chapter by Akgerman) are useful technologies for the separation and concentration of various contaminated matrices.

Although bioremedation is attractive in terms of its economics and its potential to transform *in situ* toxic materials into safe, innocuous products; most examples to date are for laboratory situations with moderately toxic substances. As the papers here confirm, major R&D efforts will be required to achieve practical and effective systems. Both the microorganisms and their environment must be optimized in terms of selectivity, efficiency, durability, and remediation rates. At best bioremediation technologies will be able to reduce the levels of some toxic materials by one or two orders of magnitude within time periods of 3-25 weeks. The challenge is to develop microorganisms with sufficient durability and specificity to achieve these goals with a single inoculation.

# PRACTICAL TECHNIQUES FOR REDUCING WASTE

Kenneth E. Nelson

Energy Conservation Manager
Dow Chemical U.S.A.
Plaquemine, LA 70765

## INTRODUCTION

The future of the chemical industry is closely tied to waste reduction. At the Dow Chemical Company, reducing waste has been an important objective for many years, and we have a wealth of knowledge and experience in this important field. This paper is designed to share some of that knowledge. Nearly all the ideas that will be discussed are based on process changes that have actually been made by Dow plants.

Since waste *reduction* (rather than treatment) is the ultimate way to lower emissions, we will focus our attention on techniques that produce fewer pounds of waste. Because waste reduction and energy conservation often occur simultaneously, many of the projects discussed will also reduce energy consumption.

## WHAT IS WASTE?

Let's start by defining waste. Waste includes streams or materials that are:

- Vented to the air
- Discharged to the water
- Sent to landfill
- Sent to an incinerator
- Sent to a flare
- Sent to a biological treatment facility

Waste reduction can be accomplished by creating less waste initially or by recycling waste products back to the process. It is also possible to reduce emissions by operating waste treatment units more efficiently, but that is not the concern of this paper. Also, we will not discuss the nature of the waste (i.e. hazardous or non-hazardous, toxic or non-toxic, flammable or non-flammable, etc.). This is not meant to be a comprehensive paper. The author has not made an exhaustive study of all possible ways of reducing

*Industrial Environmental Chemistry*, Edited by D.T. Sawyer
and A.E. Martell, Plenum Press, New York, 1992

waste. But it does contain a wealth of practical ideas and is intended to serve as the basis for discussion and brainstorming. We'll begin with ideas associated with raw materials.

## RAW MATERIALS

Raw materials are usually purchased from an outside source or transferred from an on-site plant. Each raw material needs to be studied to determine how it affects the amount of waste produced. The specifications for each raw material entering the plant should be closely examined.

### Improve Quality of Feeds

Although the percentage of undesirable impurities in a feed stream may be low, it can be a major contributor to the total waste produced by a plant. Reducing the level of impurities may involve working with the supplier of a purchased raw material, working with on-site plants that supply feed streams or installing new purification equipment. Sometimes the effects are indirect (e.g. water gradually kills the reactor catalyst causing formation of by-products, so a drying bed or column is added).

### Use Off-spec Material

Occasionally, a process can use off-spec material (that would otherwise be burned or landfilled) because the particular quality that makes the material off-spec is not important to the process.

### Improve Quality of Products

Impurities in your own products may be creating waste in your customers' plants. Not only may this be costly, it may cause customers to look elsewhere for higher quality raw materials. Take the initiative in discussing the effects of impurities with your customers.

### Use Inhibitors

Inhibitors prevent unwanted side reactions or polymer formation. A wide variety of inhibitors are commercially available. If inhibitors are already being used, check with suppliers for improved formulations and new products.

### Change Shipping Containers

If raw materials are being received in containers that can not be reused and need to be burned or landfilled, change to reusable containers or bulk shipments. Similarly, consider using alternative containers for shipping products to customers.

### Re-examine Need for each Raw Material

Sometimes the need for a particular raw material (one which ultimately ends up as waste) can be reduced or eliminated by modifying the process. A cooling tower, for example, uses inhibitors to control algae growth. In one cross-flow cooling tower, the need for algae inhibitors was cut in half by shielding the water distribution decks from sunlight.

## REACTORS

The reactor is the heart of the process, and can be a primary source of waste products. The quality of mixing in a reactor is crucial. Too often, insufficient design time is spent on scaling up this key parameter and a new production facility has disappointing yields when compared to lab or pilot plant data. Several of the ideas listed below are concerned with quality of mixing.

### Distribute Feeds Better

Here is an area that deserves more attention than it typically gets. The problem is illustrated in Figure 1 (left). Reactants entering at the top of a fixed catalyst bed are poorly distributed. Part of the feed short-circuits down through the center of the reactor and does not allow adequate time for conversion to desired products. In contrast, feed that is closer to the walls remains in the reactor too long and "over-reacts" to by-products that eventually become waste. One solution is to add some sort of distributor and the reactor inlet that causes the feed to move uniformly through all parts of the reactor as shown on the right. A special collector at the bottom of the reactor may also be necessary to prevent the flow from necking down to the outlet. Similarly, if a gas reactant is added to a liquid, the gas needs to be finely dispersed and evenly distributed throughout the liquid phase.

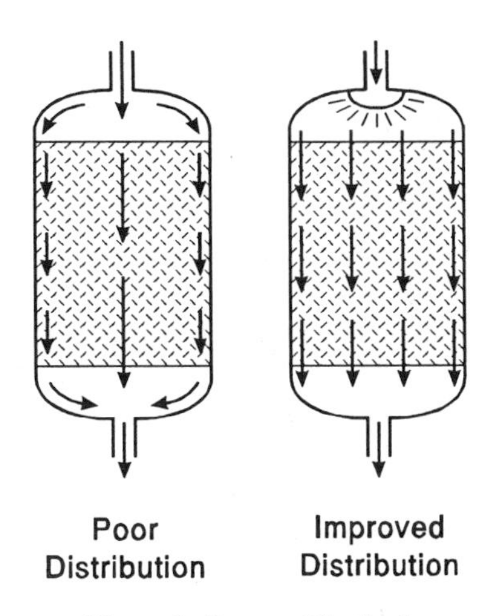


**Figure 1.** Reactor Distribution

### Improve Catalyst

Because of the significant effect a catalyst has on reactor conversion and product mix, searching for better catalysts should be an ongoing activity. Changes in the chemical makeup of a catalyst, the method by which it is prepared, or its physical characteristics (size, shape, porosity, etc.) can lead to substantial improvements in catalyst life and effectiveness.

### Improve Physical Mixing in the Reactor

Modifications to the reactor such as adding or improving baffles, installing a higher rpm motor on the agitator(s), or using a different mixer blade design (or multiple impellers) can improve mixing. Pumped recirculation can be added or increased. Two fluids going through a pump, however, do *not* necessarily mix well and a static mixer may be needed to insure good contacting.

### Improve Way in Which Reactants are Added

The idea here is to get closer to ideal reactant concentrations before the feed enters the reactor. This helps avoid secondary reactions which form unwanted by-products. The way *not* to add reactants is shown in the upper half of Figure 2 on the following page. It is doubtful that the "ideal" concentration exists anywhere in this reactor. A consumable catalyst, especially, should be diluted in one of the feed streams (one which does not react in the presence of the catalyst). The lower half of Figure 2 illustrates one approach to improving the situation by using three in-line static mixers.

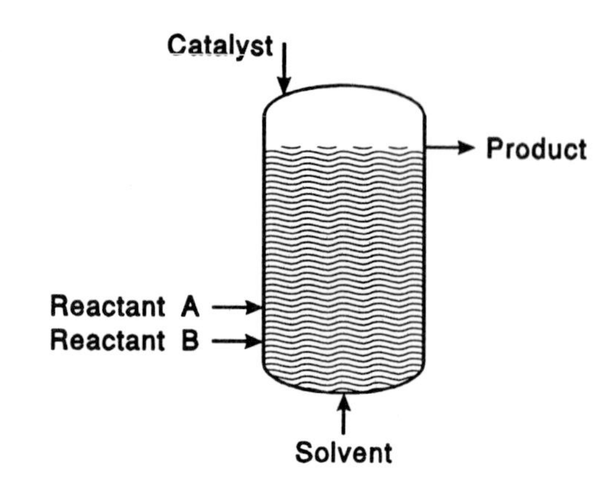


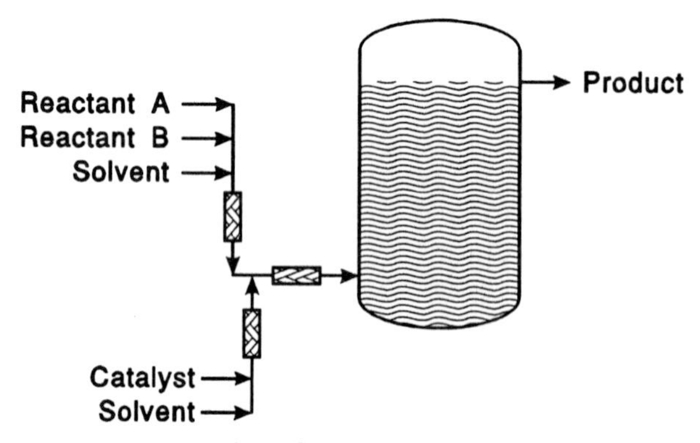


**Figure 2.** Adding Reactants

**Consider Different Reactor Design**

The classic stirred-tank backmix reactor is not necessarily the best choice. A plug flow reactor offers the advantage that it can be staged, and each stage can be run at different conditions (especially temperatures), closely controlling the reaction for optimum product mix (and minimum waste). Many innovative hybrid designs are possible.

**Provide Separate Reactor for Recycle Streams**

Recycling by-product and waste streams is an excellent technique for reducing waste, but often the ideal reactor conditions for converting recycle streams back to usable products are considerably different from conditions in the primary reactor. One solution is to provide a separate, smaller reactor for handling recycle and waste streams as illustrated in Figure 3 at the right. Temperatures, pressures and concentrations can then be optimized in both reactors to take maximum advantage of reaction kinetics.

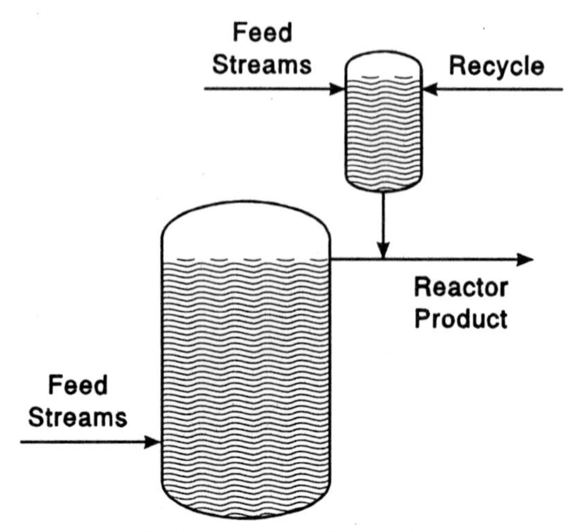


**Figure 3.** Separate reactor for recycle streams

### Improve Control

For a given reactor configuration, there is one set of operating conditions that is optimum at any given time. The control system should know that condition and make it occur, with little fluctuation. Such control may be complex, particularly in the case of batch reactors, but can yield major improvements. In less sophisticated systems, simply stabilizing reactor operation frequently reduces the formation of waste products.

Advanced computer control systems are capable of responding to process upsets and product changes swiftly and smoothly, producing a minimum of unwanted by-products (see section on **PROCESS CONTROL**).

### Examine Heating and/or Cooling Techniques

The technique for heating or cooling the reactor needs to be examined, especially to avoid hot-spots or overheated feed streams, both of which cause unwanted by-products.

## HEAT EXCHANGERS

Heat exchangers can be a source of waste, especially with products that are temperature sensitive. There are a number of techniques for minimizing the formation of waste products in heat exchangers. All reduce the amount of fouling.

### Desuperheat Plant Steam

High pressure plant steam may contain several hundred degrees of superheat. Desuperheating the steam when it enters a process (or just upstream of an exchanger) reduces tube wall temperatures and actually increases the effective surface area of the exchanger because the heat transfer coefficient of condensing steam is ten times greater than that of superheated steam.

### Use Lower Pressure Steam

When plant steam is available at fixed pressure levels, a quick expedient may be to switch to steam at a lower pressure, reducing tube wall temperatures.

### Install a Thermocompressor

Another method of reducing tube wall temperature is to install a thermocompressor, shown in Figure 4. These relatively inexpensive units work on an ejector principle, combining high and low pressure steams to produce an intermediate pressure steam. Variable throat models are available that operate like control valves, automatically mixing the correct amounts of high and low pressure steam.

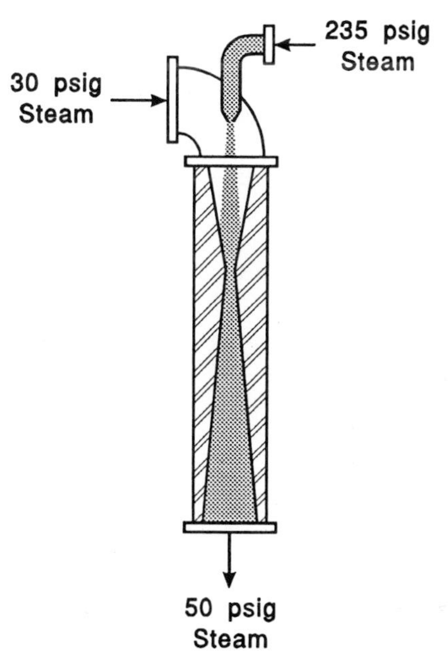


**Figure 4.** Thermocompressor

### Use On-line Cleaning Techniques

On-line cleaning devices such as recirculated sponge balls and reversing brushes have been on the market for many years. In addition to reducing exchanger maintenance, they also keep tube surfaces clean so that a lower temperature heat source can be used. Processes benefiting from close temperature approaches (e.g. refrigeration and condensing steam turbine systems) gain from on-line cleaning.

### Use Scraped-wall Exchangers

Scraped-wall heat exchangers consist of a set of rotating blades inside a vertical, cylindrical jacketed column. They can be used to recover saleable products from viscous streams. A typical application is to recover monomer from polymer tars.

### Use Staged Heating

If a heat-sensitive fluid must be heated, staged heating can lessen degradation. Begin, for example, with low level waste heat, then use low pressure steam, and finally desuperheated intermediate pressure steam. See Figure 5 below.

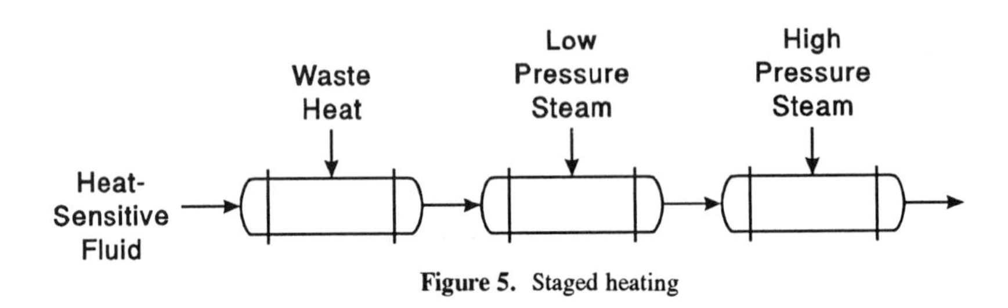


**Figure 5.** Staged heating

### Monitor Exchanger Fouling

Exchanger fouling does not necessarily occur steadily. Sometimes, an exchanger fouls because of a change in operating conditions or because of an upset elsewhere in the process. The cause of fouling can often be found by continuously monitoring the fouling factor and correlating sudden increases in fouling to operational changes.

### Use Non-corroding Tubes

Corroded tube surfaces foul more quickly than non-corroded tube surfaces. Changing to non-corroding tubes can significantly reduce fouling.

## PUMPS

Pumps don't usually contribute to waste except in two areas:

### Recover Seal Flushes and Purges

Each pump seal flush and purge needs to be examined as a possible source of waste. Most can be recycled to the process with little difficulty.

### Use Seal-less pumps

Leaking pump seals lose product and create environmental problems. Using can-type seal-less pumps or magnetically driven seal-less pumps eliminates these losses.

## FURNACES

Furnace technology is constantly evolving and represent one of the most fertile areas for efficiency improvement. Furnace manufacturers should be contacted for the latest techniques in optimizing furnace operation and reducing tube fouling. Although furnaces are often advertised as having high efficiency (e.g. 80%), it should be recognized that burning fuel is inherently inefficient, and the true thermodynamic (2nd law) efficiencies of industrial furnaces are much lower (normally 20-40%).

### Replace Furnace with Intermediate Exchanger

Another option is to eliminate direct heating in the furnace (which necessarily exposes the heated fluid to high tube wall temperatures) by using an intermediate heat transfer medium (e.g. Dowtherm®). This is illustrated in Figure 6 at the right.

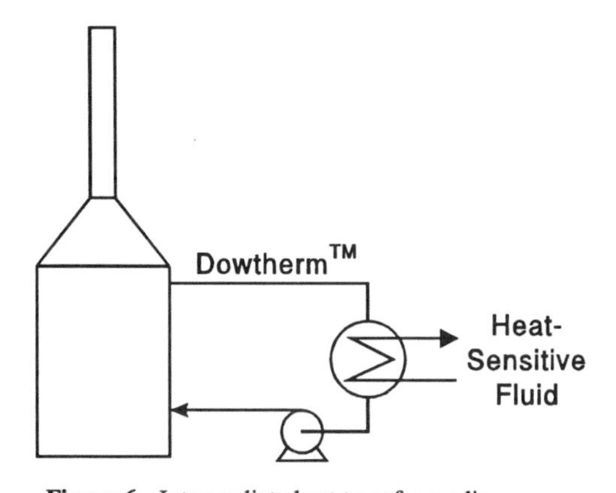

**Figure 6.** Intermediate heat transfer medium

### Replace Coil

In certain applications, significant improvements can be made by replacing the existing furnace coil with one having improved design features (e.g. tubes with low residence time or designed for split flow). Although it may not be practical to replace an undamaged coil, alternative designs should be considered whenever replacement becomes necessary.

### Use Existing Steam Superheat

Even though the temperature required may be above the *saturation* temperature of plant steam, there may be sufficient *superheat* available in the steam to totally eliminate the need for a furnace. This is illustrated in Figure 7. This may also be an efficient way to desuperheat plant steam as suggested earlier. Many people are not even aware that their plant steam contains considerable superheat.

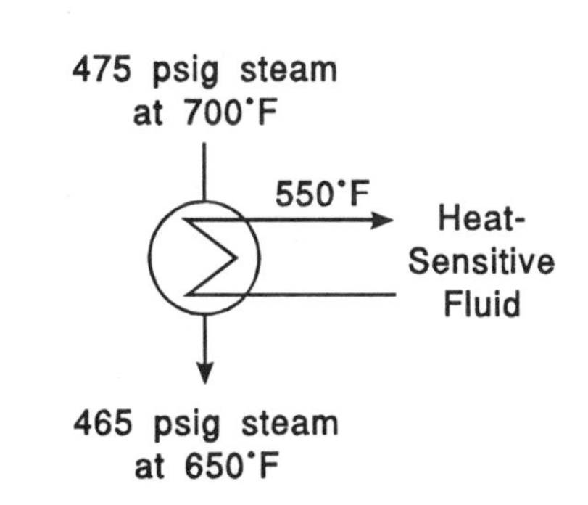

**Figure 7.** Steam superheat replaces furnace

## DISTILLATION COLUMNS

Distillation columns typically contribute to waste in three ways:

A. By allowing impurities to remain in a product. The impurities ultimately become waste. The solution: better separation. In some cases it may be desirable to exceed the normal product specifications.

B. By forming waste within the column itself, usually because of high reboiler temperatures which cause polymerization. The solution: find ways to lower column temperatures.

C. By inadequate condensing, which results in vented or flared product. The solution: improve condensing.

In the following paragraphs, we'll look at some column and process modifications that reduce waste by attacking one or more of these three problems.

### Increase Reflux Ratio

The most common way of improving separation is simply to increase the reflux ratio. This raises the pressure drop across the column and increases the reboiler temperature (using additional energy), but is probably the simplest solution if column capacity is adequate.

### Add Section to Column

If a column is operating close to flood, a new section can be added to increase capacity and improve separation. The new section can use trays, regular packing or high efficiency packing. It need not be the same as the original portion of the column.

### Retray or Repack Column

Another method of increasing separation is to retray or repack part or all of the column. High efficiency packing is available and has the added advantage that it also lowers pressure drop through the column, decreasing the reboiler temperature.

### Change Feed Tray

Don't overlook changing the feed tray. Many columns are built with multiple feed trays, but the valving is seldom changed. In general, the closer the feed conditions are to the top of the column (high lights concentration and low temperature), the higher the feed tray; the closer feed conditions are to the bottom of the column (high heavies concentration and high temperature), the lower the feed tray.

### Insulate

Good insulation is necessary to prevent heat losses. Poor insulation requires higher reboiler temperatures and allows column conditions to fluctuate with weather conditions. During a storm, inadequately insulated columns may cool rapidly, sending light components in the column bottoms stream, ultimately producing waste.

**Improve Feed Distribution**

The effectiveness of feed distributors (particularly in packed columns) needs to be analyzed to be sure that distribution anomalies are not lowering overall column efficiency. As with reactors, poor distribution in columns results in inefficiencies which become accepted with time.

**Preheat Column Feed**

Preheating the feed should improve column efficiency. Supplying heat in the feed requires lower temperatures than supplying the same heat to the reboiler, and also reduces the reboiler load. It can often be done by cross-exchange with other process streams.

**Remove Overheads Product from Tray Near Top of Column**

If the overheads product contains a light impurity, it may be possible to obtain a higher purity product from one of the trays close to the top of the column. A bleed stream from the overheads accumulator can be recycled back to the process to purge the column of lights. Another solution is to install a second column, which may be expensive but well worth doing.

**Increase Size of Vapor Line**

In low pressure or vacuum columns, pressure drop is especially critical; installing a larger vapor line reduces pressure drop and decreases reboiler temperature.

**Modify Reboiler Design**

While thermosyphon reboilers are the most common, they may subject the product to tube-wall temperatures which cause product degradation. Other designs are available such as falling film and pumped recirculation reboilers. High flux tubes can sometimes be retrofitted to an existing reboiler allowing use of lower pressure steam.

**Reduce Reboiler Temperature**

The same general temperature reduction techniques discussed earlier (in the section on heat exchangers) such as desuperheating steam, using lower pressure steam, installing a thermocompressor, using an intermediate heat transfer fluid, etc., also apply to the reboiler of a distillation column.

**Lower Column Pressure**

Reducing the column pressure will also decrease the reboiler temperature and may favorably load the trays or packing as long as the column stays below flood. The overheads temperature, however, will also be reduced, which may create a condensing problem (see following paragraph).

**Improve Overhead Condenser**

If overheads are lost because of an undersized condenser, consider retubing, replacing the condenser, or adding a supplementary vent condenser to minimize losses. It may also be possible to reroute the vent back to the process (if process pressure is stable). If a refrigerated condenser is used, be sure to keep the tubes above 32°F if there is any moisture in the stream.

### Improve Column Control

The comments in the Reactor section about closer process control and controlling at the right point apply to distillation columns as well as reactors.

### Forward Vapor Overheads to Next Column

If the overheads stream is sent to another column for further separation, it may be possible to use a partial condenser and introduce a vapor stream to the downstream column.

Before any equipment modifications are undertaken, it is recommended that a computer simulation be done and that a variety of operating conditions be examined. If the column operating temperature or pressure changes, equipment ratings should also be reexamined.

## PIPING

Something as seemingly innocuous as plant piping can sometimes cause waste, and simple piping changes can result in a major reduction of waste. Consider the following piping changes to a process:

### Recover Individual Waste Streams

In many plants, various waste streams are combined and sent to a waste treatment facility as shown in Figure 8. Each waste stream needs to be considered individually. The nature of the impurities may make it possible to recycle or otherwise reuse a particular stream *before* it is mixed with other waste streams and becomes unrecoverable. Stripping, filtering, drying or some other type of treatment may be necessary before the stream can be reused. Condensate used for seal flushes are an example of a stream that does not require a "pure" stream. Simply filtering a waste water stream to remove particulates may be adequate.

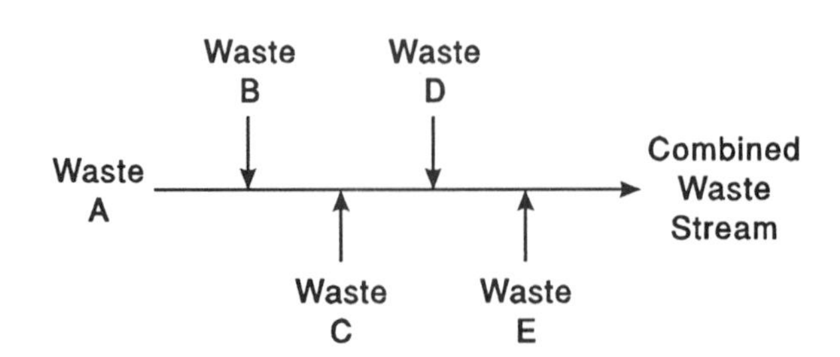


**Figure 8.** Recover individual waste streams

### Check Pipe Sizes

Undersized pipe, especially in low pressure gas service (e.g. the inlet to a compressor) can be costly. All low pressure gas piping should be checked for optimum flow and pressure drop. Also look for obstructions, restrictions or undersized valves in lines.

### Avoid Overheating Lines

If a process stream contains temperature sensitive materials, the temperature level of line and vessel tracing systems and and/or vessel jacketing needs to be reviewed. If plant steam is hot enough to cause product degradation, a recirculated warm fluid can be used to prevent freezing. Tracing systems that use a recirculated fluid heated in a central vessel consume much less energy than typical steam tracing systems. Electric tracing is also an option.

### Avoid Sending Hot Material to Storage

If a temperature sensitive material is sent to storage it should first be cooled. If this is uneconomical because the stream from storage needs to be heated when it's used, simply piping the hot stream directly into the suction of the storage tank pump as shown in Figure 9 may solve the problem adequately. Make sure that the storage tank pump can handle hot material without cavitating and that material in the storage tank does not degrade.

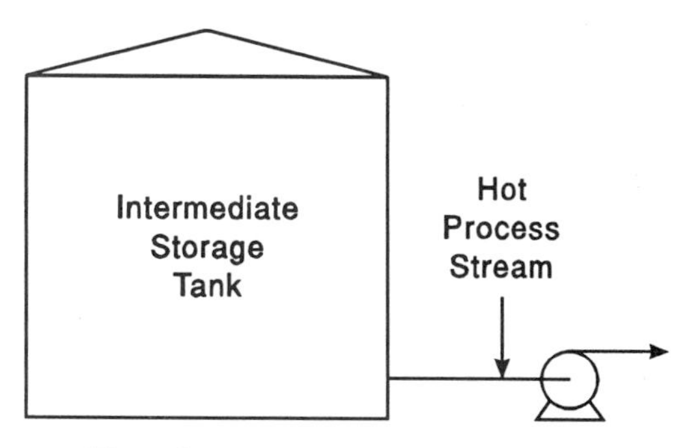


**Figure 9.** Reducing storage tank heat losses

### Eliminate Leaks

Leaks can be a major contributor to a plant's overall waste, especially if the products can not be seen or smelled. A good way to document *leaks* is to measure the quantity of *raw materials* that must be purchased to replace "lost" streams (e.g. the amount of refrigerant purchased).

### Change Metallurgy

The type of metal used for vessels or piping may be causing a color problem or be acting as a catalyst for the formation of unwanted by-products. If this is the case, change to more inert metals (see next paragraph).

### Monitor Vent And Flare Systems

The measurements need not be highly accurate, but should give a reasonable estimate of how much product is lost and when losses occur. It is often worthwhile to install whatever piping is necessary to recover products that are vented or flared. Additional purification may be needed before the recovered streams can be reused.

### Use Lined Pipes or Vessels

Using lined pipes or vessels is often a cheap alternative to using exotic metallurgy. A variety of coatings are available for different applications. FRP vessels are also an alternative.

## PROCESS CONTROL

Modern technology allows us to install highly sophisticated computer control systems that respond far more quickly and accurately than human beings. We can use that capability to reduce waste. Some suggestions are summarized on the following pages.

### Improve On-line Control

Good process control reduces waste by minimizing cycling and improving a plant's ability to handle normal changes in flows, temperatures, pressures and compositions. Statistical quality control techniques help plants analyze process variations and document improvements. Sometimes additional instrumentation or on-line stream monitors (e.g. gas chromatographs) are necessary, but good control optimizes process conditions and reduces plant trips, often a major source of waste.

### Optimize Daily Operation

If a computer is incorporated into the control scheme, it can be programmed to analyze the process continuously and optimize operating conditions. If the computer is not an integral part of the control scheme, off-line analyses can be performed and used as a guide for setting process conditions.

### Automate Startups, Shutdowns and Product Changeovers

Huge quantities of waste can be produced during plant startups, shutdowns and product changeovers, even when such events are well planned. Programming a computer to control these situations brings the plant to stable operating conditions quickly and minimizes the time spent generating off-spec product. Further, since minimum time is spent in undesirable running modes, equipment fouling and damage are also reduced. This is an excellent place to use input from experienced plant operators.

### Program Plant to Handle Unexpected Upsets and Trips

Even with the best control systems, upsets and trips occur. Not all can be anticipated, but operators who have lived with the plant for years probably remember most of the important ones and know the best ways to respond. With computer control, optimum responses can be pre-programmed. Then, when upsets and trips occur, the computer takes over, minimizing downtime, spills, equipment damage, product loss and waste.

## MISCELLANEOUS

In addition to the ideas already discussed, there are a number of other miscellaneous improvements that can be made to reduce waste:

### Avoid Unexpected Trips and Shutdowns

A good preventative maintenance program and adequate sparing of equipment are two keys to minimizing trips and unplanned shutdowns. Another key is to provide an adequate warning system for critical equipment (e.g. vibration monitors). Plant operators can be extremely helpful by reporting unusual conditions so that minor maintenance problems get corrected before they become major and cause a plant trip.

### Use Waste Streams from Other Plants

Within a chemical complex, each plant's waste streams should be clearly identified. The quantity and quality of these streams should be documented, including the presence of trace quantities of metals, halides or other impurities that render a stream useless as a raw material. This list should be reviewed by all plants in the complex to determine if any are suitable as feedstocks.

### Reduce Number and Quantity of Samples

Taking frequent and large samples can generate a surprising amount of waste. Many plants find that the quantity can be reduced or that the samples can be returned to the process after analysis.

### Recover Product from Tank Cars and Tank Trucks

Product vented or drained from a tank car or tank truck (especially those dedicated to a single service) can often be recovered and reused.

### Use Removable Insulation

When conventional insulation is removed from equipment it is typically scrapped and sent to landfill. A number of companies manufacture reusable insulation. Their products are particularly effective on equipment where the insulation is removed regularly in order to perform maintenance (e.g. heat exchanger heads, manways, valves, transmitters, etc.)

### Maintain External Painted Surfaces

Even in plants handling highly corrosive materials, *external* corrosion can be a major cause of pipe deterioration. Piping and vessels should be painted before being insulated, and all painted surfaces should be well-maintained.

### Find a Market for Waste Products

Converting a waste product to a saleable product may require some additional processing, but can be an effective means of reducing waste. Converted product, however, should not create a waste problem for the customer.

## CONCLUSIONS

Reducing waste is a never-ending activity. Each year, we are faced with new demands, new regulations, new legislation, new challenges and of course, new problems. Amidst all this activity, we never want to lose sight of one extremely important concept:

**The way you reduce waste is by installing projects that reduce waste.**

You can state corporate positions, write detailed mission statements, outline ambitious goals, and prepare detailed plans, but:

**The way you reduce waste is by installing projects that reduce waste.**

You can give dynamic speeches, publicize inspiring slogans, develop lofty mottoes, design award-winning posters, and distribute multi-colored brochures, but:

**The way you reduce waste is by installing projects that reduce waste.**

You can establish high-level committees, appoint technical task forces, organize environmental teams, and restructure entire departments, but:

**The way you reduce waste is by installing projects that reduce waste.**

You can meet with environmentalists, negotiate with regulators, confer with politicians, talk to community leaders, and address the general public, but:

**The way you reduce waste is by installing projects that reduce waste.**

You can write letters to the newspaper, be interviewed on radio, appear on TV, speak at technical meetings, lecture at symposia, and even testify before Congress, but:

**The way you reduce waste is by installing projects that reduce waste.**

By now, I hope my message is clear! A successful waste reduction program is more than talk; it must get results. It must raise employee awareness and concern about environmental issues to the point where specific ideas for projects to reduce waste are conceived, designed and implemented.

As a closing thought (and additional impetus to action), recognize that waste reduction is not synonymous with lower profits. You might be surprised to learn that nearly every idea listed in this paper had a Return on Investment—an ROI—greater than 30%! Most paid for themselves in less than a year. Some required no capital expenditure.

**Reducing waste is not just responsible management, it's good business!**

---

Note: A version of this paper was originally presented at "Pollution Prevention for the 1990's: A Chemical Engineering Challenge" December, 1989, Washington, D.C., and was subsequently published in Hydrocarbons Processing (March, 1990) and Chemtech (August, 1990).

## WARR - IMPLEMENTING A WASTE AND RELEASE REDUCTION STRATEGY

James A. Conner

Hoechst Celanese Corporation
Route 202-206
Somerville, NJ 08876

## INTRODUCTION

Hoechst Celanese Corporation is embarked upon a Corporate-wide Waste and Release Reduction program, which we have given the acronym "WARR". WARR is a comprehensive program for achievement of major reductions across the corporation of all emissions and waste materials.

The WARR program is a part of our Vision of Excellence effort. Vision of Excellence is a Hoechst Celanese initiative aimed at integration of environmental, health and safety considerations into everything we do as a Corporation. The Vision of Excellence is complementary to the CMA Responsible Care® initiative.

We have defined eight end states within Vision of Excellence - one of which is the Release Reduction/Waste Minimization End State (Table 1).

**Table 1.** Hoechst Celanese Vision of Excellence End States

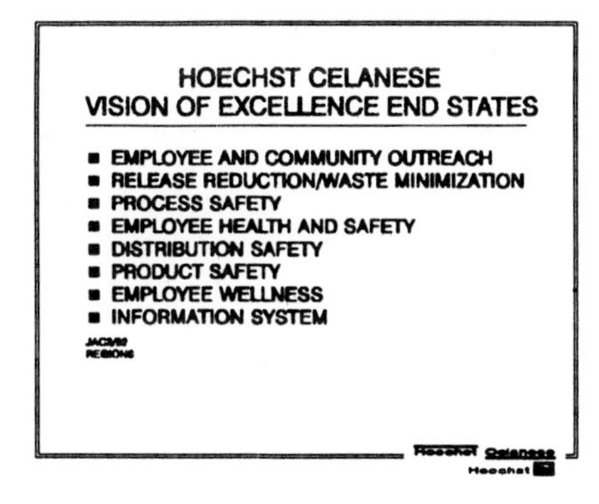

HOECHST CELANESE
VISION OF EXCELLENCE END STATES

- EMPLOYEE AND COMMUNITY OUTREACH
- RELEASE REDUCTION/WASTE MINIMIZATION
- PROCESS SAFETY
- EMPLOYEE HEALTH AND SAFETY
- DISTRIBUTION SAFETY
- PRODUCT SAFETY
- EMPLOYEE WELLNESS
- INFORMATION SYSTEM

The WARR program falls within, and is the action plan for, this Release Reduction/Waste Minimization end state.



*Industrial Environmental Chemistry*, Edited by D.T. Sawyer
and A.E. Martell, Plenum Press, New York, 1992

## OVERVIEW OF HOECHST CELANESE CORPORATION

Hoechst Celanese Corporation was formed in 1987 by the merger of the former Celanese Corporation and the former American Hoechst Corporation. It is a wholly owned subsidiary of Hoechst AG of Frankfurt, Germany. The Corporation is a major US manufacturer, with worldwide sales, of fibers, chemicals, plastics, and pharmaceuticals. At present, the company has 32 major US locations with approximately 23,000 employees. Our 1990 sales totalled roughly $6 billion. Major business units are located in Texas (Chemicals and Plastics), the Carolinas and Virginia (Fibers and Film), and New Jersey (Pharmaceuticals).

## WHAT IS WARR

WARR is a corporate-wide program focused on major reductions of emissions and waste. It addresses emission to all media - air, water, and land.

There are 3 primary goals of the WARR effort:

1. An 80% reduction (vs. 1988 base year) of all types of discharges and releases (excluding combustion products from energy production, treated process wastewater and non-contact cooling water) by the end of 1996

2. A 70% reduction (vs. 1988 base year) of releases of SARA Title III, Section 313 chemicals by the end of 1996

3. Discontinue the use of all underground injection wells for waste disposal by the end of 1996

Hoechst Celanese Corporation was also one of the early signatories to the US EPA's Industrial Toxics Program (ITP), which is also commonly referred to as the "33/50 Program". The ITP targets 17 priority chemicals (all SARA) for a 33% reduction by end of 1992 and a 50% reduction by the end of 1995 (all vs. a base year of 1988).

Hoechst Celanese has reportable emission quantities of 12 of the 17 targeted chemicals. The goals of the ITP are in concert with, and complementary to, our own WARR program goals. The projects which are planned to meet our end-of-1996 WARR goals will also assist us in meeting the 1995 50% reduction target set by EPA for these 17 priority chemicals.

The initial phase of the WARR effort focuses on nine major facilities (out of 32). There are presently in excess of 180 projects on the WARR program list.

The anticipated cost of the program is $500 to $600 million. Funding for the WARR effort is separate and in addition to the Corporation's normal annual capital budget. This WARR spending amounts to roughly 1/4 - 1/3 of our Corporation's total capital spending in the years 1991-1995. This substantial expenditure is viewed by our Corporation as an investment in the future, as defined by our Vision of Excellence.

## PRIORITIES OF THE WARR PROGRAM

Emission reduction planning and prioritizing of projects is guided by our "Waste and Release Reduction Hierarchy", a prioritization list for reduction alternatives which

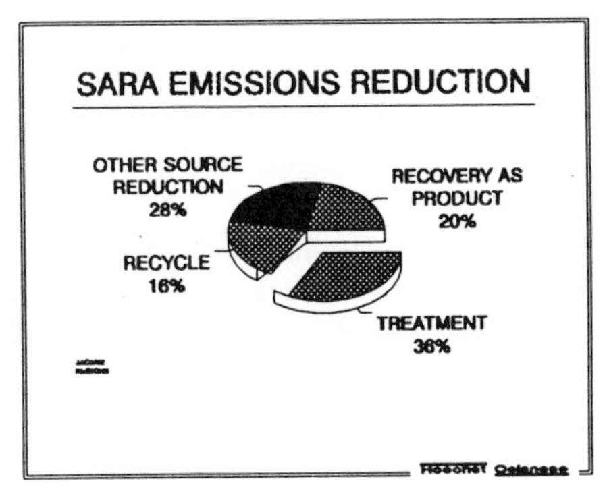


**Figure 1.** Sara Emission Reductions

**Table 2.** Release Reduction Hierarchy

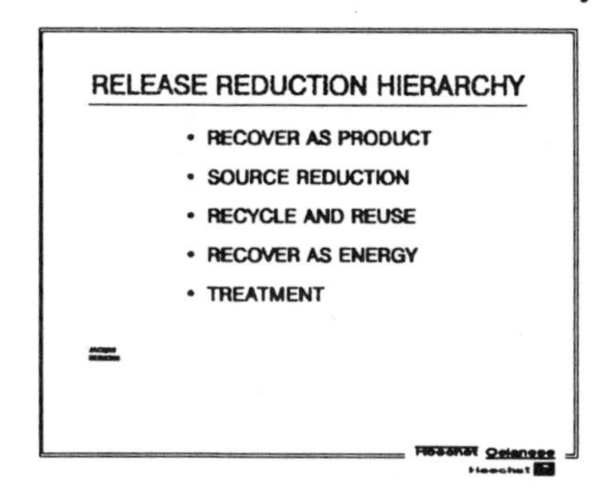


encourages source reduction and recycle/reuse over end-of-pipe treatment (Table 1).

Our organization has been challenged to develop release reduction projects and solutions which are as high on this hierarchy as can practically be achieved. This challenge has resulted in considerable success as both source reduction and recycle/reuse figure prominently in our planned reductions. Roughly half of our total program effort falls into the source reduction/recycle-reuse categories. Further, nearly 2/3 of our planned SARA reductions are source reductions or recycle/reuse, and of this total, 1/3 is through recovery of the emissions as saleable product (Figure 1).

## ORIGIN AND DEVELOPMENT OF THE PROGRAM

The WARR program is rooted in our Corporate Values, which were established at the time of the merger. We utilize the Values to help develop visions, and from those visions we establish plans for implementation.

One of the Values is "Commitment to Safety, Employee Health and Protection of the Environment". Our vision for environmental excellence was to be recognized as a leader in environmental stewardship by our toughest critics - our employees, their families, and the communities in which we operate. Implicit in this vision was recognition that leadership requires that you commit to the proactive mode of dealing with environmental issues.

The initial step in implementing the Vision was formation in 1989 of the Release Reduction Task Force. This group developed a comprehensive inventory of emissions and wastes from each of our locations. They then prioritized the issues to be addressed, and, using the 80/20 principle, the Group recommended an initial focus on nine major facilities.

A Release Reduction Strategy Committee was then appointed, led by the President of one of our Operating Groups. This committee was chartered to develop a specific strategy to achieve significant reductions in our emissions and wastes, and to identify the technology and resource needs required from the organization to achieve that strategy.

The committee challenged each of the targeted facilities to attempt to assemble a zero-discharge project list - to approach from an "emission elimination" mindset as opposed to simply "emission reduction". For reasons covered later, the zero-discharge project list was not achievable, but the value was in the approach of zero-based thinking. This represented a significant advancement beyond typical thinking, which is normally only incremental in nature. In retrospect, this approach was of vital importance to the development of our current program for several key reasons:

- It created a mind-set that this was not just an incremental program
- It forced a long-term perspective
- It forced a source-reduction perspective
- It forced creativity and original solutions

The zero-discharge list was not achievable. The rough estimate for the aggregate initial lists for the targeted facilities was in the billion of $'s, and in many cases, total elimination of the emission source would require development of unproven or unavailable technologies. However, judicious selection from these original lists yielded a project list which accomplished major reductions of emissions, emphasized source reduction over end-of-pipe treatment, and was achievable in a reasonably rapid time frame. This carefully selected strategy and project list was approved by the Corporation's Executive Committee (CEO and Group Presidents) in October 1990.

## TECHNOLOGIES

To be sure that we are installing the most effective and appropriate technology for the projects on the WARR list, we have developed mechanisms for technology transfer from a variety of sources. Interplant and interdivision committees have been set up to share and coordinate common approaches and technology applications. We have also benefitted from sharing of technologies with Hoechst in Germany. Other avenues which we utilize for sharing of technology and solutions include industry groups, such as the AIChE Center for Waste Reduction Technology, and liaisons set up with universities. '

The primary technologies for waste treatment are biological degradation and incineration. In the biotreatment arena, the technology has shifted to enclosed above-ground bioreactors, such as the Hoechst Biohoch design, to maximize energy efficiency and minimize air emissions. Anaerobic systems will work on some wastes, and offer the advantage of lower sludge yields and methane recovery. A few technology firms offer specialized micro-organism cultures which can be utilized to digest specific compounds which are otherwise toxic or refractory to mixed-bug treatment plants.

For incineration, we are seeing a trend toward more exotic stack gas cleanup systems, and toward fluidized bed or rotary kiln designs which are capable of handling wastes from a variety of media - solid, liquid, and gaseous. For some applications, we are developing catalytic incineration technology, which has the advantage of lower temperatures and reduced $NO_x$ generation compared to typical thermal incinerators.

For source reduction, the key is not so much new technologies as it is a reapplication and refocus of basic chemical process fundamentals and unit operations on the minimization of wastes. Examples are improvements in reaction selectivity or conversion, improved separations, higher purity feedstocks, solvent substitution, and in-process recycling for yield improvement. One notable exception to this is improvements in separations made possible by the rapidly evolving membrane technologies. Our Separations Products Division develops and markets membrane systems for phase contact (e.g. liquid extraction), ultrafiltration, and pervaporation (separation of liquid or vapor streams via selective permeation and evaporation across a membrane). Separations Products is working closely with our WARR effort to develop specific applications of these technologies to enhance our source reduction efforts.

For the longer term, we have increased the focus of the Hoechst Celanese R&D Community on source prevention. Pollution Prevention begins in the development lab, where the opportunity exists for toxic use reduction and the development of "green" chemistry and processes. Another leverage point for source prevention is our project teams - the process design engineers and the plant layout engineers play a key role in source prevention through process and plant designs which have fewer leak points, better separations, and plant layouts which minimize waste during equipment cleaning and maintenance.

## REFINEMENT OF ORIGINAL STRATEGY

As work began on definition and development of the original WARR project list, the technical staffs at the facilities began to find better alternatives to many of the projects. In late 1991, we completed a substantial optimization of the project list which increased our projected reductions of emissions and also shifted more of the projects into the source reduction category. Many of the original end-of-pipe projects were replaced by multiple smaller source reduction projects, thereby eliminating the need for treatment. In other cases, treatment projects dropped from the list after operational changes reduced the source and made treatment unnecessary. Further optimization of the project list remains a priority, and we expect to continue the trends of additional emissions eliminations, and movement toward greater source reduction.

## MANAGEMENT OF THE PROGRAM

Responsibility for oversight and direction of the program is vested in the Waste and Release Reduction Implementation Steering Committee. This group is chaired by the President of one of our largest operating divisions, Technical Fibers. Members are a cross-sectional group of the Corporation (plant managers, environmental directors, division VP's, etc.). The Steering Committee has responsibility for:

- Tracking vs. objectives
- Assure best technology

- Consistency with long term environmental goals
- Grow program beyond initial focus to encompass major reductions at all facilities

WARR Coordinators were named at each site. These are predominantly technical people, many with backgrounds in Operations, Process Engineering or Project Management. The strengths and skills of these people have direct bearing upon the amount of source reduction that will be achieved. To catalyze source reduction, you need to involve these people who best know the process - design engineers, process engineers and chemists, and operations personnel. The results we have achieved with the optimized project list proved the value of this approach.

At the group level, additional people were appointed to assist the plantsites to crossfeed and coordinate shared solutions. A good example is our Film and Fibers Group, where we have a Vice President of Waste and Release Reduction, a Director of WARR projects to coordinate plant activities, and a focus person for technology transfer between divisions and sites. At the Corporate level, the position of Manager, Environmental Projects was created to provide overall coordination and management across all divisions.

## RESULTS AND FORECASTS

The WARR program is on track to achieve all stated objectives by the end of 1996. Our disposal well elimination projects are on track, and we project a substantial reduction in all emissions over the next 3-4 years. We have already achieved significant reductions in SARA releases. As a Corporation, our SARA releases declined roughly 20% from 1988 to 1990. In addition, we are projecting additional SARA release reductions of approximately 10% annually for the next five years, achieving our 70% reduction target by the end of 1996.

## SUCCESS FACTORS

Unwavering management commitment from the top of the Corporation has played a major role in our program's success. This commitment held steady in spite of the economic downturn which beset the industry. In addition, champions for the program emerged in key positions within the company to help eliminate barriers and assemble the necessary resources.

Plantsite participation in development of the strategy was also key. Early input to the project list established an ownership at the plantsites to the project goals, the overall strategy, and the specific contribution needed from each facility.

Finally, the WARR effort is not perceived as simply a program from the Environmental organization. Through our Vision of Excellence, we have moved Waste and Release Reduction into the mainstream of our business objectives and plans - a way of doing our business, and an integral part of our long term planning.

# WASTE MINIMIZATION BY PROCESS MODIFICATION

J. R. Hopper, C. L. Yaws, T. C. Ho, M. Vichailak and A. Muninnimit

Chemical Engineering Department
Lamar University
Beaumont, Texas

## INTRODUCTION

Waste minimization means the reduction, to the extent feasible of hazardous waste that is generated prior to treatment, storage or disposal of the waste. The terms waste minimization and pollution prevention are becoming key initiatives of the United States Environmental Protection Agency (U.S.EPA) and the term is increasingly being used in all areas of current applied research and development [15] .

Waste minimization has been defined to consist of two basic techniques: source reduction and recycling [14]. See Figure 1. Source reduction or avoiding waste generation is the most desirable goal and should be explored first [16]. In the waste management hierarchy, source reduction is followed in order by recycling, treatment and, finally, disposal. Although such processes as incineration, stabilization and/or storage are clearly preferable to land disposal, they are not included in the definition of waste minimization. The total concept of pollution prevention is that it is much more desirable not to produce waste rather than to implement extensive treatment schemes to insure that the quality of the environment is not damaged. Under the category of source reduction , there are two sub-categories: source control and product substitution. Source control can be accomplished by: (1) good housekeeping practices, (2) input material modification, and (3) technology modification. EPA has reported that , generally, technology modifications have been found to be the most effective means of reducing waste generation [38]. Technology modifications have been categorized as follows:

- Process Modifications
- Improved Controls
- Equipment Changes
- Energy Conservation
- Water Conservation

The need for waste minimization studies has become apparent to the Chemical Processing Industries (CPI) because Congress initially declared waste minimization to be the national policy of the United States that, wherever feasible, the generation of hazardous waste is to be reduced or eliminated as expeditiously as possible [35]. The Clean Air Act

*Industrial Environmental Chemistry*, Edited by D.T. Sawyer
and A.E. Martell, Plenum Press, New York, 1992

(CAA) and the Superfund Amendments Reauthorization Act, Title 313 which includes the Toxics Releases Inventory (TRI) list will clearly impact the drive to implement this concept which has been given a high priority by Congress.

The objective of this project is to develop examples of waste minimization by process modifications which can be achieved by modification of chemical process reaction and separation parameters. These examples can be used as guidelines for process and environmental engineers to follow in suggesting process modifications which will minimize waste generation.

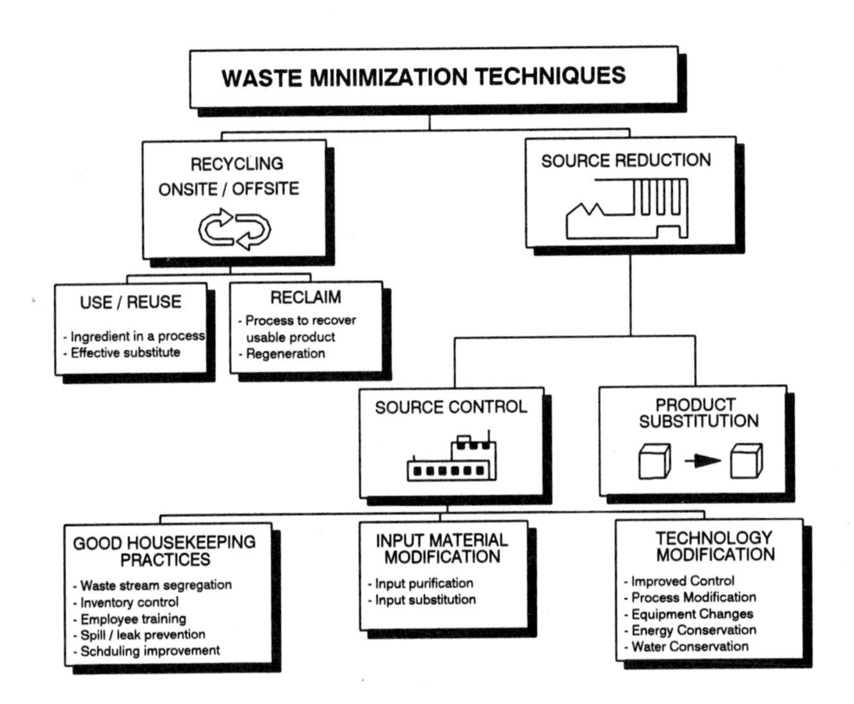


**Figure 1.** Waste Minimization Techniques Diagram

## PROCESS MODIFICATION CONCEPTS

Process modifications or changes mean: (1) the use of alternative low-waste process pathways to obtain the same product, and (2) modification of reaction or separation parameters.

Multiple chemical reactions can be classified into three general reaction types [10, 12, 20, 24, 30]: (1) series, (2) parallel and (3) combinations of series and parallel reactions. It is the many possible reactions which generate many of our waste products. Generally, we have desirable products and undesirable products. The undesirable products may become waste streams. The ratio of the desirable products to the undesirable products is generally defined as the selectivity of the reaction. This selectivity can be influenced by the choice of the reaction parameters or by modification of these reaction parameters.

Waste minimization by modification of reaction parameters is defined as changing the selectivity of the reaction so that undesirable reactions which produce waste products are minimized, while at the same time producing the desirable products. This selectivity may be affected by all of the reaction conditions. These reaction conditions include: temperature, pressure, catalyst, mixing conditions, ratio of feed concentrations and conditions, reaction type and reactor design options. Although catalyst changes are, without a doubt, the most effective method of changing a reaction selectively [26, 31], catalyst changes are very specific for each process and are in most cases highly proprietary. However, changes in the other reaction parameters have more general applications for many processes and can have an impact far greater than any of our previous expectations [29, 2].

This work summarizes the results of two studies of modification of the process parameters for pollution prevention.

## CASE STUDY FOR THE ACRYLONITRILE PROCESS

General studies of process modifications have been made in the past [7, 25, 34]and an earlier report was presented for the allyl chloride process on our preliminary results [21]. Recently two specific cases have been reported for the effect of modification of reaction parameters [29, 2] for industrial processes.

This case study is an example of how alteration of reactor design parameters can reduce waste generation for the arylonitrile process.

### Process Description

Since the commercialization of the Sohio process in the early 1960's, all acrylonitrile plants built in the world have been based on the reaction between propylene, ammonia, and air [18, 19, 22, 32, 36, 37, 38] . Acrylonitrile is the product from the gas phase catalytic air oxidation of propylene and ammonia. The reaction chemistry for the ammoxidation is given below:

$$\underset{\text{Propylene}}{CH_2{=}CH\text{-}CH_3} + \underset{\text{Ammonia}}{NH_3} + \underset{\text{Oxygen}}{3/2\ O_2} \rightarrow \underset{\text{Acrylonitrile}}{CH_2{=}CH\text{-}CN} + \underset{\text{Water}}{3H_2O} \quad (1)$$

with the combination of five others possible side reactions

$$CH_2{=}CH\text{-}CH_3 + O_2 \rightarrow \underset{\text{Acrolein}}{CH_2{=}CH\text{-}CHO} + H_2O \quad (2)$$

$$CH_2{=}CH\text{-}CH_3 + NH_3 + 9/4O_2 \rightarrow \underset{\text{Acetonitrile}}{CH_3\text{-}CN} + 1/2\ CO_2 + 1/2\ CO + 3H_2O \quad (3)$$

$$CH_2{=}CH\text{-}CHO + NH_3 + 1/2\ O_2 \rightarrow CH_2{=}CH\text{-}CN + 2H_2O \quad (4)$$

$$CH_2{=}CH\text{-}CN + 2O_2 \rightarrow CO_2 + CO + \underset{\text{Hydrocyanic acid}}{HCN} + H_2O \quad (5)$$

$$CH_3\text{-}CN + 3/2\ O_2 \rightarrow CO_2 + HCN + H_2O \quad (6)$$

The catalyst used for the reaction in the early process version was Bi-Mo-O [38]. The current commercial catalyst is of unknown composition with antimony-tellurium catalyst manufactured by Nitto Corporation being one possible catalyst. The flow diagram for this process is shown in Figure 2. Approximately stoichiometric quantities of propylene, ammonia, and oxygen (as air) are introduced to a fluidized-bed catalytic reactor at 15 psig and the contact time in the range of several seconds and a temperature of 350 - 600 °C. The use of polymerization grade propylene is unnecessary. Once through flow is used since conversion of propylene is practically complete. The reaction is exothermic, so heat-removal must be provided. The reactor effluent is neutralized to remove unconverted ammonia. It is next washed with water to yield an unabsorbed stream of inert gases and the solution of acetonitrile, acrylonitrile and HCN. This solution is stripped of the dissolved product which are fractionated to remove pure HCN and then sent to the main purification section. A main fractionator yields an overhead consisting of wet acrylonitrile which is dried and purified by azeotropic and conventional distillation. The bottoms wet acetonitrile is similarly concentrated, dried, and converted to high purity product [36, 38].

Acrolein, acetonitrile, and HCN are by products and potential wastes. This study has evaluated the reaction parameters shown in Table 1 to determine the conditions which will minimize the production of by-products while maintaining the production of acrylonitrile.

## Reaction System Analysis

The rate equation for the ammoxidation of propylene over Bi-Mo-O has a first order dependence with respect to propylene and a zero order dependence for both oxygen and ammonia when they are supplied in at least stoichiometric amounts [4, 5, 6, 9, 12, 23, 28].

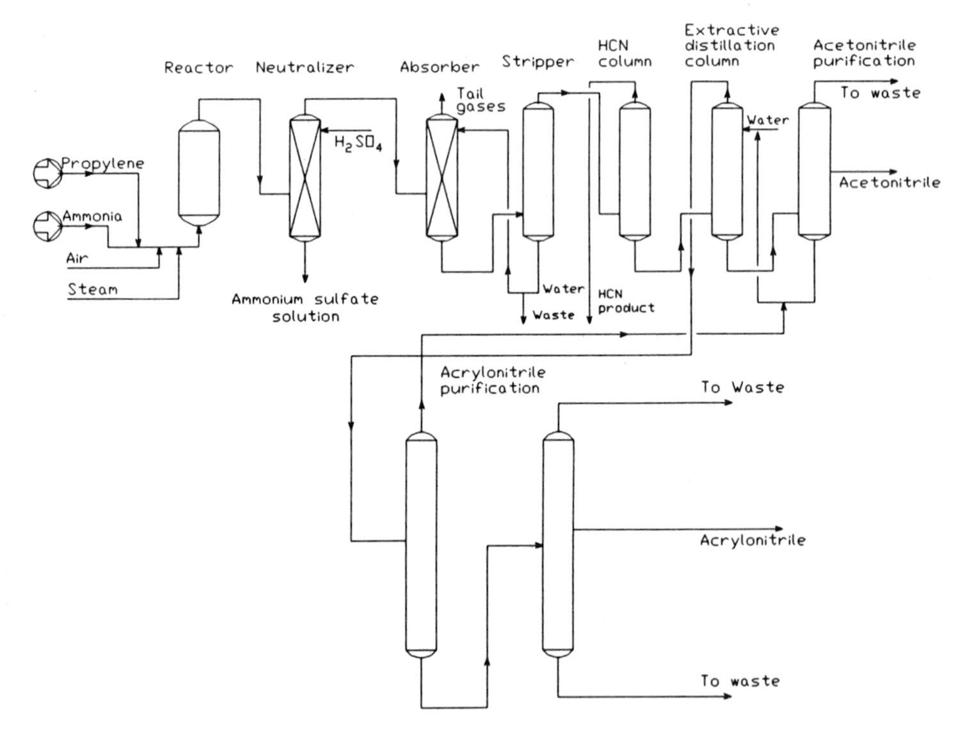


**Figure 2.** Acrylonitrile Process Flow Diagram

**Table 1.** The Reaction Parameters.

| | |
|---|---|
| Reactor Type | PFR / CSTR / FBR |
| Reactor Condition | Temperature<br>Residence Time |
| Feed Condition | Entering Feed Temperature |
| Kinetic Model | Six Reactions |

The rate equations are:

$$(-r_1) = k_1 C_{C_3H_6} \tag{7}$$

$$(-r_2) = k_2 C_{C_3H_6} \tag{8}$$

$$(-r_3) = k_3 C_{C_3H_6} \tag{9}$$

$$(-r_4) = k_4 C_{CH_2CHCHO} \tag{10}$$

$$(-r_5) = k_5 C_{CH_2CHCN} \tag{11}$$

$$(-r_6) = k_6 C_{CH_3CN} \tag{12}$$

where the rate constants are expressed in the Arrhenius form by:

$$k_{i,\,T_1} = k_{i,\,T_2}\, e^{-(\frac{E_i}{R})\,[(\frac{1}{T_1}) - \frac{1}{T_2})]} \tag{13}$$

The reaction system constants are shown in Table 2.

**Table 2.** The Reaction System Constants.

| | Reaction System Constants | | |
|---|---|---|---|
| Reaction No. | Activation Energy, $E_i$ (Cal/mol) | Rate constant, $k_i$ at 470 °C (sec$^{-1}$) | Heat of Reaction, $\Delta H_i$ at 298 K (J/mol) |
| 1 | 19,000 | 0.40556 | -515,610 |
| 2 | 19,000 | 0.00973 | -333,350 |
| 3 | 7,000 | 0.01744 | -864,990 |
| 4 | 7,000 | 6.81341 | -182,260 |
| 5 | 19,800 | 0.16222 | -800,900 |
| 6 | 7,000 | 0.07300 | -593,120 |

The heat of reaction at other temperatures can be obtained from:

$$\Delta H_{r_i}(T) = \Delta H_{r_i}(298°K) + \int_{298°K}^{T} (\Delta C_p)_i \, dT \tag{14}$$

using heat capacity values for each of the species $j$ as given by:

$$C_{Pj} = \alpha_j + \beta_j T + \gamma_j T^2 + z_j T^3 \tag{15}$$

The values for the constants are given in Table 3.

**Table 3.** The Heat Capacities Constants

| | Heat Capacities Constants | | | |
|---|---|---|---|---|
| | $\alpha_j$ | $\beta_j$ | $\gamma_j$ | $z_j$ |
| 1. Nitrogen, $N_2$ | 30.78 | -1.178E-02 | 2.390E-05 | -1.000E-08 |
| 2. Carbon Monoxide, CO | 30.87 | -1.285E-02 | 2.789E-05 | -1.272E-08 |
| 3. Oxygen, $O_2$ | 28.11 | -3.680E-06 | 1.746E-05 | -1.065E-08 |
| 4. Carbon Dioxide, $CO_2$ | 19.80 | 7.344E-02 | -5.602E-05 | 1.715E-08 |
| 5. Propylene, $C_3H_6$ | 3.710 | 2.345E-01 | -1.160E-04 | 2.205E-08 |
| 6. Ammonia, $NH_3$ | 27.31 | 2.383E-02 | 1.707E-05 | -1.185E-08 |
| 7. Hydrogen Cyanide, HCN | 21.86 | 6.062E-02 | -4.961E-05 | 1.815E-08 |
| 8. Acrolein, $C_3H_4O$ | 11.97 | 2.106E-01 | -1.071E-04 | 1.906E-08 |
| 9. Acrylonitrile, $C_3H_3N$ | 10.69 | 2.208E-01 | -1.565E-04 | 4.601E-08 |
| 10. Acetonitrile, $C_2H_3N$ | 20.48 | 1.196E-01 | -4.492E-05 | 3.203E-09 |
| 11. Water, $H_2O$ | 32.24 | 1.924E-03 | 1.055E-05 | -3.596E-09 |

$C_P \equiv$ Joule/g-mole K
$T \equiv$ K

## Stoichiometric Equation

The stoichiometric equation for each of the species $j$ for the $n$ independent reactions is:

$$F_j = F_{j0} + F_{A0} \sum_{i=1}^{n} \nu_{ij} X_i \tag{16}$$

where $\nu_{ij}$ is the stoichiometric coefficient for component $j$ in reaction $i$. $F_{j0}$, $F_j$, and $F_{A0}$ are the molar flowrates for component $j$ initially and at any point in time and the primary reactant $A$.

$X_i$ is the extent of conversion for reaction $i$ and is defined as:

$$X_i = \frac{N_{ji} - N_{j0}}{\nu_{ij} N_{A0}} = \frac{\Delta N_{ji}}{\nu_{ij} N_{A0}} \tag{17}$$

where $\Delta N_{ji}$ is moles of species $j$ consumed (formed) in reaction $i$.

## Component Mass Balance

The steady state component mass balance for the PFR and CSTR is:

$$F_{j0} - F_j + \int_0^V r_j \, dV = 0 \tag{18}$$

For $q$ simultaneous reactions, $n$ of which are independent, $r_j$ is defined as:

$$r_j = - \sum_{i=1}^{q} \nu_{ij} \, r_{ij} \tag{19}$$

where $r_{ij}$ is the rate of formation of species $j$ in reaction $i$. The bubbling bed model [8] has been used for the fluidized bed reactor. This model has a rising bubble surrounded by a cloud-wake and enclosed by the emulsion catalyst bed. Mass transfer and reaction proceeds in each of these phases according to the following component mass balance relations for reactant $A$.

$$-\frac{dC_A}{dt} = -u_b \frac{dC_A}{dl} = \gamma_b K_r C_{Ab} + (K_{bc})_b (C_{Ab} - C_{Ac}) \tag{20}$$

$$(K_{bc})_b (C_{Ab} - C_{Ac}) \cong \gamma_c K_r C_{Ac} + (K_{ce})_b (C_{Ac} - C_{Ae}) \tag{21}$$

$$(K_{ce})_b (C_{Ac} - C_{Ae}) \cong \gamma_e K_r C_{Ae} \tag{22}$$

**Total Energy Balance**

The total energy balance for $n$ independent multiple reactions involving $j$ species are: For the CSTR and for the FBR.

$$\dot{Q} - \dot{W}_s - F_{A0} \int_{T_0}^{T} \sum_{j=1}^{m} \theta_j \, C_{Pj} \, dT - F_{A0} \sum_{i=1}^{n} \Delta H_{ri}(T) \, X_i = 0 \tag{23}$$

where $\theta_j = F_{j0}/F_{A0}$ : ratio of the initial molar feed rate for each specie $j$ to the initial molar feed rate of primary reactant.

For the PFR.

$$F_{A0}\left( \sum_{j=1}^{m} \theta_j \, C_{Pj} + \sum_{i=1}^{n} X_i \, \Delta C_{Pi} \right) \frac{dT}{dV} - \sum_{i=1}^{n} r_i \, \Delta H_{ri}(T) = \dot{Q} - \dot{W}_s \tag{24}$$

where $\dot{Q} = UA(T_s - T)$ and $T_s$ = temperature of the surroundings.

These models have been used to evaluate waste minimization by reaction parameter modifications for this process.

## DISCUSSION AND RESULTS

The major by-products (potential wastes) considered for the acrylonitrile process are acetonitrile and HCN since the production of acrolein is very much less than either acetonitrile (ACE) and HCN. An analysis of the reactor type, temperature, and residence time has been done to establish the reaction conditions which will minimize the production of these by-products and maintain the production of acrylonitrile (ACN).

A plot of the mole fraction for each of the reactants and products versus residence time at 400 °C and 2 atm. for the fluidized bed reactor (FBR) is shown in Figure 3. In Figure 3 for the FBR the reactant oxygen decreases from 18% in the feed to 4.7% after a residence time of 20 seconds while the reactants ammonia and propylene decrease from 7% in the feed to 1.1% after a residence time of 20 seconds. Water is a significant product and reaches a value of 19 mole%, acrylonitrile reaches 3.6%, acetonitrile reaches 0.3%, HCN reaches 1.8%, CO2 reaches 2.1% and CO reaches 1.9% for this same residence time. A similar profile of reactants and products can be obtained for the PFR and CSTR. Similar plots can also be generated at various temperatures.

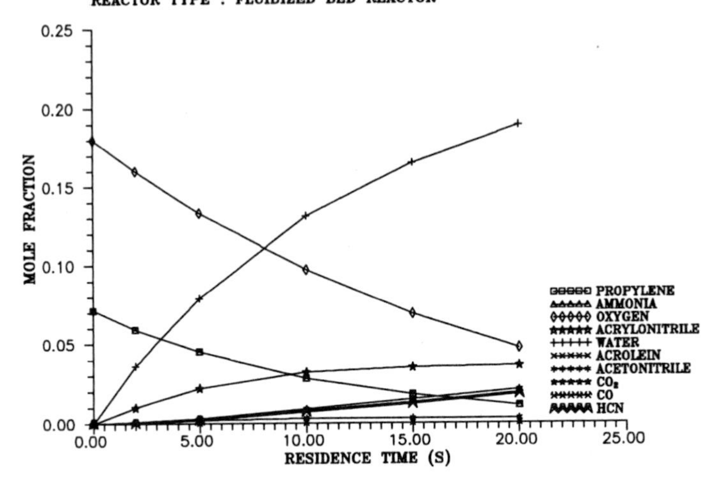


**Figure 3.** Plots of Mole Fraction versus Residence Time of Fluidized Bed Reactor.

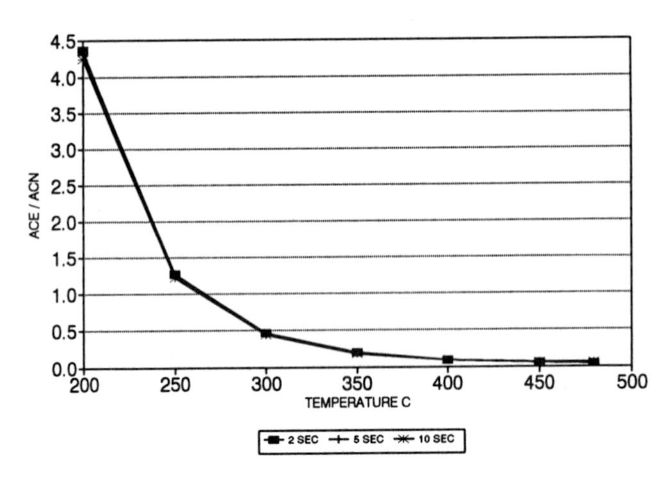


**Figure 4.** Plot of the Mole Ratio of ACE/ACN as a function of Temperature for Fluidized Bed Reactor

A comparison of waste generation with the variation in temperature and residence time can be obtained from the cross section of Figures 3 at various times and temperatures, by plotting the ratio of HCN/ACN, ACE/ACN, and the combination of both ([HCN + ACE]/ACN) with respect to the temperature and residence time for each of the reactor types.

A plot of the mole ratio of ACE/ACN as a function of temperature for the FBR is shown in Figure 4. This plot shows that the ratio of ACE/ACN decreases as the temperature and residence time increase. The change in residence time has no significant effect on this ratio. A similar profile of the plot shown in Figure 4 can be obtained for the PFR and the CSTR

A plot of the mole ratio of HCN/ACN as a function of temperature for the FBR is shown in Figure 5. The plots for each of the reactor types shows that the ratio HCN/ACN

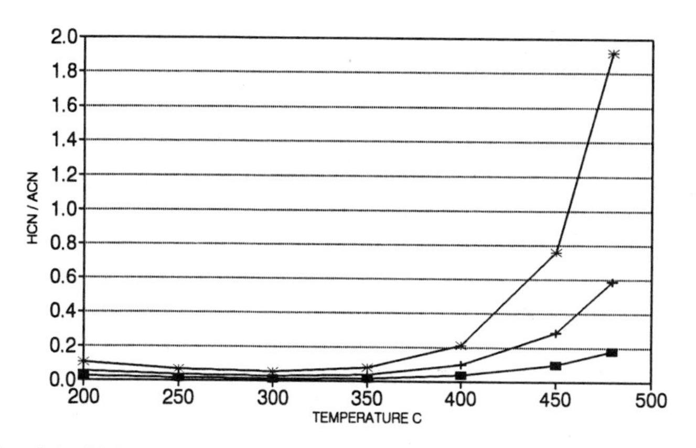


**Figure 5.** Plot of the Mole Ratio of HCN/ACN as a function of Temperature for Fluidized Bed Reactor.

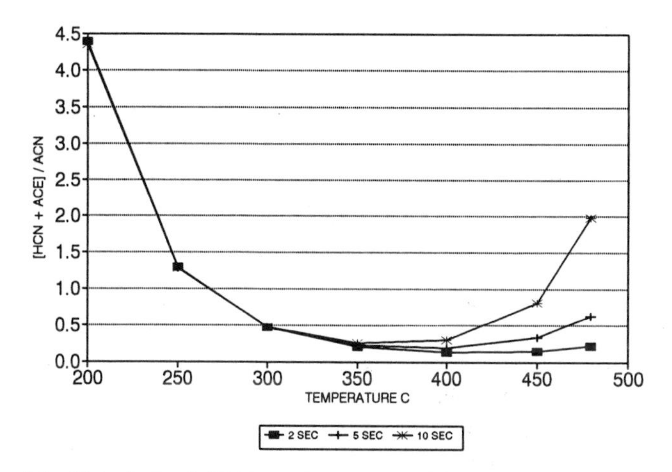


**Figure 6.** Plot of the Mole Ratio of (HCN+ACE)/ACN as a function of Temperature for Fluidized Bed Reactor.

increases as the temperature and residence time increase. The minimum ratio occurs at a temperature of about 300 °C, however the conversion of propylene at this temperature is only 12.6% at a residence time of 10 sec. This indicates that a higher temperature and/or residence time may be necessary.

A plot of total waste ([HCN+ACE]/ACN) versus temperature and residence time is shown in Figure 6 for the FBR. This plot shows that there is an optimum temperature for minimizing the waste and it is in the range of temperature of 400 - 480 °C for each type of reactor. For this temperature range, a high conversion can be obtained by selecting the correct residence time. A higher conversion can be obtained by choosing a longer residence time but the ratio of waste to acrylonitrile produced will also increase.

**Table 4.** Reactor Comparison for Waste Generation at 400 °C, 2 atm at Constant Residence Time

| Reactor type | Residence time (s) | % Conversion | Mole Ratio | |
|---|---|---|---|---|
| | | | HCN/ACN | ACE/ACN |
| PFR | 2.00 | 19.97 | 0.04356 | 0.09767 |
| | 10.00 | 66.88 | 0.26389 | 0.09565 |
| | 15.00 | 80.72 | 0.44471 | 0.09442 |
| CSTR | 2.00 | 19.29 | 0.08907 | 0.09728 |
| | 10.00 | 54.43 | 0.44404 | 0.09491 |
| | 15.00 | 64.18 | 0.66527 | 0.09400 |
| FBR | 2.00 | 17.14 | 0.03600 | 0.09647 |
| | 10.00 | 60.69 | 0.21180 | 0.09086 |
| | 15.00 | 75.11 | 0.34930 | 0.08744 |

**Table 5.** Reactor Comparison for Waste Generation at 400 °C, 2 atm at Constant Conversion

| Reactor type | Residence time (s) | % Conversion | Mole Ratio | |
|---|---|---|---|---|
| | | | HCN/ACN | ACE/ACN |
| PFR | 0.90 | 10 | 0.02050 | 0.09820 |
| | 4.25 | 40 | 0.10647 | 0.09695 |
| | 9.90 | 70 | 0.29294 | 0.09544 |
| CSTR | 0.95 | 10 | 0.04234 | 0.09789 |
| | 5.60 | 40 | 0.24901 | 0.09599 |
| | 19.60 | 70 | 0.86948 | 0.09336 |
| FBR | 1.15 | 10 | 0.02016 | 0.09723 |
| | 5.51 | 40 | 0.10590 | 0.09398 |
| | 12.99 | 70 | 0.29020 | 0.08880 |

An exact comparison of the waste generation for each type of reactor can be made by comparing the mole ratio of undesired product to the acrylonitrile (HCN/ACN, ACE/ACN) at the same residence time (as shown in Table 4.) and also at the same conversion of propylene (as shown in Table 5.). These two comparisons were made at an operating temperature of 400 °C and pressure of 2 atm.

A comparison of the mole ratio of HCN/ACN and the mole ratio of ACE/ACN for each of the reactors at constant residence times of 2, 10, and 15 seconds in Table 4 shows that the FBR has the lowest values. The greatest differences in these ratios is for HCN/ACN. The highest ratio of HCN/ACN occurs in the CSTR which also has the lowest conversion. The mole ratio of both HCN/ACN and ACE/ACN are higher with the PFR than the FBR but the conversion in the PFR is also higher.

Table 5. shows the ratio of HCN/ACN and ACE/ACN at three different conversion for three reactors. The residence time used for each reactor varies with the type of the reactor. The FBR again generates less waste than other two reactors.

## CONCLUSION

To minimize the by-products in the production of acrylonitrile by modification of the reaction parameters in the process, the following conditions would be recommended:

| | |
|---|---|
| Reactor Type : | Fluidized Bed Reactor |
| Temperature : | 400 - 480 °C |
| Residence Time : | 2 - 10 sec. |

The reactor effluent from the operation of the reactor at 450 °C and the residence time of 7 seconds is:

| Component | mole% |
|---|---|
| Propylene | 1.4 |
| Ammonia | 1.4 |
| Oxygen | 5.5 |
| Acrylonitrile | 3.6 |
| Water | 18.1 |
| Acrolein | 0.0 |
| Acetonitrile | 0.2 |
| Carbon dioxide | 1.8 |
| Carbon monoxide | 1.7 |
| Hydrogen cyanide | 1.6 |

and the conversion is 81.05%

## CASE STUDY FOR ALLYL CHLORIDE PROCESS

This example illustrates how selection of reactor and separation design parameters can reduce waste generation for the allyl chloride synthesis reaction. The production of allyl chloride is the first step in the epichlorohydrin manufacture. This example is a classical case of the combination reactions. Allyl chloride is produced from a non-catalytic reaction of propylene and chlorine according to the following reaction:

$$\underset{\text{Propylene}}{CH2{=}CH{-}CH3} + Cl2 \rightarrow \underset{\text{Allyl Chloride}}{CH2{=}CH{-}CH2Cl} + HCl \qquad (25)$$

The by-products (or wastes) of this reaction include 1,2-dichloropropane and isomers of 1,3-dichloropropene as shown in equations (26) and (27):

$$\underset{\text{Propylene}}{CH2{=}CH\text{-}CH3} + Cl2 \rightarrow \underset{\text{1,2-dichloropropane}}{CH2Cl\text{-}CH2Cl\text{-}CH3} \tag{26}$$

and

$$\underset{\text{Allyl Chloride}}{CH2{=}CH\text{-}CH2Cl} + Cl2 \rightarrow \underset{\text{1,3-dichloropropene}}{CHCL{=}CH\text{-}CH2Cl} + HCl \tag{27}$$

Chlorine is reacting with the hydrocarbons in parallel while the hydrocarbon is reacting in series. Thus selectivity to allyl chloride and minimization of 1,2-dichloropropane and 1,3-dichloropropene are desirable.

Formation of heavy and light ends in the allyl chloride synthesis is governed by these side reactions. However, the heavy and light ends were neglected because the kinetic data were not available. In principle, the desired substitution reaction to form allyl chloride is accompanied by a low temperature addition chlorination reaction, an unsaturated monochloride isomer formation , an unsaturated dichloride formation by further chlorination of allyl chloride, and by thermal degradation to tars, carbon and benzene.

In the commercial process, propylene and chlorine gas are continuously fed into the adiabatic non-catalytic reactor to produce allyl chloride. The reactor effluent is cooled and sent to a prefractionator to separate HCl and propylene. The remaining organic chloride fraction is separated in a two-step distillation. The dichloropropane is taken overhead in the first column and the heavy boiling fraction consisting of the dichloropropene and degradation products are taken off as bottom products.

This allyl chloride synthesis example is one which demonstrates that modification of the reaction parameters may have a major effect on the by-product production. An analysis of the process requires that component mass balance, kinetic rate equations and total energy balances be solved simultaneously. A complete analysis would also investigate the effects of mixing on the reaction system.

A study of this process is based on literature data [11, 17] and two other studies [1, 30]. The emphasis of this study in contrast to previous studies is directed toward the concept of waste minimization. The results are based on evaluating the selectivity to allyl chloride, the conversion of chlorine and propylene, and the minimization of 1,2-dichloropropane, 1,3-dichloropropene, and HCl. The evaluated reaction parameters are given in Table 6.

**Table 6.** The Reaction Parameters

| | |
|---|---|
| Reactor Types | PFR/CSTR |
| Reactor Conditions | Isothermal/Adiabatic/Non-Adiabatic<br>Residence Time |
| Feed Conditions | Propylene/Chlorine Ratio, $\theta$<br>Entering Feed Temperature |
| Kinetic Model | Three Reactions |

The reaction rate equation for the reactions of equation (25) and (26) were given by Smith [30] as developed from literature data [11]. Biegler and Hughers [1] did further analysis of this data and developed a rate equation for the reaction of equation (27).

The rate equations are:

$$(-r_1) = k_1 P_{C_3H_6} P_{Cl_2} \tag{28}$$

$$(-r_2) = k_2 P_{C_3H_6} P_{Cl_2} \tag{29}$$

$$(-r_3) = k_3 P_{C_3H_6} P_{Cl_2} \tag{30}$$

where the rate constants are expressed in the Arrhenius form by:

$$k_i = A_i \, e^{\frac{-E_{a\,i}}{RT}} \tag{31}$$

**Table 7.** The Reaction System Constants

| Equation | Frequency Factor, $A_i$ (lb-mol/hr-ft$^3$-atm$^2$) | Activation Energy, $E_i$ (Btu/lb-mol) | Heat of Reaction, $\Delta H_{ri}$ (298 K) (Btu/lb-mol) |
|---|---|---|---|
| 25 | 206,000 | 27,200 | -48,000 |
| 26 | 11.7 | 6,860 | -79,200 |
| 27 | 4.6E+08 | 10,650 | -91,800 |

The heat of reaction associated with each reaction $i$ is given by equation (14) and the specific heat for each specie $j$ is given by equation (15).

Values for the constants for the Arrhenius Equation and the heat of reaction are given in Table 7 [1] and Table 8 [27, 39].

The stoichiometric equation, component mass balance, and total energy balance are given by equation (16), (18), (23) and (24), respectively.

**Table 8.** Constants for Heat Capacity Equations

| Component | $\alpha_j$ | $\beta_j$ | $\gamma_j$ | $z_j$ |
|---|---|---|---|---|
| Propylene | 0.8861 | 5.601E-02 | -2.771E-05 | 5.267E-09 |
| Chlorine | 6.432 | 7.278E-03 | -9.241E-06 | 3.695E-09 |
| Allyl Chloride | 0.604 | 7.278E-02 | -5.441E-05 | 1.742E-08 |
| 1,3-dichloropropene | 2.809 | 7.700E-02 | -6.306E-05 | 2.147E-08 |
| 1,2-dichloropropane | 2.496 | 8.730E-02 | -6.220E-05 | 1.849E-08 |
| Hydrogen Chloride | 7.325 | -1.720E-03 | 2.976E-06 | -9.310E-10 |

$C_p$ ≡ Btu/(lb-mol °F)
$T$ ≡ Fahrenheit

## DISCUSSION AND RESULTS

The variation of 1,2 DCP by-product produced in the process is shown in Figure 7 as a function of the propylene/chlorine mole ratio. The figure applies for residence times of 1.0, 1.5 and 2.0 seconds. These curves show that the production of 1,2 DCP merge together at propylene/chlorine mole ratio less than 6. At propylene/chlorine mole ratios above 6, there is slightly more 1,2 DCP at the residence time of 1.0 second than there is at 1.5 seconds. The amount of 1,2 DCP at 1.5 and 2.0 seconds residence time are almost identical. This observation suggests that a residence time of about 1.5 seconds is desirable. Higher residence times corresponds to higher cost (larger reactor) and do not provide significantly less by-product. Lower residence times produce more by-product.

**Table 9.** Boiling Point Temperature

| Component | Boiling Point, K |
|---|---|
| 1. Hydrogen Chloride, HCl | 188.1 |
| 2. Propylene, $C_3H_6$ | 225.7 |
| 3. Chlorine, $Cl_2$ | 239.1 |
| 4. Allyl Chloride, $C_3H_5Cl$ (3-Chloro-1-Propene) | 318.4 |
| 5. 1,2 Dichloropropane, $C_3H_6Cl_2$ (1,2 DCP) | 370.0 |
| 6. cis 1,3 Dichloropropene, $C_3H_4Cl_2$ (1,3 DCP) | 377.0 |
| 7. trans 1,3 Dichloropropene, $C_3H_4Cl_2$ (1,3 DCP) | 385.2 |

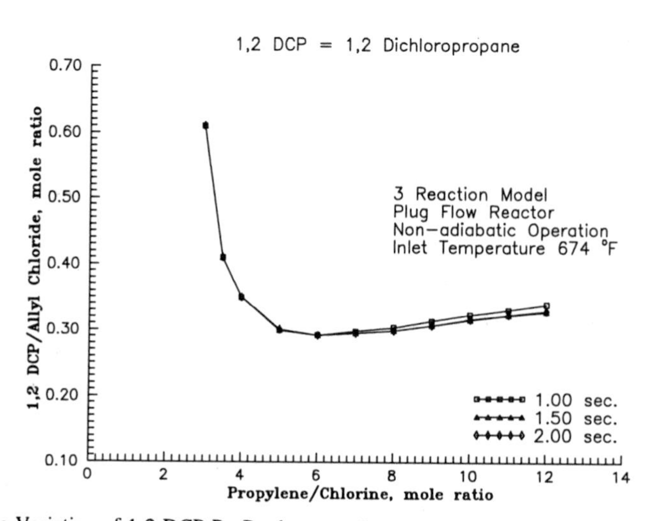


**Figure 7.** The Variation of 1,2 DCP By-Product as a function of Propylene/Chlorine Mole Ratio

At the propylene/chlorine mole ratio of 3.0, the production of 1,2 DCP is about 0.62 at the residence time of 1.5 seconds. As the propylene/chlorine mole ratio increases from 0.20 to 6, the production of 1,2 DCP decreases from about 0.62 to 0.30 as shown in the graph. This decrease is desirable since this corresponds to less production of by-product per quantity of allyl chloride produced. As the feed mole ratio of propylene to chlorine increases from 6 to 7, the by-product increases from 0.30 to 0.305 as shown. The quantity of by-product is larger and continues to increase moderately as the propylene/chlorine mole ratios increases.

The minimum production of 1,2 DCP appears to occur at a propylene/chlorine mole ratio of about 6 and at a residence time of about 1.5 seconds.

Thus the minimum production of 1,2 DCP appears to occur at the following reactor parameter conditions:

| | | |
|---|---|---|
| Reactor Type | : | Plug Flow Reactor |
| Reactor Conditions | : | Non-adiabatic Operation |
| | | Outlet Temperature, °F = 876 |
| | | Residence time, seconds = 1.5 |
| Feed Condition | : | Propylene/Chlorine, mole ratio = 6 |
| | | Inlet Temperature, °F = 674 |

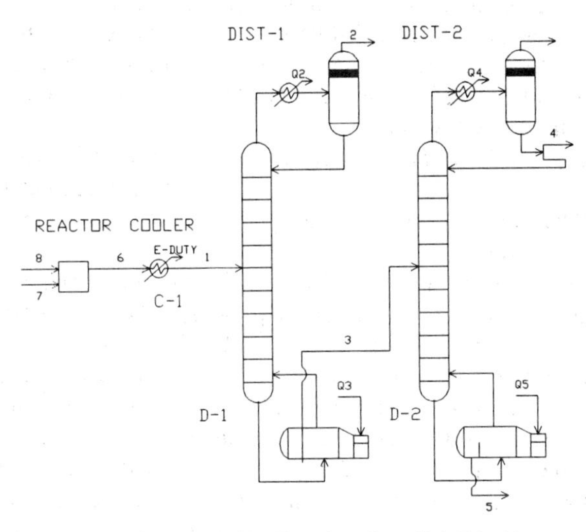


**Figure 8.** Distillation Flowchart for Allyl Chloride.

The stream exiting from the non-adiabatic plug flow reactor is a mixture containing hydrogen chloride, trace propylene, trace chlorine, allyl chloride, 1,2 dichloropropane (1,2 DCP) and 1,3 dichloropropene (1,3 DCP). Boiling Point temperatures for these compounds are provided in Table 8. Separation of allyl chloride form the mixture is required.

Inspection of the boiling points in Table 9 show that the boiling point temperature of 1,3 DCP (both cis- and trans-1,3 dichloropropane) is higher than the boiling point temperature of 1,2 DCP (377 and 385.2 K vs 370 K). Property data for 1,3 DCP was very sparse, therefore 1,2 DCP was used to represent 1,3 DCP in the distillation calculations. This representation corresponds to a conservative view since 1,3 DCP is less volatile.

A distillation flowchart to accomplish the separation and provide a purified allyl chloride product is shown in Figure 8.

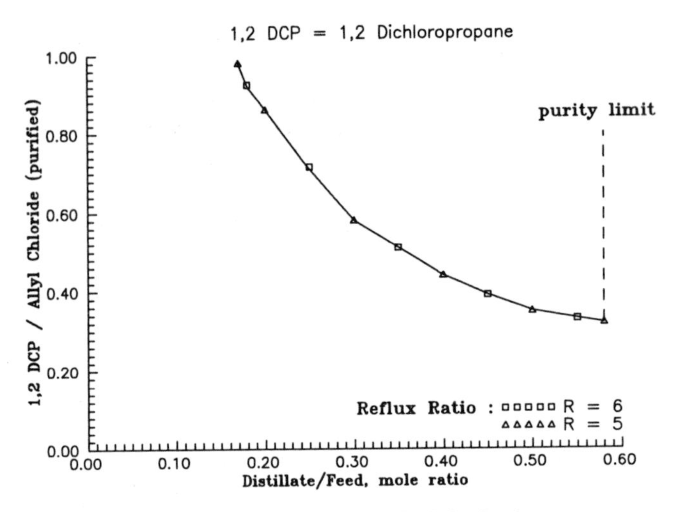


**Figure 9.** Variation of 1,2 DCP By-Product.

In the flowchart of Figure 8, the mixture issuing from the non-adiabatic plug flow, 1, reactor is cooled and sent to distillation for separation of the allyl chloride. In the first distillation column, D-1, the light boiling fraction (hydrogen chloride, trace propylene, and trace chlorine) are taken as the overhead distillate, 2. The bottoms from the first distillation, 3, are sent as the feed to the second distillation, D-2. In the second distillation, D-2, the purified allyl chloride is taken as the overhead distillate, 4. The bottoms from the second distillation comprise the heavy fraction, 5.

The quantity of 1,2 DCP by-product produced per quantity of purified allyl chloride product is influenced by the distillate/feed mole ratio for the second distillation. The variation of 1,2 DCP by-product produced in the process is shown in Figure 9 as a function of the distillate/feed mole ratio.

In preparing the figure, distillation calculations were performed using the HYSIM process simulation software and the PRSV property package (Peng-Robinson Stryjek-Vera) was used for the vapor-liquid calculations. The PRSV Equation of State is a two fold modification of the PR (Peng-Robinson) equation of state that extends the application of the original PR method to highly non-ideal systems. For this application, binary interaction coefficients were assigned a value of zero in the absence of experimental data.

In the Figure 9, the variation of 1,2 DCP by-product is shown at reflux ratios of R = 5 and 6. Calculation at lower reflux ratios did not provide a purified allyl chloride product. Inspection of the curve indicates that results at R = 5 are equivalent to results at R = 6. The same finding of equivalent results was ascertained at even higher reflux ratios (R = 8,10,12,14,...). Since such higher reflux ratios correspond to higher energy costs (more boil-up in reboiler and more condensation in condenser), the lower value for reflux ratio of R = 5 is preferred.

The effect of distillate/feed mole ratio on minimizing the 1,2 DCP by product is illustrated in the Figure 9. At the low value of distillate/feed mole ratio of 0.18, the amount of 1,2 DCP by-product per purified allyl chloride is approximately 0.98. As the distillate/feed mole ratio increases from 0.18 to 0.58, the 1,2 DCP by-product decreases from about 0.98 to about 0.30. This decrease is desirable since this corresponds to less production of by-product per quantity of purified allyl chloride product.

The minimum production of 1,2 DCP by-product appears to occur at a distillate/feed mole ratio of about 0.58 and reflux ratio of 5 or greater.

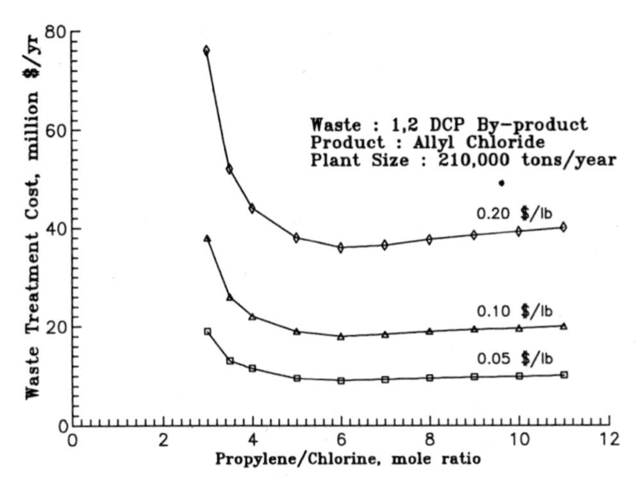


**Figure 10.** Cost of Waste Treatment

Thus the minimum production of 1,2 DCP by-product appears to occur at the following separation parameter conditions:

Distillation column = D-2
Pressure, psia = 20
Distillate/Feed, mole ratio = 0.58
Reflux ratio, R = 5

If the by-product in a process is not salable in the market or valuable in another process, there is a cost for treatment of the by-product as a waste. The cost of waste treatment is shown in Figure 10. This figure was developed for a plant size of 210,000 tons/year of allyl chloride (equivalent to large plant producing epichlorohydrin with 210,000 tons/year capacity).

For Figure 10, the cost of waste treatment for the 1,2 DCP by-product is shown as a function of propylene/chlorine mole ratio at three levels of specific treatment costs of 0.20, 0.10 and 0.05 $/lb. At a low value for the propylene/chlorine mole ratio of about 3, the waste treatment costs are about 76, 38, and 19 million dollars for the respective specific treatment costs. As the propylene/chlorine mole ratio is increased from the low value of 3 to about 6, the waste treatment cost are about 36, 18 and 9 million dollars for the respective curves. This decrease in waste treatment costs is desirable. As the propylene/chlorine mole ratio increases to higher values (6 to 7, 8, 9, 10 and 11), the waste treatment costs associated with handling the by-product increases to slightly higher values.

Thus the minimum cost associated with handling the 1,2 DCP by-product occurs at a propylene/chlorine mole ratio of about 6.

## ACKNOWLEDGMENT


This project has been funded in part with Federal Funds as part of the program of the Gulf Coast Hazardous Substance Research Center which is supported under cooperative agreement R815197 with the United States Environmental Protection Agency. The contents do not necessarily reflect the views and policies of the U.S.EPA nor does the mention of trade names or commercial products constitute endorsement or recommendation for use.

## REFERENCES


1. Biegler, L.T. and R.R. Hughes, *Comput. Chem. Eng.*, 7(5), 645-661 (1983).
2. Bundick, H.L., T.B. Kopecky, B.R. Allen, and J.R. Hopper, "Continuous Lube Oil Additive Process : Waste Reduction By Process Modification and Control," AIChE 1992 Spring National Meeting, New Orleans, LA, March 29 - April 2 (1992).
3. Callahan, H.L., B. Gertisser, and J.J. Szaba, "Process for the Manufacture of Unsaturated Nitriles from Olefins and Ammonia," U.S. Patent 3,354,197 (1967).
4. Callahan, J.L., R.K. Grasselli, E.C. Milberger, and H.A. Strecker, "Oxidation and Ammoxidation of Propylene over Bismuth Molybdate Catalyst.," *Ind. Eng. Chem. Prod. Res. Develop.*, 9(2), p. 134-142 (1970).
5. Callahan, J.L., R.K. Grasselli, E.C. Milberger, and H.A. Strecker, "Oxidation and Ammoxidation of Propylene over Bismuth Molybdate Catalyst," Division of Peteroleum Chemistry Inc. ACS., p. C13-C27, September (1969).
6. Cathala, M. and J.E. Germin, "Ammoxidation Catalytique des hydrocarbures et reactions apparentees.," *Bulletin de la Societe Chimique de France.*, No.11, p. 4114-4119 (1970).
7. CMA Conference Proceedings November, 1987, "Waste Minimization," *Chem. Eng.* (NY), 95 (No. 11) 37, August 15 (1988).
8. Daizo, K. and O. Levenspiel, Fluidization Engineering, John Wiley & Sons, New York, NY (1969).
9. Dalin et al., *Azerb Khim. Zhur.*, No.5, p. 99-102 (1963).
10. Denbigh, K.G., and J.C.R. Turner, Chemical Reactor Theory, 2nd ed., Cambridge University Press (1971).
11. Fairbairn, A.W., H.A. Cheney, and A.J. Cherniavsky, Chem. Eng. Prog., 43, 280 1947.
12. Farkas, A., "What You Should Know ... Catalytic Hydrocarbon Oxidation.," *Hydrocarbon Processing*, p. 121, July (1970).
13. Fogler, H.S., Elememts of Chemical Reaction Engineering, Prentice-Hall, Engelwood Cliffs, NJ (1986).
14. Freeman, H.M., "Hazardous Waste Minimization," JAPCA 38, No. 1, 59, January (1988).
15. Freeman, H.M., Hazardous Waste Minimization, McGraw-Hill, New York, NY (1990).
16. Fromm, C.H., M.S. Callahan, H.M. Freeman, and M. Drabkin, "Succeeding at Waste Minimization," *Chem. Eng.* (NY), 91-94, September 14 (1987).
17. Groll, H.P.A. and G. Hearne, *Ind. Eng. Chem.*, 31, 1530 (1939).
18. Hadley D.J., in H.G. Hancock, ed., Propylene and its Derivatives, Ernest Benn Ltd., London, p. 416-497 (1973).
19. Hanh, A.V., The Petrochemical Industry., McGraw-Hill, New York, NY (1970).
20. Holland, C.D. and R.G. Anthony, Fundamentals of Chemical Reaction Engineering, Prentice-Hall, Englewood Cliffs, NJ (1979).
21. Hopper, J.R. and A. Muninnimit, "Waste Minimization by Process Modification.", Preceedings of the National AIChE Conference, Washington D.C., December (1989).
22. Kirk-Othmer., Encyclopedia of Chemical Technology, 3rd ed., A Wiley-Inter-science Publication, John Wiley & Sons, New York, NY (1978).
23. Lankhuyzen, S.P., P.M. Florack, and H.S. van der Baan, "The Catalytic Ammoxidation of Propylene over Bismuth Molybdate Catalyst.," *Journal of Catalyst*, 42, p. 20-28 (1976).

24. Levenspiel, O., Chemical Reaction Engineering, John Wiley & Sons, New York, NY (1962).
25. Overcash, M., Technique for Industrial Pollution Prevention, Lewis Publishing Co., Chelsea, MI (1986).
26. Perkinson, G. and E. Johnson, "Designer Catalysts Make Their Debut," *Chem. Eng.* (N.Y.), 96 (No. 9), 30, September (1989).
27. Reid, R.C., J.M. Pransnitz, and B.E. Poling, The Properties of Gases & Liquids., 4th ed., McGraw-Hill, New York, NY (1987).
28. Shelstad, K.A. and T.C. Chong, "Kinetics of Ammoxidation of Propylene over Bistmuth-Molybdate Catalyst.," *Can. J. Chem. Eng.*, 47, p. 597-602 (1969).
29. Shirtum, R.P., "Kinetic Control of By-Product Reaction with In Situ Distillation In A Continuous Flow Device," AIChE 2nd Topical Pollution Prevention Conference, Pittsburg, PA, August 20 - 21 (1991).
30. Smith, J.M., Chemical Engineering Kinetics, 3rd ed., McGraw-Hill, New York, NY (1981).
31. Smyth, L.C., and J.R. Hopper, "Waste Control by Reaction Design and Catalyst Choice.," Pollution Prevention for the 1990's, AIChE Conference, Washington D.C., December 4-5 (1989).
32. Stobaugh, R.B., S.G. McH. Clark, and G.D. Camirand, "Acrylonitrile : How, Where, Who-Future.," *Hydrocarbon Processing.*, p.109-120, January (1971).
33. Stull, D.R., E.F. Westrum, Jr., and G.C. Sinke, The Chemical Thermodynamics of Organic Compounds., John Wiley & Sons, Inc., New York, NY (1969).
34. Tavalerides, L.L., Process Modification for Industrial Pollution, Industrial Waste Management Service, J.W. Patterson, Editor, Lewis Publishing Co., Chelsea, MI (1985).
35. USEPA Report to Congress : Minimization of Hazardous Waste, EPA/530 SW-86-033A Office of Solid Waste, U.S. EPA, Washington, DC 20460, October (1986).
36. Vectch, F., J.L. Callahan, J.D. Idol, and E.C. Milberger, "New Route to Acrylonitrile.," Chem. Eng. Prog., 56(10), p. 65-67 (1960).
37. Vectch, F., J.L. Callahan, J.D. Idol, and E.C. Milberger, "New Data on Sohio's Acrylo Process.," *Hydrocarbon Processing & Petroleum Refinery.*, 41(11), p. 187-190 (1962).
38. Waste Minimization: Issues and Options Vol I, II USEPA Office of Solid Wastes and Emerging Response, EPA/530 SW-86-041 October (1986).
39. Yaws, C.L., P.Y. Chiang, "Enthalpy of Formation for 700 Major Organic Compounds.," *Chem. Eng.* (NY), p. 81-88, September 26 (1988).

## KINETIC MODELS TO PREDICT AND CONTROL MINOR CONSTITUENTS IN PROCESS REACTIONS

Selim Senkam

Department of Chemical Engineering, University of California

Los Angeles, California 90024 U.S.A.

### Introduction and Background

Detailed Chemical Kinetic Mechanisms (DCKMs) are comprehensive descriptions of chemical processes expressed in term of irreducible chemical events or elementary reactions. That is, they provide descriptions of chemical processes at the molecular level, and as a consequence they are of considerable utility to predict and control the formation of both the major and minor products. In principle, detailed mechanisms comprise large, putatively complete sets of elementary reactions for which independent rate coefficient parameters, frequently expressed in the following form:

$$k = AT^n \exp(-E/RT)$$

are available from direct measurements or estimable from theoretical considerations, together with the thermochemistry of the associated species participating in the reaction. Estimations are based on the judicious application of theories of rate processes in a conventional manner, as well as in conjunction with modern quantum chemical calculations, and by using analogies between similar types of reactions. At present we relý on all of these complementary methods. Although DCKMs have primarily been used to describe gas phase reaction processes, its application to condensed systems, e.g. catalysis and chemical vapor deposition, is also possible. Nevertheless, we shall describe the DCKM approach to gas phase processes in this paper for convenience.

Our ability to develop detailed chemical kinetic models has improved considerably because of simultaneous developments in a number of interrelated areas. First, with the availability of fast computers together with efficient numerical algorithms to handle stiff differential equations, an increasing complexity of problems can now be coded and solved rapidly without making any fundamental assumptions. For example, it is no longer necessary to the invoke pseudo-steady-state assumption to treat reactive radicals or intermediates in modeling the kinetics of complex chemical processes in an effort to obtain analytical expressions. In fact, conclusions based on the assumption of pseudo-steady-state conditions or through the neglect of reactions presumed to be unimportant in a process should be viewed questionable in the absence of a detailed modeling work that proves the validity of these assumptions.

Second, the quality and quantity of data on the thermochemistry of species and on the kinetics and mechanisms of individual elementary reactions, especially those in the gas-phase, have improved significantly over the past decade due to advances made in experimental methods. In particular, the thermochemistry of a large number of species are now available in convenient tabulations (see for example Pedley et al. 1985, Chase et al. 1985). Similarly, the rates and mechanisms of a large number of elementary reactions have also been evaluated and

*Industrial Environmental Chemistry*, Edited by D.T. Sawyer
and A.E. Martell, Plenum Press, New York, 1992

documented (see for example Demore et a. 1990, Berces and Marta 1988, Tsang and Hampson 1986, Kerr and Drew 1987, Basevich 1987, Westbrook and Dryer 1984, Hanson and Salimian 1984, Warnatz 1984, Kondratiev 1972, Kerr and Moss 1981, Baulch et al. 1981, Westley 1976, 1980, Benson and O'Neal 1970). In addition, mature empirical methods are now in place to estimate thermochemistry and reaction rate parameters from theoretical considerations (see for example Gaffney and Bull 1988, Benson 1976, Senkan 1992). Although considerable progress also has been made in heterogeneous reactions, the available data are more limited, and the theoretical framework to estimate thermochemical and rate parameters is less well developed.

Third, with recent advances made in theoretical and computational quantum mechanics, it is possible to estimate thermochemical information via electronic structure calculations (Dunning et al. 1988, Senkan 1992). The development of this capability obviously is crucial for the development of DCKMs to predict the formation of trace by-products in view of obvious experimental and cost considerations. Electronic structure calculations, together with the Transition State Theory (TST) also allows the determination of the rate parameters of elementary reactions from first principles. Our ability to estimate activation energy barriers is particularly significant because, until the advent of computational quantum mechanics no fundamentally-based methods were available for the determination of this important variable.

In order to illustrate the utility of quantum mechanical calculations consider the reaction of $C_2H_5Cl$ with H atom, which can lead to 3 distinct products: $CH_3CH_2+HCl$, $CH_3CHCl+H_2$, and $CH_2ClCH_2+H_2$. At present the available data on the kinetics of each of these pathways are quite uncertain. Consequently, we undertook semi-empirical quantum chemical calculations to assess the relative importance of each channel. In Figure 1 the potential energy diagrams and geometries of the reactants, products and the associated transition states in each reaction are shown. As evident from this figure, we expect the ordering of the rates of reactions to be: $CH_3CH_2+HCl > CH_3CHCl+H_2 > CH_2ClCH_2+H_2$. Although the results of such calculations may not be very accurate in an absolute sense, the trends that arise are crucial for the development of sound DCKMs.

Finally, although DCKMs comprise of a large number of reactions, to preserve the elementary nature of these mechanisms, only a subset of these reactions are important at any given time and application. That is, different reactions often dominate the process behavior under different conditions. Consequently, parts of DCKMs can be developed and tested independently for the evolution of these models. This is an important consideration in DCKMs as we can start exploiting the advantages of this type of modeling today.

Since detailed chemical kinetic mechanisms involve the participation of a large number of species in a large number of elementary reactions, sensitivity and reaction path analyses are also essential elements of DCKM. Sensitivity analysis provides a means to assess the limits of confidence we must put on our model predictions in view of uncertainties that exist in reaction rate parameters, thermochemical and thermophysical data utilized, as well as the initial and boundary conditions used in the modeling work. Through reaction path analysis, major reaction pathways responsible for the production and consumption of each species can be identified. Powerful formalisms have been developed to undertake sensitivity analysis in detailed chemical kinetic modeling (see for example Tilden et al. 1981, Siegneur et al. 1982, Rabitz et al. 1983, Hwang 1983). In addition, efficient and transportable computer codes to implement these techniques have also been developed (Lutz et al. 1988, Caracotsios and Steward 1985).

In Figure 2 various elements involved with the development of detailed chemical kinetic mechanisms are illustrated and we shall illustrate the application of DCKM in the following example.

## An Example

We have recently studied the high temperature oxidation and pyrolysis of $C_2HCl_3$, a prototypical chlorinated hydrocarbon, in order to better understand the identities and concentrations of its reaction products and by-products (Chang and Senkan 1988, Chang and

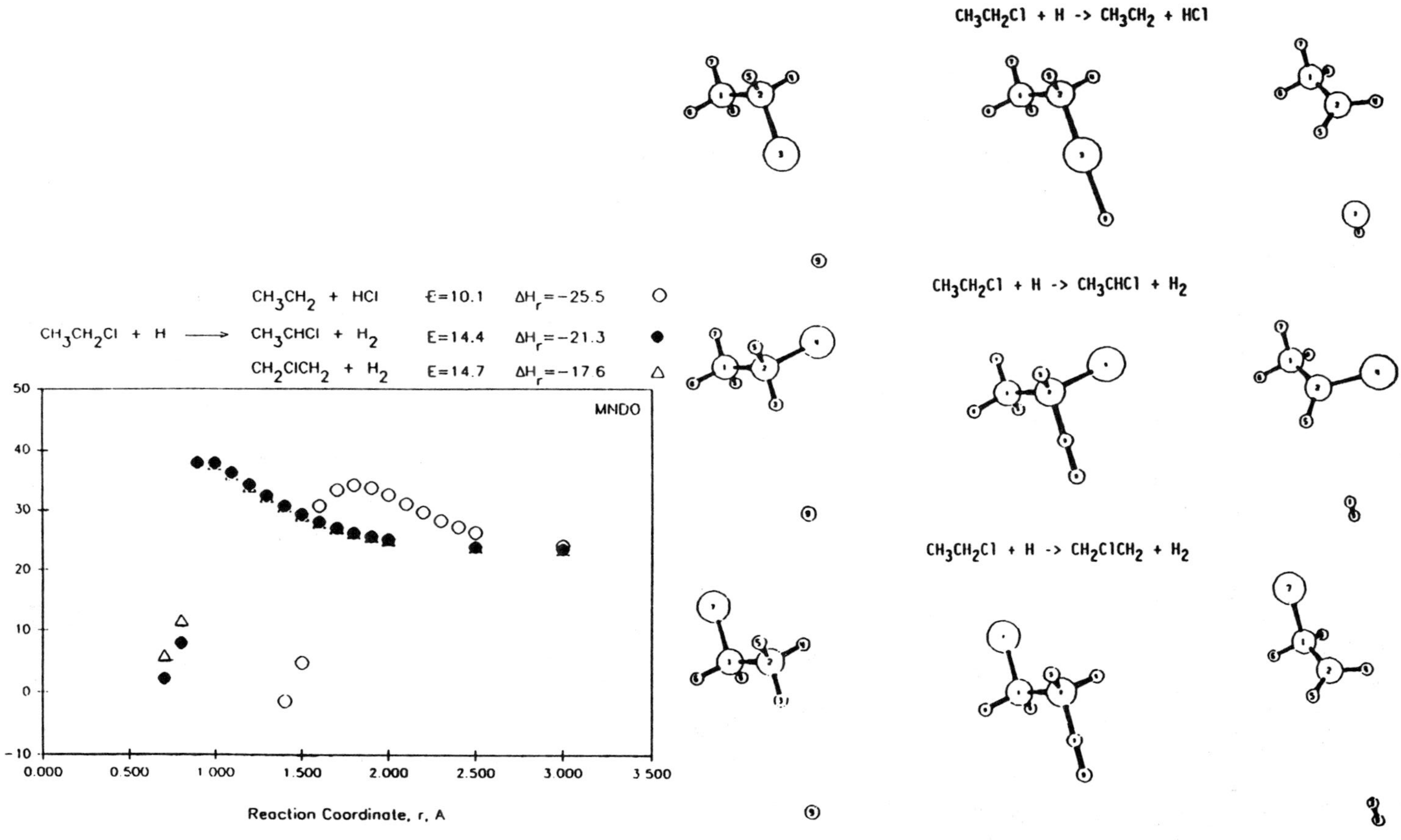


Figure 1. Reaction Pathways and Energy Diagrams for the $C_2H_5Cl + H$

Senkan 1989). The development of better insights for the high temperature reactions of chlorinated hydrocarbons (CHC) is important because of the growing need to better understand and control pollutant emissions into the environment from combustion systems burning hazardous materials that frequently contain CHCs For example, CHCs are associated with the formation of chlorinated aromatics, di-benzo-dioxins, and di-benzo-furans in incinerators, some of which are extremely toxic (Junk and Ford 1980, Schaub and Tsang 1983, Oberg et al. 1985). Second, the manufacture of useful chemicals by the controlled high temperature reactions of CHCs also is a promising enterprise. This was shown recently by the development of the chlorine-catalyzed oxidative-pyrolysis (CCOP) process to convert methane, the major component in natural gas and an important product in the anaerobic digestion of organic matter, into more valuable products such as ethylene, acetylene and vinyl chloride (Granada et al. 1987).

Reaction Mechanism

An elementary reaction set describing the high-temperature fuel-rich combustion of $C_2HCl_3$ is presented in Table I together with the rate parameters for the forward reaction paths (Chang and Senkan 1989). This mechanism was constructed by systematically considering all

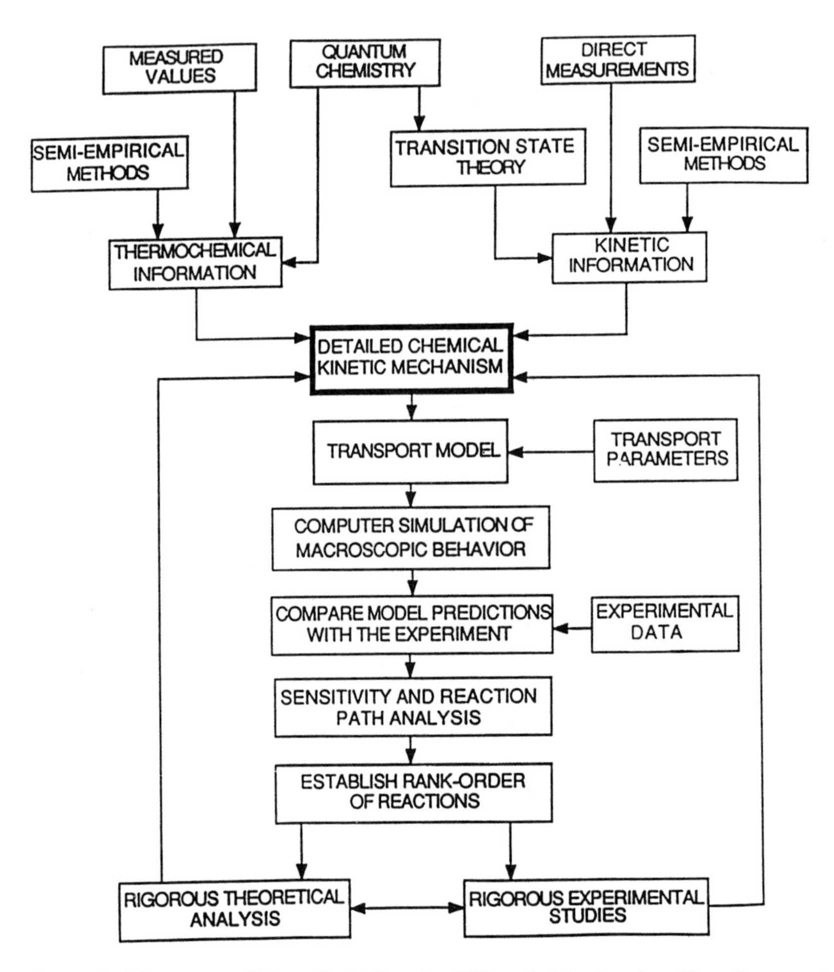


Figure 2. Elements of Detailed Chemical Kinetic Mechanism Development

Table 1. Detailed Chemical Kinetic Mechanism for $C_2HCl_3$ Combustion ($k = AT^n \exp(-E/RT)$, in $cm^3$, kcal, mole, s units)

| Reaction | A | n | E |
|---|---|---|---|
| **$C_2$ Kinetics** | | | |
| 1. $C_2HCl_3 \rightleftharpoons C_2HCl_2 + Cl$ | 3.02E33 | -5.88 | 88300. |
| 2. $C_2HCl_3 + Cl \rightleftharpoons C_2HCl_4$ | 3.66E32 | -6.52 | 5520. |
| * 3. $C_2HCl_3 \rightleftharpoons C_2Cl_2 + HCl$ | 1.07E26 | -3.38 | 61560. |
| 4. $C_2HCl_3 \rightleftharpoons C_2HCl + Cl_2$ | 9.55E35 | -7.06 | 83500. |
| 5. $C_2HCl_3 + Cl \rightleftharpoons C_2HCl_2 + Cl_2$ | 3.16E13 | 0. | 20000. |
| ** 6. $C_2HCl_3 + Cl \rightleftharpoons C_2Cl_3 + HCl$ | 2.00E12 | 0. | 5000. |
| 7. $C_2HCl_3 + O \rightleftharpoons CHOCl + CCl_2$ | 1.00E13 | 0. | 2000. |
| 8. $C_2HCl_3 + OH \rightleftharpoons CHCl_2COCl + H$ | 3.19E11 | 0. | -880. |
| * 9. $C_2HCl_3 + ClO \rightleftharpoons CHCl_2COCl + Cl$ | 1.00E11 | 0. | 2000. |
| 10. $C_2HCl_3 + CCl_3 \rightleftharpoons CCl_4 + C_2HCl_2$ | 3.16E11 | 0. | 8000. |
| 11. $C_2HCl_3 + CHCl_2 \rightleftharpoons CHCl_3 + C_2HCl_2$ | 2.51E12 | 0. | 5000. |
| 12. $C_2HCl_3 + C_2HCl_4 \rightleftharpoons C_2HCl_5 + C_2HCl_2$ | 1.00E11 | 0. | 8000. |
| 13. $C_2Cl_4 + OH \rightleftharpoons CHCl_2COCl + Cl$ | 5.66E12 | 0. | 2400. |
| 14. $C_2Cl_4 + O \rightleftharpoons COCl_2 + CCl_2$ | 1.00E13 | 0. | 5000. |
| 15. $C_2Cl_4 + ClO \rightleftharpoons CCl_3COCl + Cl$ | 1.00E11 | 0. | 5000. |
| 16. $C_2Cl_4 + Cl \rightleftharpoons C_2Cl_5$ | 2.62E35 | -7.71 | 5280. |
| ** 17. $C_2Cl_3 + Cl_2 \rightleftharpoons C_2Cl_4 + Cl$ | 2.51E12 | 0. | 3000. |
| 18. $C_2HCl_5 \rightleftharpoons HCl + C_2Cl_4$ | 1.66E33 | -5.79 | 68050. |
| 19. $C_2HCl_5 \rightleftharpoons Cl + C_2HCl_4$ | 2.19E29 | -4.14 | 72000. |
| 20. $C_2HCl_5 + Cl \rightleftharpoons HCl + C_2Cl_5$ | 2.00E12 | 0. | 3300. |
| 21. $C_2HCl_5 + Cl \rightleftharpoons Cl_2 + C_2HCl_4$ | 1.00E13 | 0. | 16600. |
| 22. $C_2Cl_6 \rightleftharpoons Cl_2 + C_2Cl_4$ | 1.62E35 | -6.53 | 63200. |
| 23. $C_2Cl_6 \rightleftharpoons Cl + C_2Cl_5$ | 1.35E36 | -6.48 | 74430. |
| 24. $C_2Cl_6 + Cl \rightleftharpoons Cl_2 + C_2Cl_5$ | 6.31E13 | 0. | 18300. |
| 25. $C_2HCl_2 + M \rightleftharpoons C_2HCl + Cl + M$ | 8.00E14 | 0. | 28000. |
| 26. $C_2HCl_2 + Cl \rightleftharpoons C_2Cl_2 + HCl$ | 7.26E15 | -0.97 | 1940. |
| 27. $C_2HCl_2 + Cl \rightleftharpoons C_2HCl + Cl_2$ | 5.30E05 | 1.65 | 1140. |
| 28. $C_2HCl_2 + O_2 \rightleftharpoons CHOCl + COCl$ | 1.00E11 | 0. | 5000. |
| 29. $C_2Cl_3 + Cl \rightleftharpoons C_2Cl_4$ | 4.42E33 | -7.21 | 5710. |
| 30. $C_2Cl_3 + Cl \rightleftharpoons C_2Cl_2 + Cl_2$ | 5.26E17 | -1.66 | 5090. |
| ** 31. $C_2Cl_3 + O_2 \rightleftharpoons COCl_2 + COCl$ | 3.16E11 | 0. | 5000. |
| ** 32. $C_2Cl_3 + O_2 \rightleftharpoons C_2Cl_2O + ClO$ | 3.16E11 | 0. | 5000. |
| ** 33. $C_2Cl_2O + Cl \rightleftharpoons CO + CCl_3$ | 1.00E13 | 0. | 0. |
| 34. $C_2Cl_3 + O \rightleftharpoons CO + CCl_3$ | 1.00E13 | 0. | 0. |
| 35. $C_2Cl_3 + ClO \rightleftharpoons CO + CCl_4$ | 1.00E12 | 0. | 0. |
| 36. $C_2HCl_4 + O_2 \rightleftharpoons CHCl_2COCl + ClO$ | 1.00E11 | 0. | 5000. |
| * 37. $CHCl_2COCl + Cl \rightleftharpoons CCl_2COCl + HCl$ | 1.00E13 | 0. | 5000. |
| 38. $CHCl_2COCl + Cl \rightleftharpoons CHCl_2CO + Cl_2$ | 1.00E14 | 0. | 17600. |
| * 39. $CCl_2COCl \rightleftharpoons \dot{C}Cl_3 + CO$ | 1.00E12 | 0. | 5000. |
| 40. $CHCl_2CO \rightleftharpoons CHCl_2 + CO$ | 3.00E13 | 0. | 13500. |
| 41. $C_2HCl_4 + Cl \rightleftharpoons CHCl_2 + CCl_3$ | 6.15E20 | -1.75 | 16510. |
| 42. $C_2Cl_5 + Cl \rightleftharpoons CCl_3 + CCl_3$ | 1.70E27 | -4.01 | 12120. |
| 43. $C_2Cl_5 + Cl \rightleftharpoons C_2Cl_4 + Cl_2$ | 1.27E27 | -4.73 | 8900. |
| 44. $C_2Cl_5 + O_2 \rightleftharpoons CCl_3COCl + ClO$ | 1.00E12 | 0. | 12000. |
| 45. $CCl_3COCl + Cl \rightleftharpoons CCl_3CO + Cl_2$ | 1.00E14 | 0. | 17600. |
| 46. $CCl_3CO \rightleftharpoons CCl_3 + CO$ | 3.00E13 | 0. | 8000. |

Table 1. (continued)

| Reaction | A | n | E | Reaction | A | n | E |
|---|---|---|---|---|---|---|---|
| 47. $C_2Cl_2 + O \rightleftharpoons CCl_2 + CO$ | 7.00E13 | 0. | 0. | 70. $CCl_2 + Cl_2 \rightleftharpoons CCl_3 + Cl$ | 5.00E12 | 0. | 3000. |
| ** 48. $C_2Cl_2 + O_2 \rightleftharpoons COCl + COCl$ | 1.00E11 | 0. | 5000. | 71. $CHOCl + M \rightleftharpoons CO + HCl + M$ | 1.00E17 | 0. | 40000. |
| 49. $C_2Cl_2 + Cl_2 \rightleftharpoons C_2Cl_4$ | 2.63E36 | -7.17 | 40860. | 72. $CHOCl + Cl \rightleftharpoons COCl + HCl$ | 2.00E13 | 0. | 3000. |
| * 50. $C_2Cl_3 + M \rightleftharpoons C_2Cl_2 + Cl + M$ | 8.00E14 | 0. | 28000. | $C_3 - C_6$ **Kinetics** | | | |
| 51. $C_2Cl_2 + ClO \rightleftharpoons CO + CCl_3$ | 1.00E11 | 0. | 0. | * 73. $CCl_3 + C_2Cl_2 \rightleftharpoons C_3Cl_5$ | 3.16E11 | 0. | 5300. |
| 52. $C_2Cl_2 + OH \rightleftharpoons CO + CHCl_2$ | 1.00E12 | 0. | 0. | * 74. $C_3HCl_5 + Cl \rightleftharpoons C_3Cl_5 + HCl$ | 5.00E12 | 0. | 5000. |
| $C_1$ **Kinetics** | | | | * 75. $C_3Cl_6 + Cl \rightleftharpoons C_3Cl_5 + Cl_2$ | 3.16E13 | 0. | 20000. |
| 53. $CHCl_3 \rightleftharpoons CHCl_2 + Cl$ | 2.51E27 | -4.02 | 79700. | * 76. $CCl_3 + C_2HCl_3 \rightleftharpoons C_3HCl_6$ | 5.77E11 | 0. | 5280. |
| 54. $CHCl_3 + O \rightleftharpoons COCl_2 + HCl$ | 1.00E11 | 0. | 4000. | 77. $C_3HCl_6 + O_2 \rightleftharpoons C_3Cl_6 + HO_2$ | 1.00E12 | 0. | 5000. |
| 55. $CHCl_3 + O \rightleftharpoons CCl_3 + OH$ | 2.90E12 | 0. | 5000. | 78. $C_3HCl_6 \rightleftharpoons C_3HCl_5 + Cl$ | 1.00E13 | 0. | 22000. |
| 56. $CHCl_3 + Cl \rightleftharpoons CHCl_2 + Cl_2$ | 1.00E14 | 0. | 21000. | * 79. $C_3Cl_7 \rightleftharpoons C_3Cl_6 + Cl$ | 1.00E13 | 0. | 18600. |
| * 57. $CHCl_3 + Cl \rightleftharpoons CCl_3 + HCl$ | 6.92E12 | 0. | 3340. | 80. $CCl_3 + C_2Cl_4 \rightleftharpoons C_3Cl_7$ | 3.16E11 | 0. | 4300. |
| 58. $CCl_4 \rightleftharpoons CCl_3 + Cl$ | 7.41E35 | -6.52 | 75360. | * 81. $C_2Cl_3 + C_2Cl_2 \rightleftharpoons C_4Cl_5$ | 6.31E11 | 0. | 4200. |
| 59. $CCl_4 + O \rightleftharpoons CCl_3 + ClO$ | 2.50E10 | 0. | 2270. | * 82. $C_2Cl_2 + C_4Cl_5 \rightleftharpoons C_6Cl_7$ | 2.00E12 | 0. | 4000. |
| * 60. $CCl_3 + O_2 \rightleftharpoons COCl_2 + ClO$ | 1.00E13 | 0. | 28000. | * 83. $C_6Cl_7 \rightarrow C_6Cl_6 + Cl$ | 1.00E10 | 0. | 0. |
| 61. $CCl_3 + O \rightleftharpoons COCl_2 + Cl$ | 1.00E14 | 0. | 0. | * 84. $C_4Cl_6 + Cl \rightleftharpoons C_4Cl_5 + Cl_2$ | 5.00E12 | 0. | 20000. |
| * 62. $CCl_3 + Cl_2 \rightleftharpoons CCl_4 + Cl$ | 2.51E12 | 0. | 6000. | * 85. $C_4HCl_5 + Cl \rightleftharpoons C_4Cl_5 + HCl$ | 2.00E13 | 0. | 3000. |
| 63. $CCl_3 + CCl_3 \rightleftharpoons C_2Cl_6$ | 1.41E36 | -7.48 | 6680. | 86. $C_2Cl_3 + C_2Cl_4 \rightleftharpoons C_4Cl_7$ | 1.81E11 | 0. | 4100. |
| 64. $CCl_3 + CCl_3 \rightleftharpoons C_2Cl_4 + Cl_2$ | 2.22E26 | -4.43 | 8980. | 87. $C_4Cl_7 \rightleftharpoons C_4Cl_6 + Cl$ | 1.00E13 | 0. | 18000. |
| 65. $CCl_3 + CHCl_2 \rightleftharpoons C_2HCl_5$ | 1.66E34 | -6.79 | 6010. | 88. $C_4HCl_6 + O_2 \rightleftharpoons C_4Cl_6 + HO_2$ | 1.00E12 | 0. | 5000. |
| 66. $CCl_3 + CHCl_2 \rightleftharpoons C_2Cl_4 + HCl$ | 2.36E20 | -2.45 | 6380. | * 89. $C_2Cl_3 + C_2HCl_3 \rightleftharpoons C_4HCl_6$ | 5.00E11 | 0. | 3000. |
| 67. $CHCl_2 + O_2 \rightleftharpoons CHOCl + ClO$ | 1.00E13 | 0. | 28000. | * 90. $C_4HCl_6 \rightleftharpoons C_4HCl_5 + Cl$ | 1.00E13 | 0. | 16000. |
| 68. $CHCl_2 + O \rightleftharpoons CHOCl + Cl$ | 1.00E14 | 0. | 0. | * 91. $C_4Cl_6 + C_2Cl_3 \rightleftharpoons C_6Cl_8 + Cl$ | 5.00E11 | 0. | 1000. |
| 69. $CCl_2 + O_2 \rightleftharpoons ClO + COCl$ | 1.00E13 | 0. | 1000. | * 92. $C_6Cl_8 \rightarrow C_6Cl_6 + Cl_2$ | 1.00E10 | 0. | 0. |

| Reaction | $A$ | $n$ | $E$ | Reaction | $A$ | $n$ | $E$ |
|---|---|---|---|---|---|---|---|
| **$CO$ and $H_2$ Kinetics** | | | | 116. $COCl + OH \rightleftharpoons CO + HOCl$ | 1.00E14 | 0. | 0. |
| 93. $CO + O + M \rightleftharpoons CO_2 + M$ | 5.30E13 | 0. | -4538. | 117. $COCl + O \rightleftharpoons CO + ClO$ | 1.00E14 | 0. | 0. |
| 94. $CO + OH \rightleftharpoons CO_2 + H$ | 4.40E06 | 1.5 | -740. | 118. $COCl + O \rightleftharpoons CO_2 + Cl$ | 1.00E13 | 0. | 0. |
| * 95. $CO + O_2 \rightleftharpoons CO_2 + O$ | 1.60E13 | 0. | 41000. | 119. $COCl + O_2 \rightleftharpoons CO_2 + ClO$ | 7.94E10 | 0. | 3300. |
| 96. $CO + HO_2 \rightleftharpoons CO_2 + OH$ | 1.50E14 | 0. | 23600. | *120. $COCl + Cl \rightleftharpoons CO + Cl_2$ | 1.26E13 | 0.5 | 500. |
| 97. $H_2 + O_2 \rightleftharpoons OH + OH$ | 1.70E13 | 0. | 48100. | 121. $Cl_2 + M \rightleftharpoons Cl + Cl + M$ | 2.32E13 | 0. | 46960. |
| 98. $OH + H_2 \rightleftharpoons H_2O + H$ | 1.00E08 | 1.6 | 3300. | 122. $HCl + M \rightleftharpoons H + Cl + M$ | 2.75E13 | 0. | 81760. |
| 99. $H + O_2 \rightleftharpoons OH + O$ | 1.20E17 | -0.91 | 16510. | 123. $HCl + H \rightleftharpoons H_2 + Cl$ | 7.94E12 | 0. | 3400. |
| 100. $O + H_2 \rightleftharpoons OH + H$ | 1.50E07 | 2.0 | 7550. | 124. $H + Cl_2 \rightleftharpoons HCl + Cl$ | 8.51E13 | 0. | 1170. |
| 101. $H + O_2 + M \rightleftharpoons HO_2 + M$ | 7.00E17 | -0.8 | 0. | 125. $O + HCl \rightleftharpoons OH + Cl$ | 3.16E13 | 0. | 6700. |
| 102. $OH + HO_2 \rightleftharpoons H_2O + O_2$ | 2.00E13 | 0. | 0. | 126. $OH + HCl \rightleftharpoons Cl + H_2O$ | 2.25E12 | 0. | 1020. |
| 103. $H + HO_2 \rightleftharpoons OH + OH$ | 1.50E14 | 0. | 1003. | 127. $O + Cl_2 \rightleftharpoons ClO + Cl$ | 1.26E13 | 0. | 2800. |
| 104. $O + HO_2 \rightleftharpoons O_2 + OH$ | 2.00E13 | 0. | 0. | 128. $O + ClO \rightleftharpoons Cl + O_2$ | 5.70E13 | 0. | 364. |
| 105. $OH + OH \rightleftharpoons O + H_2O$ | 1.50E09 | 1.14 | 0. | 129. $Cl + HO_2 \rightleftharpoons HCl + O_2$ | 7.94E12 | 0. | 0. |
| 106. $H + H + M \rightleftharpoons H_2 + M$ | 1.00E18 | -1.0 | 0. | 130. $Cl + HO_2 \rightleftharpoons OH + ClO$ | 6.31E13 | 0. | 1700. |
| 107. $H + H + H_2 \rightleftharpoons H_2 + H_2$ | 9.20E16 | -0.6 | 0. | 131. $ClO + H_2 \rightleftharpoons HOCl + H$ | 1.00E13 | 0. | 13500. |
| 108. $H + H + H_2O \rightleftharpoons H_2 + H_2O$ | 6.00E19 | -1.25 | 0. | 132. $H + HOCl \rightleftharpoons HCl + OH$ | 1.00E13 | 0. | 1000. |
| 109. $H + H + CO_2 \rightleftharpoons H_2 + CO_2$ | 5.49E20 | -2.0 | 0. | 133. $Cl + HOCl \rightleftharpoons HCl + ClO$ | 1.00E13 | 0. | 2000. |
| 110. $H + OH + M \rightleftharpoons H_2O + M$ | 1.60E22 | -2.0 | 0. | 134. $Cl + HOCl \rightleftharpoons Cl_2 + OH$ | 1.26E13 | 0. | 6000. |
| 111. $H + O + M \rightleftharpoons OH + M$ | 6.20E16 | -0.6 | 0. | 135. $O + HOCl \rightleftharpoons OH + ClO$ | 5.00E13 | 0. | 1500. |
| 112. $H + HO_2 \rightleftharpoons H_2 + O_2$ | 2.50E13 | 0. | 693. | 136. $OH + HOCl \rightleftharpoons H_2O + ClO$ | 1.80E12 | 0. | 3000. |
| *113. $ClO + CO \rightleftharpoons CO_2 + Cl$ | 6.02E11 | 0. | 7400. | 137. $HOCl + M \rightleftharpoons OH + Cl + M$ | 1.00E18 | 0. | 56000. |
| **114. $COCl + M \rightleftharpoons CO + Cl + M$ | 2.00E14 | 0. | 6500 | 138. $ClOO + CO \rightleftharpoons CO_2 + ClO$ | 1.00E14 | 0. | 20000. |
| 115. $COCl + H \rightleftharpoons CO + HCl$ | 1.00E14 | 0. | 0. | 139. $ClOO + M \rightleftharpoons Cl + O_2 + M$ | 1.00E15 | 0. | 7000. |

Table 1. (concluded)

| Reaction | $A$ | $n$ | $E$ |
|---|---|---|---|
| 140. $Cl + ClOO \rightleftharpoons Cl_2 + O_2$ | 7.94E13 | 0. | 500. |
| 141. $Cl + ClOO \rightleftharpoons ClO + ClO$ | 5.01E12 | 0. | 500. |
| 142. $O + ClOO \rightleftharpoons ClO + O_2$ | 3.16E13 | 0. | 500. |
| 143. $COCl_2 + M \rightleftharpoons COCl + Cl + M$ | 1.00E16 | 0. | 76000. |
| 144. $COCl_2 + M \rightleftharpoons CO + Cl_2 + M$ | 1.00E16 | 0. | 50000. |
| **145. $COCl_2 + Cl \rightleftharpoons COCl + Cl_2$ | 3.16E13 | 0.5 | 20000. |
| 146. $COCl_2 + OH \rightleftharpoons COCl + HOCl$ | 1.00E12 | 0. | 10000. |
| 147. $COCl_2 + H \rightleftharpoons COCl + HCl$ | 1.00E13 | 0. | 2000. |

plausible elementary reactions of $C_2HCl_3$ and $O_2$, and their daughter species consistent with the principles of physical organic chemistry and experimental data, and by eliminating those reactions that did not contribute to reaction rates and were determined to be unimportant by the sensitivity analysis. The thermochemistry of species were obtained from standard sources when available, and were estimated by group additivity (Benson 1976) and quantum chemical methods (Senkan 1992). Similarly, reaction rate parameters were obtained from published compilations and other sources when available or estimated via quantum chemical methods described in Senkan 1992.

The reaction mechanism was then used to simulate the chemical structure of a one-dimensional flat flame with the following precombustion composition: $C_2HCl_3$=22.6%, $O_2$=33.1%, and Ar=44.3% (Chang and Senkan 1988).

In Figure 3 calculated mole fraction profiles (indicated by lines) for the major species $C_2HCl_3$, $O_2$, CO, HCl, $Cl_2$, and $CO_2$ are compared to those determined experimentally (indicated by symbols). As evident from this figure, the agreement between the model and experimental data is generally reasonable, suggesting that the major features of the flame chemistry of $C_2HCl_3$ have been described reasonably well by the model.

However, what is significant is that the DCKMs also provide insights in predicting the fates of the minor products in $C_2HCl_3$ combustion. For example in Figure 4 DCKM predictions are compared to experiments for $COCl_2$, $C_2Cl_2$, $C_2Cl_4$ and $CCl_4$. Similar comparisons are provided for $C_2Cl_6$, $C_3Cl_6$, $C_4Cl_6$, in Figure 5, for $CHCl_2COCl$ and $CCl_3COCl$ in Figure 6, and for $CHCl_3$, $C_2HCl_5$ and $C_3HCl_5$ in Figure 7. As can be seen from these figures, the DCKM developed for $C_2HCl_3$ combustion provide a reasonable description of the molecular events leading to the formation and destruction minor constituents under the experimental conditions investigated. It is also significant to note that because DCKMs are fundamental models, they can be used to predict product distributions under different conditions, including conditions under which the conduct of experiments can be difficult or impossible, thus they allow for the a priori establishment of optimum operating conditions to minimize or maximize the formation of by-products.

The reaction set presented in Table I can also be subjected to a sensitivity analysis to determine the rank order of the reactions in the mechanism. For this, normalized first order sensitivity gradients ($S_{ij}$) can be calculated along the flame using the following definition:

$$S_{ij} = (A_j/C_i)[dC_i/dA_j] = d[\ln C_i]/d[\ln A_j]$$

where $C_i$ is the molar concentration for species i, and $A_j$ is the pre-exponential factor for the j'th

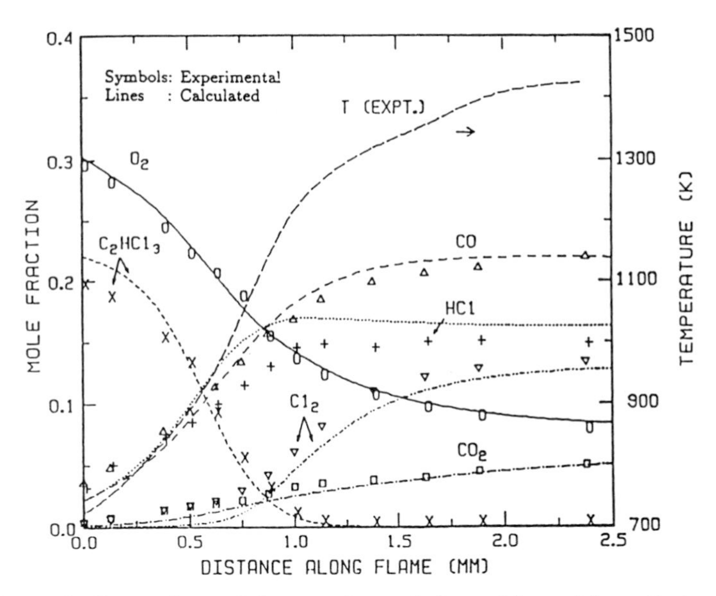


Figure 3. Comparison of the experimental data with model predictions for $C_2HCl_3$, $O_2$, CO, $CO_2$, HCl, and $Cl_2$.

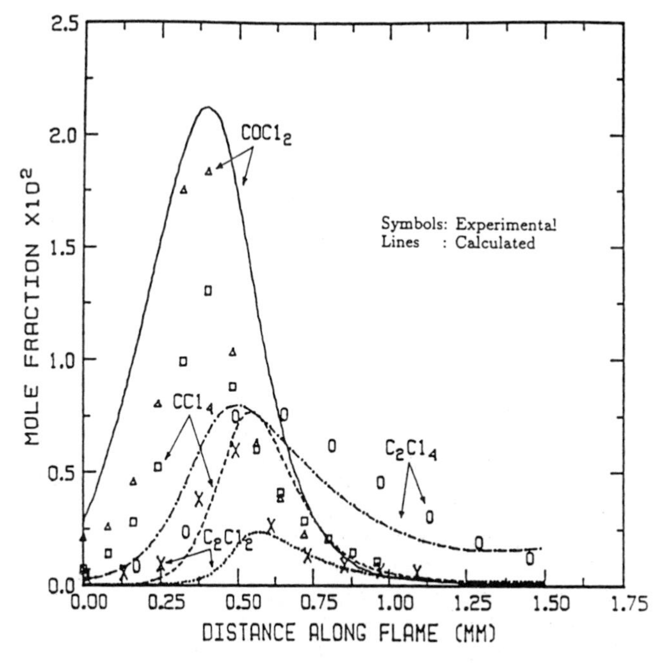


Figure 4. Comparison of the experimental data with model predictions for $COCl_2$, $C_2Cl_2$, $C_2Cl_4$, and $CCl_4$.

reaction at a particular location in the flame. This definition is particularly useful because influential reactions in the mechanism can be identified even without the explicit presence of a particular species in that reaction. In addition, the net rates of individual elementary reactions in the mechanism can be calculated to determine important reaction pathways. From such considerations the following conclusions can be reached concerning the formation and destruction of various species in fuel-rich $C_2HCl_3$ combustion.

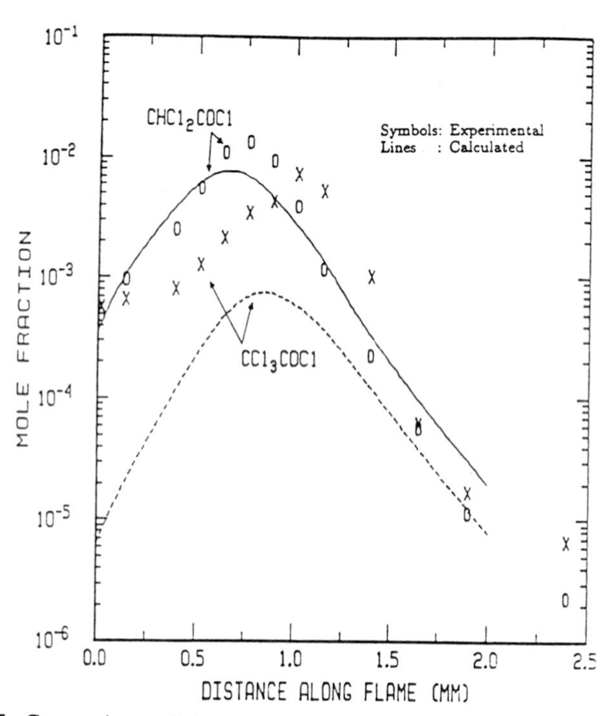


**Figure 5. Comparison of the experimental data with model predictions for $CHCl_2COCl$ and $CCl_3COCl$.**

$C_2HCl_3$

In Figure 8, rate profiles of reactions involving $C_2HCl_3$ that have the highest rates are presented. As evident from these profiles the major reaction pathway for $C_2HCl_3$ in the flame was by Cl radical attack, i.e. $C_2HCl_3+Cl=C_2Cl_3+HCl$, which also was the major route for the formation of HCl. In the earlier, i.e. cooler, parts of this fuel-rich flame the addition of $C_2Cl_3$ and ClO radicals to $C_2HCl_3$ were also significant reactions. In contrast, in fuel-lean flames $C_2Cl_3$ would have been consumed rapidly by its reaction with $O_2$, i.e. $C_2Cl_3+O_2=COCl_2+COCl$. The addition of $C_2Cl_3$ to $C_2HCl_3$ results in the formation of $C_4Cl_6$ via the sequence: $C_2HCl_3+C_2Cl_3=C_4HCl_6$ and $C_4HCl_6+O_2=C_4Cl_6+HO_2$. The addition of ClO to $C_2HCl_3$ produces $CHCl_2COCl$ by $C_2HCl_3+ClO=CHCl_2COCl+Cl$. Both $C_4Cl_6$ and $CHCl_2COCl$ form in significant levels in fuel-rich flames of $C_2HCl_3$ (see Figures 5 and 6, respectively).

## $O_2$, CO and $CO_2$

From the analysis of net rates, the major reactions responsible for the consumption of $O_2$ were determined to be: $C_2Cl_3+O_2=COCl_2+COCl$, $C_2Cl_3+O_2=C_2Cl_2O+ClO$ and $C_2Cl_2+O_2=COCl+COCl$. CO forms as a consequence of the unimolecular decomposition of COCl, i.e. $COCl+M=CO+Cl+M$, in which COCl forms via $C_2Cl_3+O_2=COCl_2+COCl$ and $C_2Cl_2+O_2=COCl+COCl$. The major reaction responsible for the consumption of CO was $CO+ClO=CO_2+Cl$, which also was the primary reaction path for the formation of $CO_2$. Because of the hydrogen lean nature of this mixture, the concentration of OH would be low, thus the $CO+OH=CO_2+H$ reaction, which is the major reaction pathway for CO oxidation in normal hydrocarbon systems, contributes very little to the formation of $CO_2$ in $C_2HCl_3$ flames.

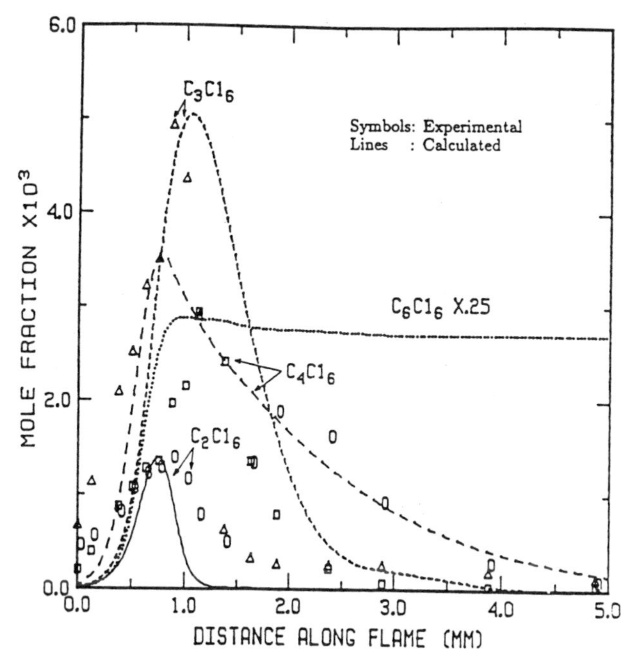


**Figure 6. Comparison of the experimental data with model predictions for $C_2Cl_6$, $C_3Cl_6$, and $C_4Cl_6$.**

## HCl and $Cl_2$

HCl forms primarily as a consequence of Cl radical attack on $C_2HCl_3$, i.e. $C_2HCl_3+Cl=C_2Cl_3+HCl$, thus its appearance early in the flame zone is expected (see Figures 3 and 9). In addition, HCl forms by the reaction $C_4HCl_5+Cl=C_4Cl_5+HCl$, in which $C_4HCl_5$ forms by the decomposition of the adduct in reaction $C_2Cl_3+C_2HCl_3=C_4HCl_6$.

According to the mechanism, major reaction pathways responsible for the formation of $Cl_2$ were: $COCl_2+Cl=COCl+Cl_2$ and $COCl+Cl=CO+Cl_2$ (Figure 10). Consumption of $Cl_2$, on the other hand occurs via $C_2Cl_3+Cl_2=C_2Cl_4+Cl$. Since $C_2Cl_3$ forms very early in the flame zone, i.e. by Cl attack on $C_2HCl_3$, it prevents the buildup of $Cl_2$ in the system, thus $Cl_2$ mole fraction profile exhibits an induction period (see Figures 3 and 10).

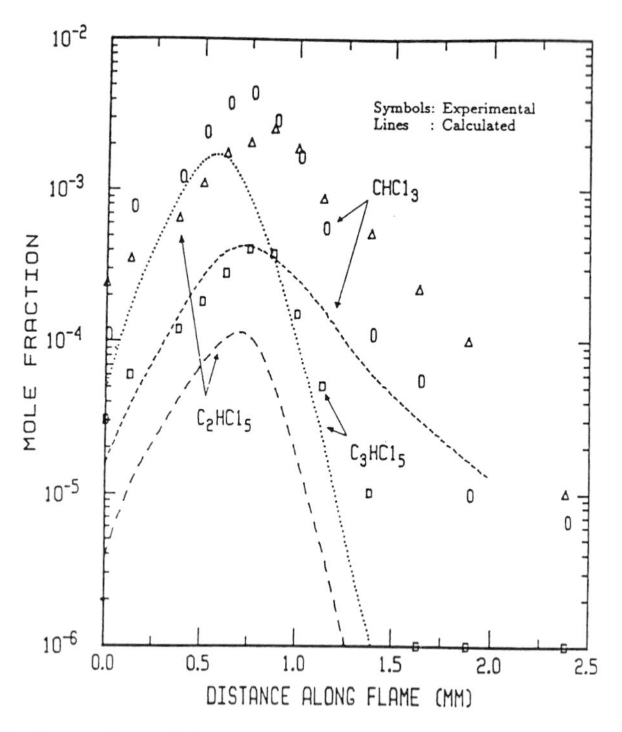


Figure 7. Comparison of the experimental data with model predictions for $CHCl_3$, $C_2HCl_5$, $C_3HCl_5$.

$COCl_2$, $C_2Cl_4$, $CCl_4$

Phosgene ($COCl_2$) is an important intermediate in $C_2HCl_3$ combustion, both unc fuel-rich and fuel-lean conditions, and the model predictions are in reasonable agreement wi the experimental data (see Figure 4). As seen from Figure 11, $COCl_2$ forms early in the fla primarily as a consequence of reaction $C_2Cl_3+O_2=COCl_2+COCl$, and to a lesser extent $CHCl_2COCl+Cl= CHCl_2+COCl_2$. The destruction of $COCl_2$ occurs essentially due to Cl radic attack, i.e. $COCl_2+Cl=COCl+Cl_2$, which also is one of the major pathways responsible the formation of $Cl_2$ and CO.

$C_2Cl_4$ also is an important chlorocarbon intermediate both in fuel-lean and fuel-ric flames of $C_2HCl_3$, and the mechanism predicts its behavior along the flame zone with reasonab accuracy (see Figure 4). The reaction $C_2Cl_3+Cl_2=C_2Cl_4+Cl$ is majorly responsible for th formation of $C_2Cl_4$, with minor contributions from $C_2Cl_5=C_2Cl_4+Cl$ where $C_2Cl_5$ forms b $C_2HCl_5+Cl=C_2Cl_5+HCl$. The destruction of $C_2Cl_4$ occurs primarily by the reversal of its maj formation reaction, i.e. $C_2Cl_4+Cl=C_2Cl_3+Cl_2$ later in the flame. Addifionally $C_2Cl_4$ reacts vi $C_2Cl_3+C_2Cl_4= C_4Cl_7$, which also represents the major path for the formation of $C_4Cl_6$.

$C_2Cl_2$ forms in substantial quantities in flames of $C_2HCl_3$ both under fuel-lean an fuel-rich conditions. As seen in Figure 4, considerable discrepancy exists between mode predictions and experimental profiles for $C_2Cl_2$. This, however, is not surprising in view c uncertainties that exist both in the experiments and in the mechanism. Uncertainties in th experimental data are substantial because $C_2Cl_2$ mole fraction profiles were determined in th absence of a calibration standard. The elementary reactions of $C_2Cl_2$ are also virtually unknown thus its reactions presented in Table I correspond to plausible steps in the absence of othe

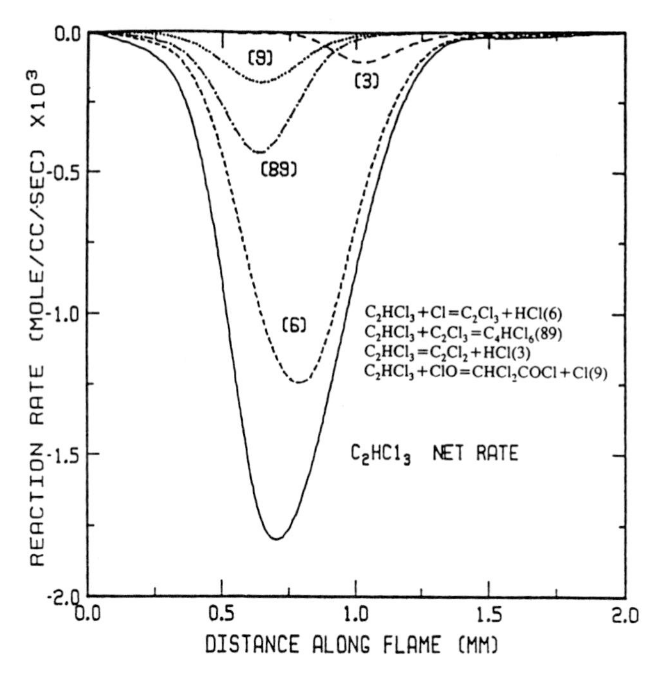


**Figure 8. Rate profiles for reactions having the highest net rates that involve $C_2HCl_3$.**

evidence. According to the mechanism, the formation of $C_2Cl_2$ is due to the decomposition of $C_4Cl_5$, $C_2Cl_3$, and $C_2HCl_3$ in an approximate order of importance, i.e. $C_4Cl_5=C_2Cl_3+C_2Cl_2$, $C_2Cl_3+M= C_2Cl_2+Cl+M$, and $C_2HCl_3=C_2Cl_2+HCl$. $C_2Cl_2$ destruction was then proposed to occur via the complex process $C_2Cl_2+O_2= COCl+COCl$, although it is unlikely to correspond to a simple elementary chemical reaction.

Carbon tetrachloride also forms in substantial quantities in fuel-rich flames of $C_2HCl_3$. In the proposed mechanism major reaction sequences responsible for the formation of $CCl_4$ were: $C_2Cl_3+O_2=C_2Cl_2O+ClO$, $C_2Cl_2O+Cl=CCl_3+CO$ followed by the chlorination of the $CCl_3$ radical by $CCl_3+Cl_2=CCl_4+Cl$ and $CCl_3+C_2HCl_3=CCl_4+C_2HCl_2$, and directly by the reaction $C_2Cl_3+ClO=CCl_4+CO$. Minor routes for $CCl_4$ formation also exists, and they include $CCl_3COCl+Cl=CCl_3CO+Cl_2$ followed by $CCl_3CO= CCl_3+CO$, and $CHCl_3+Cl=CCl_3+HCl$, again followed by the chlorination of $CCl_3$.

## $CHCl_2COCl$ and $CCl_3COCl$

For dichloroacetyl chloride ($CHCl_2COCl$) and trichloroacetyl chloride ($CCl_3COCl$), there exists discrepancies between the model predictions and the experiment (Figure 5). The model correctly predicts the relative ordering and the peak locations in $CHCl_2COCl$ and $CCl_3COCl$ profiles. However, the model underpredicts the absolute mole fractions for these species by a factor of 2. In this case, the major uncertainty rests in the reaction mechanism, as the experimental data have been estimated to be accurate within 20% (Chang and Senkan 1988). The reactions of acyl chlorides in flames are poorly understood, and the reactions presented in Table I represent plausible steps in the absence of direct experimental evidence. According to the proposed mechanism the major reaction paths responsible for the formation of acyl chlorides

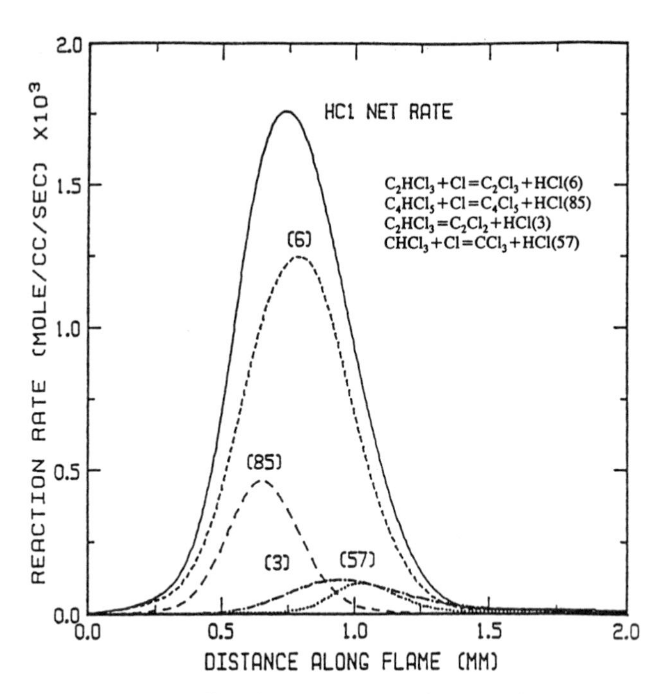


Figure 9. Rate profiles for reactions having the highest net rates that involve HCl.

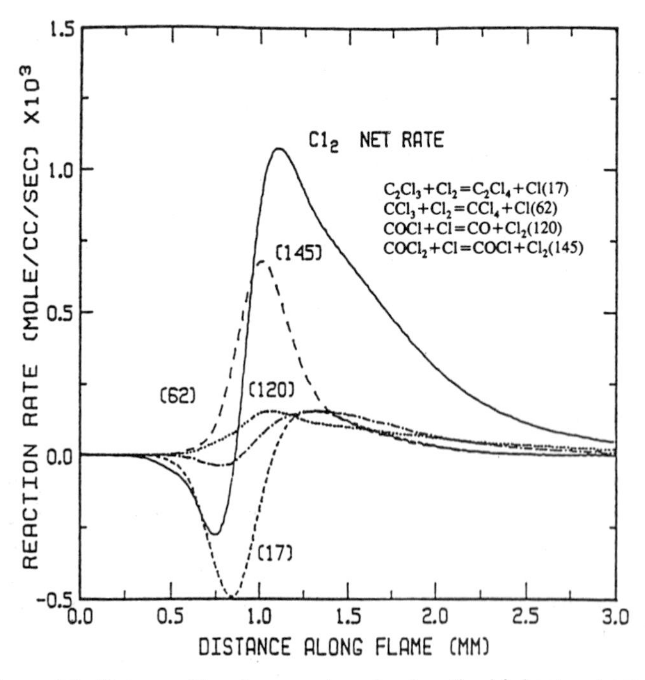


Figure 10. Rate profiles for reactions having the highest net rates that involve $Cl_2$.

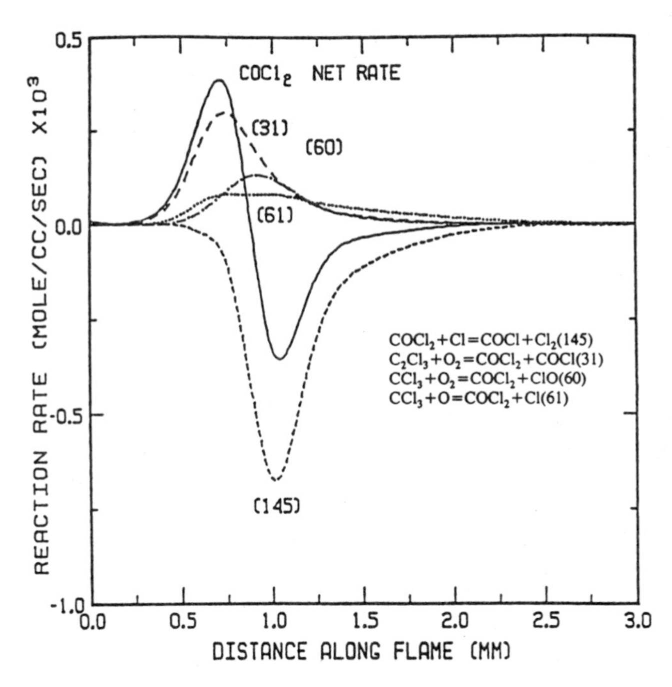


**Figure 11.** Rate profiles for reactions having the highest net rates that involve $COCl_2$.

are due to ClO addition to $C_2HCl_3$ and $C_2Cl_4$, i.e. $C_2HCl_3+ClO=CHCl_2COCl+Cl$ and $C_2Cl_4+ClO=CCl_3COCl+Cl$. Because acyl chlorides are very unstable at high temperatures, they are expected to decompose, for example by the following sequence of reactions: $CHCl_2COCl+Cl=CHCl_2CO+Cl_2$ and $CCl_3COCl+Cl=CCl_3CO+Cl_2$, followed by $CHCl_2CO=CHCl_2+CO$ and $CCl_3CO=CCl_3+CO$, respectively. These reactions also constitute pathways which subsequently lead to the formation of $CHCl_3$ and $CCl_4$ by the chlorination of the $CHCl_2$ and $CCl_3$, respectively.

$C_2Cl_6$, $C_3Cl_6$, $C_4Cl_6$:

As seen in Figure 6 major discrepancies exist between DCKM predictions and the experimental data. Specifically the second maxima in the mole percent profile for $C_2Cl_6$ was not reproduced at all by the present model, and this clearly manifests the existence of a deficiency in the mechanism. According to the model, the earlier formation of $C_2Cl_6$ is majorly due to $CCl_3$ recombination, i.e. $CCl_3+CCl_3=C_2Cl_6$, and not via the chlorination of $C_2Cl_4$, since the conditions do not favor the reaction $Cl+C_2Cl_4=C_2Cl_5$ in the forward direction in the flame. The destruction of $C_2Cl_6$ occurs via its unimolecular decomposition back to $CCl_3$ and by C-Cl bond fission, i.e. $C_2Cl_6=CCl_3+CCl_3$ and $C_2Cl_6=C_2Cl_5+Cl$, and by complex molecular elimination process, $C_2Cl_6=C_2Cl_4+Cl_2$. The increase in $C_2Cl_6$ mole fraction for the second time would likely to be caused by the fragmentation of higher molecular weight chlorohydrocarbons, which form early in the flame zone. The formation of $C_3Cl_6$ starts by the addition of $CCl_3$ to $C_2HCl_3$, $C_2Cl_4$ or $C_2Cl_2$, i.e. $CCl_3+C_2HCl_3=C_3HCl_6$, $CCl_3+C_2Cl_4=C_3Cl_7$ and $CCl_3+C_2Cl_2=C_3Cl_5$, all of which are chemically activated processes, followed by $C_3HCl_6+Cl=C_3Cl_6+HCl$, $C_3Cl_7=C_3Cl_6+Cl$ and $C_3Cl_5+Cl_2=C_3Cl_6+Cl$. The destruction of $C_3Cl_6$ occurs primarily through Cl radical attack, i.e. $C_3Cl_6+Cl=C_3Cl_5+Cl_2$. The major route for the formation of

$C_4Cl_6$ begins with the addition of $C_2Cl_3$ to unsaturated $C_2$ chlorohydrocarbons, i.e. $C_2Cl_3+C_2HCl_3=C_4HCl_6$, $C_2Cl_3+C_2Cl_4=C_4Cl_7$ and $C_2Cl_3+C_2Cl_2=C_4Cl_5$, followed by $C_4HCl_6+Cl=C_4Cl_6+HCl$, $C_4Cl_7=C_4Cl_6+Cl$ and $C_4Cl_5+Cl_2=C_4Cl_6+Cl$. In the current mechanism, the $C_4Cl_6$ formed undergoes further addition reactions via $C_2Cl_3+C_4Cl_6=C_6Cl_8+Cl$ or $C_4Cl_6+Cl=C_4Cl_5+Cl_2$ followed by $C_2Cl_2+C_4Cl_5=C_6Cl_7$, and the $C_6Cl_8$ and $C_6Cl_7$ formed subsequently are transformed into $C_6Cl_6$ via $C_6Cl_8\text{-}{>}C_6Cl_6+Cl_2$ and $C_6Cl_7\text{-}{>}C_6Cl_6+Cl$, respectively.

$CHCl_3$, $C_2HCl_5$, $C_3HCl_5$

The major reaction paths for the formation of $CHCl_3$ were $CHCl_2+Cl_2=CHCl_3+Cl$ and $C_2HCl_3+CHCl_2=CHCl_3+C_2HCl_2$, in which the $CHCl_2$ radical forms by $CHCl_2COCl+Cl=CHCl_2CO+Cl_2$, followed by $CHCl_2CO=CHCl_2+CO$. The destruction of $CHCl_3$ occurs primarily by Cl radical attack, i.e. $CHCl_3+Cl=CCl_3+HCl$. For $C_2HCl_5$, its major formation paths were the radical recombination reactions $CCl_3+CHCl_2=C_2HCl_5$ and $Cl+C_2HCl_4=C_2HCl_5$, both of which are chemically activated processes. Its major destruction path also involves Cl attack, i.e. $C_2HCl_5+Cl=C_2Cl_5+HCl$. The $C_3HCl_5$ forms via $C_3HCl_6=C_3HCl_5+Cl$, in which the $C_3HCl_6$ radical is produced from $CCl_3+C_2HCl_3=C_3HCl_6$, again a chemically activated process. The destruction of $C_3HCl_5$ again proceeds through Cl radical attack, i.e. $C_3HCl_5+Cl=C_3Cl_5+HCl$.

In Figures 12 and 13, the major reaction pathways responsible for the formation and destruction of species discussed above, as well as for several key radical species, are presented.

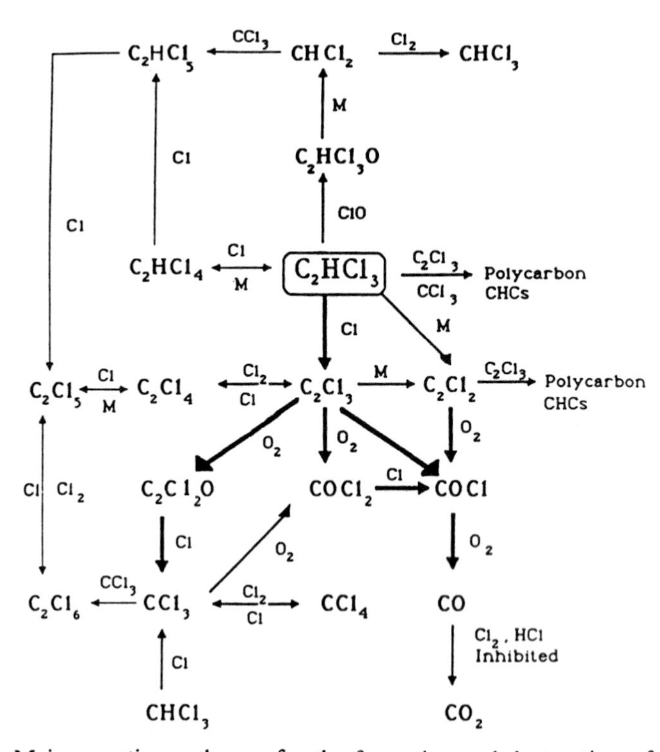


Figure 12. Major reaction pathways for the formation and destruction of $C_1$ and $C_2$ species.

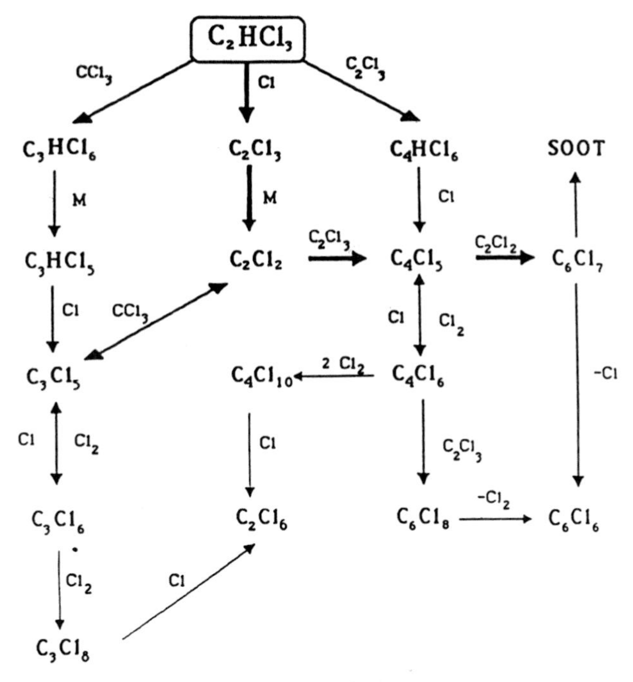


Figure 13. Major reaction pathways for the formation and destruction of $C_3$ to $C_4$ species.

As seen from these figures, the high temperature reactions of $C_2HCl_3$ are considerable complex but not intractable. The reaction rate and sensitivity analysis provided crucial information on the identities of important reactions, which in return can be used to construct simpler, albeit with reduced range of applicability, reaction mechanisms.

## Concluding remarks

As shown in the above discussion and example for $C_2HCl_3$ combustion, detailed chemical kinetic mechanisms (DCKM) provides useful insights into reaction processes, especially with regard to the trace constituents. Since DCKMs involve the participation of a large number of species in very large number of reactions, to preserve the elementary nature of these models, they represent a convenient framework to describe the formation and destruction of major as well as minor products in manufacturing processes. In addition, DCKMs, because of their fundamental bases and predictive features represent numerical tools for the rational design of clean manufacturing processes, i.e. processes with minimal formation of toxic by-products.


## Acknowledgements

This research was supported, in part, by funds from the U.S. Environmental Protection Agency, Grant No:R815136-01-0, and the UCLA Center for Clean Technologies.

## REFERENCES

Basevich, V.Ya.,"Chemical kinetics in the combustion processes: A detailed kinetic mechanism and its implementation", Prog. Energy. Combust. Sci., 13, 199 (1987).

Baulch,D.L., Duxbury,J., Grant,S.J., Montague,D.C.," Evaluated Kinetic Data for High Temperature Reactions: Homogeneous Gas Phase Reactions of Halogen and Cyanide Containing Species", 4, J.Phys.Chem.Ref.Data, 10,Supplement No.1.(1981).

Benson,S.W., Thermochemical Kinetics, John Wiley, N.Y. (1976).

Benson, S.W., and O'Neal, H.E., "Kinetic data on gas-phase unimolecular reactions", NSRDS-NBS 21, 1970.

Berces,T., and Marta,F., "Activation Energies for Metathesis Reactions of Radicals", in Chemical Kinetics of Small Organic Molecules: IV Reactions in Special Systems, Alfassi, Z.B., Ed., CRC Press, Boca Raton, Florida 1988.

Caracotsios,M., and Steward,W.E., "Sensitivity Analysis of Initial Value Problems Including ODEs and Algebraic Equations", Comp. Chem. Eng., 9, 359 (1985).

Chang,W.D., Karra,S.B., and Senkan,S.M.," A Detailed Mechanism of the High Temperature Oxidation of C2HCL3", Combust.Sci.Tech, v.49. p.107 (1986).

Chang,W.D., and Senkan,S.M.," Chemical Structure of Fuel-Rich, Premixed, Laminar Flames of Trichloroethylene", 22nd Symposium (Int'l) on Combustion, 1453 (1988).

Chang, W.D., and Senkan, S.M., "Detailed Chemical Kinetic Modeling of the Fuel-Rich Flames of Trichloroethylene", Enviro. Sci. Tech., 23, 442 (1989).

Chase, M.W., Davies, C.A., Downey, J.R., Frurip, D.J., McDonald, R.A., and Syverud, A.N., "JANAF Thermochemical Tables", J. Phys. Chem. Ref. Data, 14, Supplement No.1, (1985).

DeMore, W.B., Molina, M.J., Watson, R.T., Golden, D.M., Hampson, R.F., Kurylo, M.J., Howard, C.J., and Ravishankara, A.R. "Chemical Kinetics and Photochemical Data for Use in Stratospheric Modeling", Evaluation No. 8, JPL Publication 90-37, 1990.

Dunning, T.H., Harding, L.B., Wagner, A.F., Schatz, G.C., and Bowman, J.M., "Theoretical Studies of the Energetics and Dynamics of Chemical Reactions", Science, 240, 453 (1988).

Gaffney,J.S., Bull,K., "Prediction of the Rate Constants for Radical Reactions Using Correlational Tools", in Chemical Kinetics of Small Organic Molecules: IV Reactions in Special Systems, Alfassi, Z.B., Ed., CRC Press, Boca Raton, Florida 1988.

Granada, A., Karra, S.B., and Senkan, S.M., "Conversion of $CH_4$ into $C_2H_2$ and $C_2H_4$ by the Chlorine-Catalyzed Oxidative-Pyrolysis (CCOP) process: Oxidative pyrolysis of $CH_3Cl$", Ind. Chem. Eng. Res., 26, p.1901 (1987).

Hanson,R.K., Salimian,S., "Survey of Rate Constants in the N/H/O System", in Combustion Chemistry, (W.C. Gardiner, Jr., Ed.), Springer-Verlag, New York, 1984.

Hwang, J-T., "Sensitivity Analysis in Chemical Kinetics by the Method of Polynomial Approximations", Int. J. Chem. Kinetics, 15, 959 (1983).

Junk,.G.A., and Ford,C.S., "A Review of Organic Emissions from Selected Combustion Processes", Chemosphere, v.9, p.187 (1980).

Kee,R.J., Grear,J.F., Smooke,M.D., and Miller,J.A., "A Fortran program for modeling steady state laminar one-dimensional premixed flames", Sandia Report (1985).

Kerr,A.J., Moss,S.J., Handbook of Bimolecular and Termolecular Gas Reactions, Volumes I and II, CRC Press, Boca Raton, FL 1981.

Kerr,A.J., Drew,R.M., Handbook of Bimolecular and Termolecular Gas Reactions, Volumes III and IV, CRC Press, Boca Raton, FL 1987.

Kondratiev, V.N. "Rate Constants of Gas Phase Reactions, COM-72-10014, National Bureau of Standards, Washington, 1972.

Laidler, K.J., Chemical Kinetics, 3rd Ed., Harper and Row, New York 1987.

Lutz, A.E., Kee,R.J., and Miller,J.A., "SENKIN: A Fortran Program for Predicting Homogeneous Gas Phase Kinetics with Sensitivity Analysis", Sandia National Laboratories, Livermore, SAND87-8248, February 1988.

Oberg,T., Aittola,J-P., and Bergstrom,J.G.T., "Chlorinated Aromatics from the Combustion of Hazardous Waste", Chemosphere, v.14, p.215 (1985).

Pedley, J.B., Naylor, R.D., and Kirby, S.P., Thermochemical Data of Organic Compounds, Chapman and Hall, London 1986.

Rabitz, H., Kramer, M.A., and Dacol, D., "Sensitivity Analysis in Chemical Kinetics", Ann. Rev. Phys. Chem., 34, 419 (1983).

Senkan, S.M., "Detailed Chemical Kinetic Modeling: Chemical Reaction Engineering of the Future", Advances in Chemical Engineering, in press 1992.

Schaub, W.M., and Tsang, W., "Dioxin Formation in Incinerators", Enviro. Sci. Tech., v.17, p.721 (1983).

Siegneur, C., Stephanopoulos, G.,and Carr,Jr.,R.W., "Dynamic Sensitivity Analysis of Chemical Reaction Systems: A Variational Method", Chem. Eng. Sci., 37, 845 (1982).

Tilden, J.W., Costanza, V., McRae,G.J., and Seinfeld, J.H., "Sensitivity Analysis of Chemically Reacting Systems", in Modeling of Chemical Reaction Systems, K.H. Ebert, P. Deuflhard, and W. Jaeger, Eds., Springer-Verlag, New York, 1981.

Tsang, W., and Hampson, R.F., "Chemical Kinetic Data Base for Combustion Chemistry. Part I. Methane and Related Compounds", J Phys. Chem. Ref. Data, 15, 1087 (1986).

Warnatz, J., "Rate coefficients in the C/H/O system", in Combustion Chemistry, (W.C. Gardiner, Jr., Ed.), Springer-Verlag, New York, 1984.

Westbrook, C.K., and Dryer, F.A., "Chemical Kinetic Modeling of Hydrocarbon Combustion", Prog. Energy Combust. Sci., 10, 1 (1984).

Westley, F., "Table of Recommended Rate Constants for Chemical Reactions Occurring in Combustion", NBS Report, NSRDS-NBS 67, 1980.

Westley,F., "Chemical Kinetics of the Gas Phase Combustion of Fuels", NBS Special Publication No:449, 1976.

# KINETIC CONTROL OF BY-PRODUCT REACTIONS WITH *IN SITU* DISTILLATION IN A CONTINUOUS FLOW DEVICE

Robert P. Shirtum

Dow Plastics
The Dow Chemical Company
Freeport, TX 77541

*A detailed kinetic study of a complex, commercially important, organic synthesis was combined with math models of several reactor configurations to yield a design which proved viable for large-scale production. The combined characteristics of in situ distillation, plug-flow behavior, and continuous reagent addition were proven in pilot plant operations.*

## INTRODUCTION

A review of the volumes of SRI PEP [1] reports indicates, in general, that many industrial processes, to produce bulk building block commodity chemicals have yields to desired products of about 90%. (Certain hydrocarbons, specialties, and plastics products not included.) Our own experience and information from other major chemical companies supports this average of ten percent yield loss to by-products.

Many of these processes require substantially more capital, per unit of production, to destroy the by-products than the allocated capital required to produce them initially.

The cost of waste destruction has recently escalated so rapidly as environmental regulations and concerns have received unprecedented emphasis, that many petro chemical companies are utilizing much of their capital resources to address waste treatment.

To eliminate a substantial portion of this large volume of by-products by the research and design of more efficient processes, rather than focus on destroying these materials, would justify the expenditure of a very large effort.

This work is a successful example of the development of technology which increases the process yield of a major chemical product from about 90% yield, of one of the main reactants, to over 99% yield.

Other positive attributes of the new process were increased yield of a secondary feedstock from 55% to almost 100%, improved ability to recover and recycle spent catalyst, and several product quality improvements. Slight increases in conversion energy and capital, due to increased process complexity, were insignificant relative to waste treatment cost reduction and raw material yield improvement.

Consider the industrial synthesis of an important organic chemical product where the reaction sequence can be simplified in the following manner:

*Industrial Environmental Chemistry*, Edited by D.T. Sawyer and A.E. Martell, Plenum Press, New York, 1992

$$A + B \xrightarrow{C,S_2} D \qquad (1)$$

$$D + C \longrightarrow C_s + P_1 + S\uparrow \qquad (2)$$

$$P_1 + B \longrightarrow E \qquad (3)$$

$$E + C \longrightarrow C_s + P_2 + S\uparrow \qquad (4)$$

These main reactions are accompanied by the following by-product reactions:

$$A + S \xrightarrow{C} F \qquad (5)$$

$$A + C \longrightarrow G \qquad (6)$$

$$A + D \longrightarrow H \qquad (7)$$

$$P_1 + C \longrightarrow D_t \qquad (8)$$

The effects on product quality of these series of reactions are as follows:

$P_1/P_2$ = constant

(D,E,G,H) < 2500 ppm by weight.

Additional process desirability conditions include:

F = O (high yield to products from A) $C_s$ = 99.5+% purity (for ease of regeneration)

## EXPERIMENTAL APPROACH

The development of the kinetic model, as simplified above, began with a computer model based on stoichiometry and estimates of Arrhenius constants from previous studies. We used an approach in which the model was used to design subsequent laboratory reactions. Data from the experiments were then compared to the model and sensitivity analysis focused on weaknesses in model predictions. The model was then refined in an iterative fashion until acceptable prediction over a wide range of conditions were established. We believe that it is significant that a model preceded the first experiment and we refer to this methodology as model based experimental design.

The initial laboratory experiments were conducted in a semi-batch fashion with a conventional glass apparatus as illustrated in Figure 1. The reactor is fitted with a rectification column, condenser, and reflux reservoir to facilitate the removal of catalyst solvent and volatile by-products.

## RESULTS

Model predictions and semi-batch laboratory data at simulation conditions are shown for the major reaction components in Figure 2. Products of trace chemistry and model predictions are illustrated in Figure 3.

The model is then solved numerically using the design equation for a serial continuous stirred tank reactor (CSTR) configuration and the economically optimal number of CSTRs is determined [2]. A graphical representation of this approach is illustrated in Figure 4, where the critical ratio of $P_1/P_2$ is plotted as the dependent variable for the design basis.

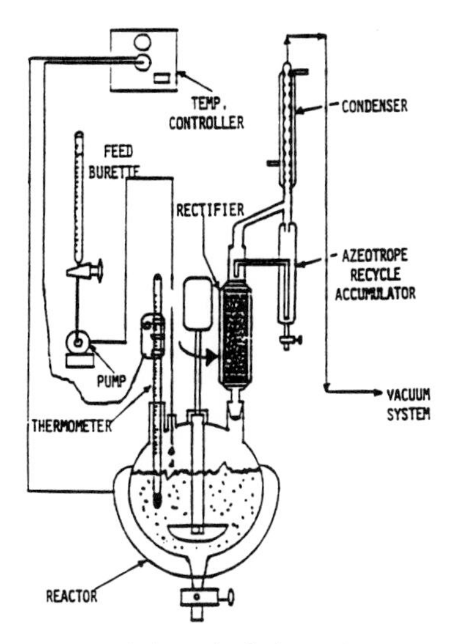


Figure 1. Semi-batch laboratory reactor.

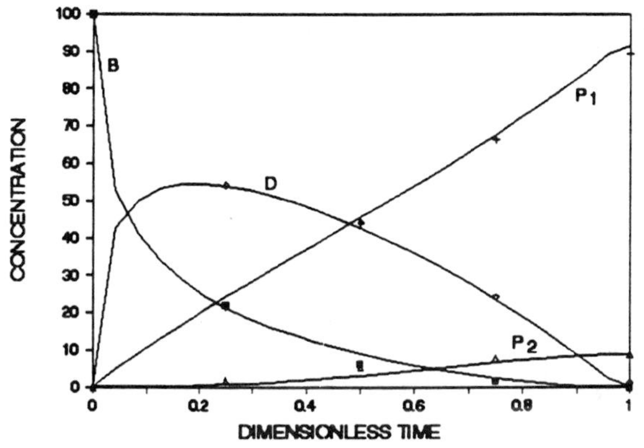


Figure 2. Concentration vs. time for semi-batch laboratory and computer model data.

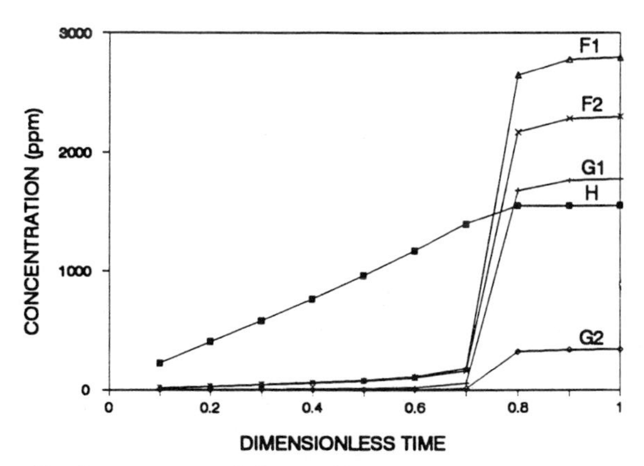


Figure 3. Computer model predictions of concentration vs. time for side reaction trace chemistry.

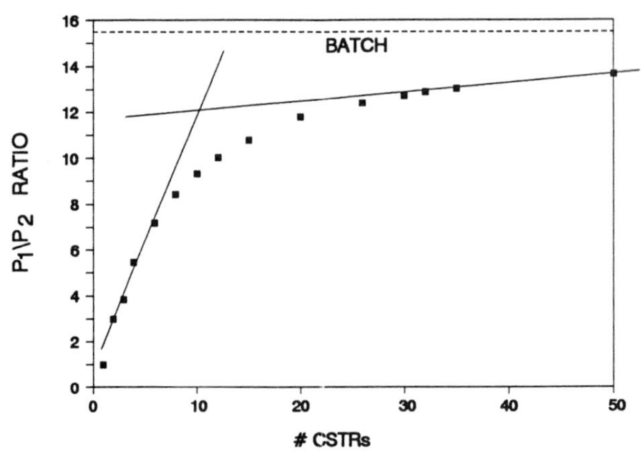


Figure 4. Isomer ratio $P_1\backslash P_2$ vs. number of CSTR stages.

This comparison between serial CSTRs and batch reactor results is not unlike the efficiency comparison made by Octave Levenspiel [3], except that the conversion of the limiting reactant of interest to us is above 99.9% and the $P_1/P_2$ ratio is of much more pratical interest than volumetric efficiency.

Reaction strategy learned by using a simplified math model:

- Maintain a low catalyst concentration, [C], by adding it slowly to the reaction mixture in concert with the rate of catalyst depletion in Reaction 2.
- Removal of the by-product, S, as its azeotrope with A via in situ distillation.
- Add an unreactive solvent, $S_2$, to insure miscibility of the catalyst.
- Control the product isomer ratio of $P_1/P_2$, using the initial stoichiometric ratio of A/B.

Later studies were conducted in a horizontal glass pipe with baffled partitions allowing a multi-stage CSTR configuration within one apparatus containing a vapor space in common communication with each stage. Reflux, rectification and condensation equipment were similar to the batch apparatus.

## REACTOR DESIGN

The descriptive model derived from semi-batch experiments was solved on the computer by Runge-Kutta [4] techniques. The equations for reaction velocities, and heat and mass balances were then solved by a Quasi-Newton [5] method to estimate the number of CSTR stages necessary for control of the important ratio of $P_1$ to $P_2$ in the product. Catalyst and solvent were added inter-stage in the continuous reactor in an optimized profile to aid in achieving the correct $P_1$ to $P_2$ ratio with a minimal number of compartments, to minimize reactor cost. Continuous addition of the catalyst greatly helped control the very fast exothermic reactions so that area requirements for enthalpy transfer were much easier than in the semi-batch case. Although a simplified model, illustrated in Equations (1-8) is presented, the model we used required over 30 chemical reactions, reversible and irreversible, and over 35 anionic and chemical species to accurately describe the trace chemistry (21 equations) along with the seven major reaction pathways. The reactor system which evolved is illustrated in Figure 5.

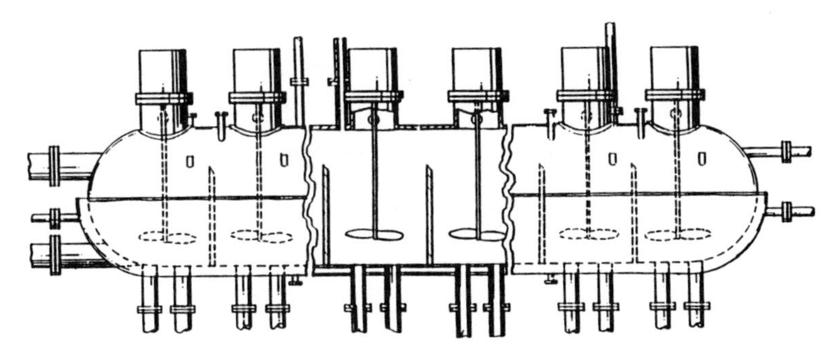

Figure 5. Continuous horizontal staged CSTR reactor.

This reactor allows for the staged addition of catalyst and co-reactants and the removal of solvents and volatile products of reaction just as the laboratory semi-batch reactor did. This is the design that was pilot plant tested to produce 1000 tons per year of product. The manufacturing scale-up from this pilot plant will be approximately two orders of magnitude.

The start-up philosophy, which proved successful, was to begin reaction by introducing catalyst while operating the fully charged reactor on total recycle. At high conversion of B, catalyst addition ceased and the desired steady-state forward flowrate was begun with catalyst profile initiated sequentially to each CSTR stage after the appropriate residence time had elapsed. The reaction model is a great aid in implementing start-up and shut down control schemes.

## CONCLUSIONS AND DISCUSSION

An environmental solution to the production of a chemical product can be approached in several ways:

1. Cease production of the product.
2. Add process unit operations to destroy process wastes.
3. Modify the process to reduce or eliminate the production of wastes.

We believe the options are listed in order of increasing desirability. One exception to this order may be where an alternate product, which fulfills market needs and is produced by a more environmentally favorable process allows the replacement of the original product.

Resources spent to add unit operations to destroy process wastes are necessarily less productive toward the mission of production of the product than resources spent toward making the process more efficient. The magnitude of resources may be less; however, if applied toward the destruction of wastes. This is especially true if short-term economics are the decision driving forces. As so often occurs, environmental goals are not so well defined and they are constantly becoming more restrictive.

Dr. W. Edwards Deming [6], the noted statistician and quality pioneer, has said that there are three things a manufacturer, interested in continued profitability, must do to be competitive. They are:

1. Innovation of product
2. Innovation of process
3. Improvement of existing process

It is interesting to notice the balance implied between product and process work, perhaps one-third and two-thirds respectively. A recent report by Suzanne Berger (et. al.) [7], which tries to compare American and Japanese industry, states that a marked difference between R&D expenditures exists. American companies devote one-third of their R&D resources against process technology while product related activity consumes the other two-thirds. In Japan these roles in R&D are reversed, this is not surprising as Dr. Deming created the systems model in practice in industrial Japan today.

We believe that the age old philosophy of "if it ain't broke, don't fix it" will no longer serve well in our new global competitive environment. The new philosophy should be to ask "Where is our process least perfect?" and then to apply prioritized resources to judiciously improve continuously.

The cost to adopt a renewed expenditure toward process development may not prove popular in time of corporate belt tightening but we believe it is necessary for long-term business viability.

## REFERENCES


1. SRI International, "Process Economics Program (PEP)," Reports: 2-2D, 11-11C, 22-22B, 33-33B, and 38-38A.
2. J.M. Smith, Chemical Engineering Kinetics, 2nd Ed., McGraw-Hill, New York, 1970, pgs. 110-111, 184-193, and 236-239.
3. Octave Levenspiel, Chemical Reaction Engineering, John Wiley and Sons, New York, 1967, pgs. 140-141.
4. B. Carnahan, H.A. Luther, and J.O. Wilkes, Applied Numerical Methods, John Wiley and Sons, New York, 1969, pg. 363.
5. Raghu Raman, Chemical Process Computations, Elsevier Applied Science Publishers Ltd., New York, 1985, pgs. 539-540.
6. Dr. W.E. Deming, Out of the Crisis, Massachusetts Institute of Technology, 1986, pgs. 5-26.
7. S. Berger, M.L. Dertouzos, R.K. Lester, R.M. Solow, and L.C. Thurow, "Scientific American," Vol. 260, Issue 6, June 1989, pgs. 39-47.

# WASTE MINIMIZATION AND REMEDIATION IN THE METAL FINISHING INDUSTRIES

Clifford W. Walton, Kevin S. Briggs, and Kevin J. Loos

University of Nebraska-Lincoln
Department of Chemical Engineering
236 Avery Laboratory
Lincoln, NE 68588-0126

## INTRODUCTION

An overview of waste assessment procedures and a knowledge of process options for remediation and recycle are necessary for a successful waste minimization program. Factors to consider when constructing a waste minimization program include U.S. EPA priorities, constructing an efficient assessment program, elements of previously successful programs, and opportunities relating to both the process and procedure of the particular system. This work describes the basic steps necessary to fulfill all the factors mentioned in order to create a successful and enduring waste minimization program in the metal finishing industries. Existing and potential technologies are discussed. Available references are presented that assist in the process of analyzing and constructing a suitable program for the particular situation, including descriptions of selected case studies related to surface finishing.

There is a need for more efficient methods for minimization of hazardous waste brought on by the increasing public concern about waste management. Various federal and state hazardous and solid waste regulations have made it more difficult and expensive to dispose of unwanted by-products, prompting prevention techniques. Some undesirable wastes eventually will be banned from land disposal completely, while other disposal methods will be restricted considerably. These restrictions have forced a shortage of approved hazardous waste disposal facilities, causing rapidly rising disposal costs. Due to the financial and legal incentives of reducing or completely eliminating unwanted by-products, and that generators of these by-products are required to certify that they have instituted a waste minimization program, it has become necessary and beneficial to implement a waste minimization program[1].

Figure 1 shows a simplified view of a chemical or product manufacturing process with special attention to discharges[2]. By developing a waste minimization program a manufacturer can save money by reducing waste treatment and disposal costs, raw material purchases, and other operating costs. Also, successful programs are necessary

*Industrial Environmental Chemistry*, Edited by D.T. Sawyer
and A.E. Martell, Plenum Press, New York, 1992

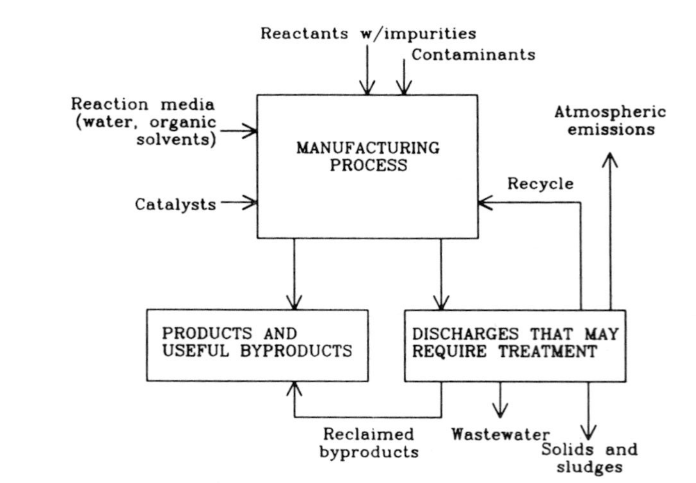


Figure 1. Manufacturing process from the viewpoint of discharges and waste minimization.

Table 1. U.S. EPA Priorities for Waste Management

| Rank | Procedure/Process |
|---|---|
| 1 | Source Reduction |
| 2 | Recycling |
| 3 | Incineration/Treatment |
| 4 | Secure Land Disposal |

to meet national, state and local waste minimization policy goals, reduce potential environmental liabilities, protect public and worker health and safety, and protect the environment[3], as well as promote a better corporate image to the public.

An important step in eliminating or reducing undesirable by-products is to first identify them. Some common by-products generated by metal finishers include industrial treatment sludge containing toxic metals such as cadmium, copper, chromium and nickel; spent plating and process baths, including various cleaners and contaminated solvents; and miscellaneous solid wastes such as absorbents, filters and empty containers[4].

## U.S. EPA PRIORITIES

The former U.S. Environmental Protection Agency (EPA) Administrator, William Reilly, announced pollution prevention as the environmental goal for the 1990's[5]. Table 1 shows the ranking of priorities for waste management established by the U.S. EPA[1,6,7], with particular emphasis on source reduction.

Source reduction is the concept of preventing or minimizing the production of material which must be treated as waste by altering the actual production process. Measures include process modifications, feedstock substitutions, improvements in feedstock purity, improvements in housekeeping and management practices, increases in efficiency of machinery, recycling within a process, and production of valuable by-products.

Recycling is the use or reuse of undesirable by-products as an effective substitute for a commercial product or as an ingredient or feedstock in an industrial process. It can occur on- or off-site and includes the reclamation of useful components from a waste material or the removal of contaminants from a waste to allow it to be reused. It is often less expensive to recycle a chemical than it is to purchase new material and pay for disposal costs. Though a material may no longer meet the specifications for a process in which it is being used, the material may still be suitable for other uses at the facility. The ability to reuse is aided by adequate waste separation and concentration. This also increases possible waste exchange possibilities, where one industry's unwanted by-products are another's raw material.

Waste treatment is any method, technique, or process that changes the physical, chemical, or biological character of any hazardous waste in a way that neutralizes the waste, or renders such waste nonhazardous, less hazardous, safer to manage, able to be recovered, able to be stored, or reduced in volume. This includes recovery of energy content by using a waste as a fuel supplement, particularly for organic solvents and oils. Incineration is a last resort for a material, such as dioxane, that can only be made non-hazardous by complete oxidation at high temperatures.

Disposal is the discharging, depositing, injecting, or placing of hazardous waste into or on any land. If a waste has hazardous characteristics, then disposal in a regulated hazardous waste disposal facility is required. This secure land disposal is the least desirable option. If the waste is treated to eliminate these hazardous characteristics, it can be disposed of in a non-hazardous waste landfill, which is considerably less expensive.

The use of the various options depends on extensive federal and state regulations. Table 2 presents a brief list of the prominent laws and regulations that have an impact on the surface finishing industry. Note that the list cannot be construed as comprehensive and any interested party should contact the appropriate federal, state, or local authority regarding the specifics of these and current regulations.

## ESSENTIAL ELEMENTS AND COMMON BARRIERS OF SUCCESSFUL MINIMIZATION PROJECTS

Various studies of waste minimization programs have identified characteristics of successful programs. Three key elements found were[1]

- Management's active involvement and encouragement at all levels,
- Production personnel strongly motivated to implement and maintain necessary changes, and
- Technologies used were elegant in their simplicity.

At least one of these elements were missing in each minimization project failure[1].

People involved in implementing waste reduction programs have noted several inhibiting factors which must be identified by those attempting to develop a waste reduction strategy. This identification is necessary so that the resistance can be recognized and overcome. Barriers to waste reduction include[4]

- Lack of information about available waste reduction techniques and the benefits that can be achieved.

  This can be overcome by seeking information on other's successes. Examples can be found in the case studies published by the U.S. EPA[8,9].

- Concern for upsetting product quality.

  Both operators and management must be convinced that pollution prevention changes will not adversely affect quality. This requires a commitment to and effective application of a new process or technology, including complete training of operators.

- The "if it ain't broke–don't fix it" attitude.
- A reluctance to develop innovative ideas because of the fear of failure.

  Management must take the initiative and absorb all effects of failures, only then will a commitment from operators and middle managers be made.

- The attitude that a new technology will not succeed because it is outside the company's normal range of expertise.

  This attitude problem can be solved by seeking outside views and help. State programs, consultants, and professional organizations such as the American Electroplaters and Surface Finishers Society can assist in answering a company's questions.

Table 2. Some Environmental Laws and Regulations Pertinent to the Surface Finishing Industry[4] (1989)

| Regulation | Description |
| --- | --- |
| 40 CFR 122, NPDES | Federal regulations governing the discharge of wastewaters to surface waters of the United States. |
| 40 CFR 413, 433 | Federal regulations specifying effluent limitations, pretreatment standards, and new source performance standards for the electroplating and metal finishing industries. |
| 40 CFR 268 | Federal regulations that restrict the disposal of spent solvents and solvent-containing wastes. |
| CCR Title 23 Subchapter 9 | State regulations governing the discharge of wastewaters to surface waters. Includes provisions for issuance of permits and setting effluent limitations. |
| Local municipal codes addressing discharges to POTWs | Discharge requirements set by local POTWs restricting the concentrations of pollutants in wastewaters discharged to sanitary sewers. |
| CCR Title 22 Sections 66900 and 66905 | Restricts the land disposal of certain liquid and solid hazardous wastes and sets time schedules for implementing the restrictions. |
| CHSC Chapter 6.95 | Requires local government agencies to implement hazardous material management programs requiring local businesses to submit applications for the storage and handling of hazardous materials. |
| CCR Title 22, Division 4 Chapter 30, Article 6 | Sets requirements for generators of hazardous wastes, including restrictions on how long wastes can be accumulated without the storage facility being permitted. |
| CHSC 25202.9 | Requires certification, by waste generators permitted as TSD facilities, that a waste minimization program is in operation and that the treatment, storage, and disposal methods minimize the threat to human health and the environment |

CFR - Code of Federal Regulations

CCR - California Code of Regulations

CHSC - California Health and Safety Code

NPDES - National Pollution Discharge Elimination System

When developing and implementing a waste reduction program it is essential that management be committed to pursuing waste reduction, be willing to experiment with various ideas, and be prepared to experience failure, as well as success[4].

## WASTE MINIMIZATION ASSESSMENT PROCEDURE

A waste minimization assessment procedure begins with the recognized need to minimize waste. The U.S. EPA has published a detailed description of such a procedure[3], which can be broken into a

1. Planning and organization phase,
2. Assessment phase,
3. Feasibility analysis phase, and
4. Implementation phase.

The Planning and Organization Phase of developing a successful waste minimization (WM) program begins with achieving support from the management level. Management support of a program will occur only if it is convinced that the benefits obtained, such as economic advantages, compliance with regulations, reduction in liabilities, improved public image, and reduced environmental impact, will outweigh the costs. A formal policy is the best way to convey to the company's objectives to the employees. A policy allows the employees to focus on what they should be achieving and is a starting point to getting the employees involved. Commitment throughout an entire organization, not just at the management level, is necessary for the implementation of successful programs. Since employees are most often responsible for the generation of wastes, they can be very important to the overall success of a program.

Task forces are one of the first steps a company should take when implementing a WM program. This group should include employees from every area that the program will affect. The top priority of a task force is to set goals for the WM program. These goals should remain consistent with the company's objectives and often it is best to develop quantitative goals, or goals which can be easily measured. Periodically, the goals should be reviewed in order to reflect any change in the company's objectives. Barriers to a specific WM program must also be analyzed in order to understand what can be achieved by a company's particular program.

After setting goals for the WM program, the Assessment Phase begins with a task force given the responsibility to identify and characterize the facility waste streams. Information on the waste streams can come from a variety of sources, including completed hazardous waste manifests, biennial reports, NPDES (National Pollutant Discharge Elimination System) monitoring reports, and the "right to know" provisions of the Superfund Amendment and Reauthorization Act (SARA). Once all the waste streams have been assessed, it is necessary to prioritize these waste streams and/or operations in order of greatest potential risk. Assessment teams can then be selected to focus on a particular waste stream or process. These teams should include people that are knowledgeable about the stream or process, and are often made up of internal staff and outside people. Outside consultants can add experience about a stream or process, especially if there are no experts in a particular area within the company.

After visiting the site, a comprehensive set of options can be generated by the various task forces. Most of these options will come from education and on-the-job expertise, however technical literature can also be helpful. Once all the options have been generated, screening of these options must occur in order to decide which options are real possibilities in minimizing waste and reducing cost. A list of particular questions should be answered in relation to each option. The goal of the screening is to decide which options are suitable for a technical and economic feasibility analysis.

The next step in developing the WM program is the Feasibility Analysis Phase, which includes a technical and economic evaluation. Now that an initial set of options have been selected it is necessary to evaluate them from a technical standpoint, in order determine whether an option will work in a specific application, and from an economic standpoint, such as payback period, return on investment, and net present value.

The technical evaluation can include visiting sites to observe in-use equipment and processes, requesting both vendor and operator comments on a particular process. Also included in the evaluation are facility constraints and product requirements. If an option does not appear feasible after the technical evaluation, it should be dropped from the list.

The economic evaluation includes analysis of capital costs, operating costs, and savings. Once these have been evaluated a profitability analysis is performed. A project's profitability is measured using net cash flows for each year of the project's life. If a project contains no significant capital costs, the profitability of the project can be decided on whether an operating cost savings occurs or not. If it is found that a project reduces operating cost, it should be implemented as soon as possible.

Implementation of the WM option(s) is the last step of a WM program. If the option(s) are approved by the management, it is then necessary to obtain funding from an appropriate level within the organization. A beneficial step is to have an evaluation team made up of financial and technical personnel. This will assure the project funders that the sponsor is being objective, and not just enthusiastic. Also, this team will be able to eliminate funders' attitudes of "it can't be done" and "if it ain't broke, don't fix it." Once the funding barrier has been overcome, installation of the option can occur. A demonstration of the option is then performed and its performance is evaluated. A measure of effectiveness of a particular WM option is its ability to pay for itself through reduced waste management costs and reduced raw materials costs.

Once a WM program is implemented, it is necessary to continually re-evaluate the changes, rather than making WM a one-time effort. After the highest priority streams or processes have been evaluated, a company should continue down the list of priorities that were formulated in the beginning. If necessary, the priorities should be re-evaluated in order to ensure that each stream or process is prioritized correctly and is still a significant part of the WM program. An ultimate goal is to maximize the reduction of waste, and if this has been achieved, companies should evaluate reduction of other processes, such as industrial wastewater discharges, air emissions, and solid wastes.

## WASTE REDUCTION OPPORTUNITIES

A waste reduction program can be broken down into two broad areas: procedure and process. The procedure includes employee training, changing methods, preventing spills, and inventory control, while the process includes optimizing solutions, substituting products, concentrating waste, and changing equipment.

The most important aspect of the procedure is employee training. Although management commitment and direction are fundamental to a waste minimization program, commitment throughout an organization is crucial to resolve conflicts and overcome barriers to the program. Employees often can increase the generation of waste, so they can be pivotal in contributing to the overall success of a waste minimization program. Bonuses, awards, plaques, and other forms of recognition often times are the best way to provide motivation and to boost employee cooperation[3].

Higgins[1] gives the example of a chemical engineer at a major consulting company who spent the early part of her career as an environmental coordinator for a major aircraft manufacturer. Solvents kept turning up in the storm drains and the source was traced to several floor drains in the shop area where workers were pouring used solvent. She then posted signs that read "NEXT STOP IS YOUR FAUCET AT HOME." The violations stopped. When the workers asked her what they could do to reduce waste, she responded, "Act as though your mother were looking over your shoulder[1]."

Some control methods that can be implemented include

- Minimizing the number of different raw materials and supplies used, such as cleaning fluids, oils, etc. This helps clear-up shelf-life problems and reduces the number of partially filled containers requiring disposal.
- Purchasing container sizes appropriate to the actual use. It can be less expensive to buy smaller containers of perishable materials than to buy bulk quantities, at cheaper prices, that require disposal of the expired portion.
- Reducing the inventory of hazardous materials to a minimum. It must be ensured that containers are being rotated on the shelves so that the oldest materials are used first. This will reduce the disposal of materials whose shelf-life has expired.

These steps will reduce spending on raw materials and waste disposal, as well as the investment tied up in inventory. Such areas as preventing spills and inventory control are simple procedures which can cut significant amounts of waste.

After preparing an inventory of chemicals being used and wastes produced from all processes, an evaluation is needed to determine if the number and volume of chemicals being used can be reduced. Optimizing solutions, taking the waste of one process and adapting it to be used in another, substituting products, less-hazardous for hazardous materials, and concentrating waste are possible modifications[1].

When dealing with a change of equipment, simplicity is best. Typically, on successful projects, off-the-shelf equipment is adapted to a new application, while complex or specialized equipment is avoided. The greater number of modifications attempted at the same time, or the more experimental the equipment is, the greater the chance that problems will occur and the process will be abandoned for more reliable and proven methods[1].

In the surface finishing industry, there are numerous simple approaches. For example, several improved rinsing techniques that can reduce the amount of process fluid losses include[1,4,10,11] extended drip times to allow complete drainage, workpiece positioning to minimize holdup of process fluids, spray/fog rinsing to reduce the amount of rinse water to be treated, countercurrent rinses that can drastically reduce required flow rates of rinsewater, use of drain boards or drip tanks to catch and return process fluids to their origin tanks, air knives to remove process fluids without dilution, and use of turbulence or agitation to aid rinsing effectiveness.

## WASTE REDUCTION TECHNOLOGIES

Evaporation, ion exchange (IX), reverse osmosis (RO), electrolytic recovery, and electrodialysis (ED) have been used to recover chemicals and metals from rinsewaters.

Table 3. Summary of Recovery Technology Applications[4]

| Plating Bath | Technology | | | | |
|---|---|---|---|---|---|
| | Evaporation | Reverse Osmosis | Ion Exchange | Electrolysis | Electrodialysis |
| Decorative Chromium | x | | x | | |
| Nickel | x | x | x | | x |
| Electroless Nickel | | | x | | |
| Cadmium | x | | | x | x |
| Zinc (CN) | x | x | | x | x |
| Zinc (Cl) | x | | | x | x |
| Copper (CN) | x | x | | x | x |
| Tin ($BF_4$) | x | | | x | x |
| Silver | x | | x | x | x |

General and site-specific factors must be evaluated to determine the best recovery process for a specific situation. Factors include the type of metal to be recovered, drag-out rates, rinsewater concentrations and flows, space requirements, staffing requirements, availability of utilities (such as steam and electricity), and cost for water and wastewater treatment and sludge disposal[1]. Table 3 shows typical recovery technologies related to specific types of plating baths.

Evaporation is the oldest method used to recover chemicals from electroplating rinse streams. In this process, enough rinsewater is evaporated to concentrate the solution sufficiently to allow its return to the particular process. Because of their high-energy use, evaporators are most cost-effective in concentrating rinsewaters that are to be returned to hot baths, such as in a chromium plating process[1].

Ion exchange (IX) uses charged sites on solid resin to selectively remove either cations or anions from the solution. The ions removed are then replaced by equivalent charged ions displaced from the resin. In general, ion exchange systems are suitable for recovery applications where the rinsewater has a relatively dilute concentration of plating chemicals. Ion-exchange systems are not cost-effective when drag-out rates are low and concentrations high[1].

Reverse osmosis (RO) is a demineralization process in which water is separated from dissolved metal salts by forcing the water through a semipermeable membrane at high pressures (400 to 800 psig). RO use is limited to moderately concentrated rinsewaters. The cost-effectiveness of RO metal recovery systems varies depending on several factors[1].

Electrodialysis (ED) concentrates or separates ionic species in a water solution through use of an electric field and semipermeable ion-selective membranes. ED has been used effectively to recover cationic metals from plating rinsewaters. The maxi-

Table 4. Some Material Recovery Technology Costs (1985)[4]

| Technology | Materials Recovered | Equipment Costs |
|---|---|---|
| Evaporation Unit: Capacity of approx. 20 gph. | Rinse water and Chromic acid | $47,000 |
| Reverse Osmosis Unit: Capacity of approx. 100 gph. | Nickel plating chemicals | $27,000 |
| Ion Exchange Unit: Capacity of approx. 20 gph. | Rinse water and Chromic acid | $38,000 |
| Electrolytic Unit: Capacity of approx. 15 gph. | Rinse water and Copper | $25,000 |

mum concentration of an ED unit is limited only by the solubility of the compounds in solution, which allows ED to produce a more concentrated solution than IX or RO. ED units are easy to use, economical, require little space, and can be operated continuously with little maintenance. However, ED cannot selectively remove ions, resulting in the requirement for periodic plating bath treatments to remove impurities. Along with evaporation, ED has the quickest payback period of a metal recovery process[1].

Table 4 lists some typical costs of applying the described technologies to recover specific materials from a single source wastewater. To achieve savings that justify the purchase of recovery technology requires the waste stream to be fairly concentrated and continuous. Information necessary to determine economic feasibility includes waste stream generation rates, chemical concentrations, and the value of materials to be recovered[4].

## SPECIFIC OPTIONS AND CASES

### A Case Study at Pratt and Whitney[12]

In 1986, Pratt and Whitney's (P&W) North Haven, Connecticut facility began to plan conceptually for a "zero discharge" metal finishing capability. At that time P&W discharged about 1 million gallons a day of treated wastewater, of which about 400,000 gal/day were generated by metal finishing operations. It was very clear that a waste management plan was needed. The decision was made for a company wide reduction of hazardous wastes by 40% and toxic air by 50% by 1996. Their program involved:

- identifying waste reduction opportunities which encompass good operating practices and proven technologies,

- working with the chemical processing industry to investigate emerging waste reduction and treatment technologies which could be piloted and evaluated, and

- determining new hazardous waste reduction areas for research that can meet specific corporate needs.

P&W's North Haven facility has already taken a jump and "closed the loop" on such key processes as their Woods nickel strike, sulfamate nickel plating, hard chromium plating, cadmium plating, chromating, and cadmium, chromium and nickel stripping. The 40% of wastewater generated by plating operations of 1986 now accounts for about 4% of the discharge water.

P&W's program has been organized around the following implementation hierarchy: **Phase One** defined good operating practices, including the decision to

- Define minimum water quality standards. All water now used is either deionized or softened.
- Use countercurrent rinses to reduce water usage.
- Use continuous process purification, as opposed to batch purification, to maintain consistent process quality. This includes dummy plating and carbon and particulate filtration.
- Alter on-line process solutions to control drag-out (for example, reduce concentration and increase temperature).
- Optimize preplate rinsing to control drag-in contaminants.
- Install automatic level controls on all heated processes.
- Train operators to understand proper rinsing and work transfer techniques to reduce drag-in and drag-out.
- Treat small concentrated batches versus high volume dilute waste streams.

**Phase Two** implemented procedural changes based upon established good operating procedures.

**Phase Three** verified closed loop technology on a single process. For this phase, a modularized approach has allowed the fine tuning of designs, while disruption of production has been minimized.

**Phase Four** incorporated good operating practices and closed-loop technologies in the design of planned and appropriated new plating lines. Designs for new plating lines encompassing nickel and chromium plating, cadmium, chromium, and nickel stripping, and titanium descaling were already under way. Initial plans were revised to incorporate

- Countercurrent rinses,
- Ion exchange for nickel strike, and nickel, cadmium and chromium stripping operations,
- Atmospheric evaporation for hard chromium and sulfamate nickel lines, and
- Deionized water in all critical rinses and softened water in all non-critical rinses and non-critical evaporation make-up.

**Phase Five** involved installing new plating lines, which were completed in October 1990.

**Phase Six** will be the renovation of remaining existing processes, including cadmium cyanide plating and chromating.

Through these waste minimization efforts P&W's raw material costs have been reduced by 80% and water usage by 95%. The costs of transportation and disposal have been reduced proportionately and so have the liabilities. In the future P&W plans on implementation of continuous process purification, implementation of closed-loop technologies on preplate rinses, and implementation of on-line monitoring of all systems[12].

**Water Use Reduction Techniques**

There are many methods for reducing the overall amount of water used in a plating process and there are corresponding benefits for these reductions. These may include reduced water treatment costs, disposal of waste costs, reduced purchasing costs for chemicals and water, and overall cost reduction for an operating day. Some of these reduction methods are:

- Increase drip and rinse times on all automated lines by reprogramming computer times, or changing gears in the hoist system to increase lift time. By increasing the drip time more of the solution will stay in the tank, thus increasing the life of the bath and not contaminating the following bath. By increasing the lift time, better rinsing is accomplished because of the slower rate of breaking the surface tension. More water is then pulled off the parts.

- Drip shields should be installed on all tanks. This technology has been around for many years and can drastically reduce the amount of solution lost to drag-out. The lives of baths are again lengthened due to reduced contamination from drag-in and chemical loss due to drag-out, and much less waste treatment of diluted streams has to be performed on the discharge effluent.

- Air agitation should be employed in all process tanks. This is another technique for breaking the surface tension, which, in turn, allows for better rinsing. This technique also promotes the elimination of temperature gradients, and brings solids and impurities to the surface where they can be filtered easily.

- A condensate return system may be installed on the plant's boilers which will reduce overall water consumption. This source of water effluent should not be mixed with the discharge of the plating operations if at all possible. This stream will most likely be clean enough to discharge without treatment and will only dilute the waste treatment stream.

- Atmospheric evaporators can be installed on plating lines. The evaporators take the rinsewater from the drag-out and concentrate it. The remaining solution (a concentrated chemical) can be put back into the plating cell to replenish the bath. Recovering and recycling chemicals from the evaporators can slow the need to make large additions of chemicals to the process tanks. This provides a savings not only on the waste treatment of such chemicals, but also on the initial purchase of the chemicals.

- Continuous filtration should be used on plating cells to clean solids from the baths. This extends the life of the process solution and fewer chemicals have to be added because there is a lower level of impurities in the solution. The net result is less water going to the waste treatment system, and a reduced amount of chemicals having to be purchased. Filtration can be done either in-tank or out-of-tank. Most current systems are out-of-tank, but the current trend is towards in-tank methods[13].

- Continuous in-tank filtration versus out-of-tank. Conventional filter systems typically recirculate solution through a hose extending from the bath to be filtered to a chamber containing the filter media. This provides many sites for leaks and spills. The in-tank filtration method greatly reduces the chances of these leaks or spills because it uses no hoses, clamps, seals, or any out-of-tank apparatus. The filter is attached to the pump body and is immersed in the solution to be filtered. The pump body is secured to the tank lip with an adjustable bracket. Solution is continuously drawn by a vacuum through the filter cartridge, after which it travels about three inches before being expelled through a discharge port on the pump body. The solution is therefore never out of the tank, and is not subject to the normal leaks or spills that can occur.

  In addition to these benefits the in-tank unit creates enough force from it's discharge port into the bath to produce a clean, oil free source of agitation that suspends particles for easy filtration. This force also gives uniform temperature and chemical gradients. Frequently, the discharge port is sufficient to replace mechanical or air agitation. If continuous filtration is not required, a single unit may be used in multiple tanks. The in-tank filtration units available today are designed for ease of cleaning, filter changing, and no use of floor space[14].

- Reverse Osmosis (RO) may be used to reduce the volume of water in waste streams. RO technology has been used for this purpose in the past but today's advancements have made smaller, more efficient, units than could have been imagined in the past.

- Ion Exchange may be used to recycle rinsewater from anodization processes. The ion exchange units can be designed and purchased for use on individual compounds, or the units may be placed in series to remove virtually all of the effluent metals. The solutions to be used in this type of process must be somewhat segregated and the costs can be high unless sufficient amounts of solution are to be treated[14].

## Recovery of Metals From Baths

There are many incentives for recovering metals from waste solutions and to reach these incentives many methods of recovery have been established. By far the most significant reason for recovery of metals is their intrinsic value. The tremendous savings in initial purchase costs of most metals should be enough to warrant a recovery system. The costs of disposal and the tremendous waste of metals are other reasons that recovery of metals is becoming a process option in more and more plating operations. Some of these process options include

- Precious metals recovery using cathodic plating. Silver and gold may be plated out of the rinsewater solution onto a cathode. These cathodes may later be sent to a precious metals refiner for credit, either being sold for scrap or recycled into anodes and reused at a reduced cost in the plating lines[13]. Precious metals may also be recovered using filtration[15].

- Electroless copper deactivation by removing formaldehyde from the waste solution. This is done by adding hydrogen peroxide to the solution. This deactivated copper will not plate out in drain lines, etc., and can therefore be almost fully removed. Aluminum is now being used to plate the deactivated copper out of solution[16].

- Electrowinning of metals such as copper, cadmium, and zinc. The advantage to this method is that only the metals in solution are plated out, the other contaminants are left unchanged. The factors affecting this method are cathode surface area, temperature, and agitation. The overall efficiency of this method is excellent, but the costs can be high unless a large amount of metal is being recovered[15].

## Specific Methods for Reduction of Hazardous Wastes

Hazardous wastes come in all shapes and forms, liquids, solids and sludges. They can be either by-products of manufacturing processes or maintenance operations or discarded commercial products. Simply burying the wastes, regardless of their form or source, is no longer an acceptable means of disposal. Effective treatment, rather than land disposal is now a necessity. Several methods of hazardous waste minimization are available, some of the methods available are

- The use of inventory management to reduce the overall amount of wastes generated as mentioned earlier. All raw materials should be inventoried and only the minimum amount should be purchased. This can eliminate excess materials and chemicals from being thrown away if a process change occurs, or if the shelf life expires. Raw materials should be purchased in a container that most closely matches the expected use. This can result in less raw materials being thrown away if a lid is not resealed on a half empty container. Reusable containers should be purchased if at all possible, thus eliminating triple rinsing of an almost empty container, or disposing of a container that contains enough residue to be considered hazardous waste.

- Raw material substitution may be used to reduce the volume of hazardous waste or its toxicity. The initial raw materials purchased should be as pure as possible to reduce hazardous by-products. Non-hazardous or less hazardous materials should be investigated and purchased. Lower volatility solvents and cleaners can be used to reduce VOC air emissions. Phosphate based corrosion inhibitor chemicals should be used instead of chromates. The use of ferrous sulfate-sodium sulfate processes instead of lime for heavy metal precipitation can result in up to 90% weight reductions in waste sludges. A terpine-water emulsion has also now been developed as a substitute for ozone-damaging chlorofluorocarbons in certain degreasing operations.

- Process design and operation changes may be implemented to reduce the volume of wastes generated. Equipment which produces the minimum amount of waste should be purchased and existing equipment should be modified towards the same end. Automatic metering pumps should be used to monitor chemical additions. Consideration of redesigning or reformulating the end products to be less hazardous can also contribute to an overall reduction. Electrostatic paint equipment may be used to reduce solvent emissions. High solids paint and water-based paint are both alternates to solvent-based liquid paints. If solvent-based liquid paint must be used, high volume, low pressure guns should be used.

- Overall volume reduction can also be achieved through simple segregation of streams. Non-hazardous streams should not be mixed with hazardous streams if at all possible. Streams containing different solvents or chemicals should be kept separate so that the chemicals may be more easily recycled. The waste streams which must be treated should be reduced with such techniques as reverse osmosis, evaporation, ion exchange, precipitation, centrifugation, gravity separation, and distillation.

- Hazardous waste can also be chemically altered to reduce its toxicity or create a non-hazardous product. The waste may be encapsulated so that it is sealed with a material that makes the overall product non-hazardous. Waste oil may be chemically treated and turned into a salable aromatic distillate. Hydrochloric acid may be converted into calcium chloride and used in manufacturing polyamide fibers or into calcium sulfate and used as a soil stabilizer[17].

- The use of liquid cleaners instead of powder. Until recently most platers have relied on powder cleaners for their pretreatment of the pieces to be plated. The problems associated with powder cleaners can be safety, handling, mixing and solubility, control of concentration, useful life, and waste treatment. The use of liquid cleaners eliminates worker exposure to irritating and corrosive dusts associated with powder blends, and reduces airborne particles to help meet OSHA exposure regulations. Powder cleaners may sometimes mix non-uniformly in the solution or may not mix at all if the temperature and agitation are not adequate. This can result in powder settling on the bottom of the tank, creating even more problems. There is no guarantee of the concentration and mixing with a powder because it is impossible to tell how much has gone into solution. The use of liquid cleaners can eliminate all of these problems. In addition, liquid cleaners can be fed by a metering pump into the solution giving excellent concentration control and requiring minimal worker exposure. The total solids of liquid cleaners are typically 5 to 10 percent, as opposed to 90 to 98 percent for powder cleaners, meaning that less sludge is generated during waste treatment. Due to manufacturing improvements, liquid cleaners now out perform most powders and can be designed for almost any use[18].

- An electrodeposited $PbO_2$ anode may be used for the decomposition of low concentrations of cyanide. Titanium anodes electroplated with lead dioxide in nitrate solutions containing a surfactant or an organic additive are more effective than the traditional graphite anode for decomposing cyanide in solutions containing a low cyanide concentration and sodium chloride. This is due to the porosity of the $PbO_2$ surface layer and the electrode potential for the reaction[19].

- Microcomputers can be very useful in waste management performing such tasks as modeling processes, using material balances. The greatest asset of these computers is that they can be put on the desk of the environmental engineer at very little cost. Many of the microcomputers of today can be purchased very inexpensively and can pay for themselves in a short amount of time. FORTRAN and Basic, as well as many other languages, spreadsheets and equation solving software, can now be used on these computers. This gives the engineer in charge of locating, reducing, and treating the effluent waste a very good tool for fast, effective, and accurate modeling and calculations[20,21].

## SUMMARY

When developing and implementing a waste reduction program it is essential that management be committed to pursuing waste reduction, be willing to experiment with various ideas, and be prepared to experience failure, as well as success. Within a successful waste reduction program three essential elements must be present: active involvement and encouragement of management at all levels, application of the least complex technologies, and motivated production personnel to implement and maintain necessary changes. The second element is especially important because the greater the number of modifications attempted at the same time, or the more experimental the equipment is, the greater the chance that problems will occur and the process will be abandoned for more reliable and proven methods.

Developing a waste minimization program has become a necessity in industry. As U.S. EPA regulations on waste disposal become stricter, newer, more efficient minimization methods of waste disposal must be developed. Waste management methods that reduce or eliminate unwanted by-products will prove to be more cost-efficient and environmentally sound, traits that will be in high demand in the future.

## ACKNOWLEDGEMENTS

Although the review described in this paper has been funded in part by the U.S. EPA under assistance agreement number R-815709, to the University of Nebraska-Lincoln through the Great Plains/Rocky Mountain Hazardous Substance Research Center, headquartered at Kansas State University, it has not been subjected to the Agency's peer and administrative review and, therefore, may not necessarily reflect the views of the Agency and no official endorsement should be inferred.

The research was partially supported by the U.S. DOE, Office of Environmental Restoration and Waste Management, Office of Technology Development and the University of Nebraska–Lincoln Water Center.

## REFERENCES

1. T.E. Higgins, "Hazardous Waste Minimization Handbook," Lewis Publishers, Chelsea, Michigan (1989).
2. S.E. Manahan, "Environmental Chemistry," 5th ed, Lewis Publishers, Chelsea, Michigan, 453 (1991).
3. "Waste Minimization Opportunity Assessment Manual," U.S. EPA, Washington, DC, EPA/625/7-88/003 (July 1988).
4. PRC Environmental Management, Inc., "Hazardous Waste Reduction in the Metal Finishing Industry," Noyes Data Corp., Park Ridge, NJ (1989).
5. J.J. Segna and R.K. Raghavan, Approaches to reducing environmental risk through pollution prevention, *in*: "The Environmental Challenge of the 1990's," U.S. EPA, Washington, DC, EPA/600/9-90/039, 613-23 (Sep 1990).
6. R.W. Rittmeyer, Waste minimization–part 1: prepare an effective pollution-prevention program," *Chemical Engineering Progress*, **87**:5 (May 1991).
7. R.J. Avendt, Waste minimization–approaches and techniques, *in*: "The Environmental Challenge of the 1990's," U.S. EPA, Washington, DC, EPA/600/9-90/039, 32-40 (Sep 1990).

8. "Seminar Publication: Meeting Hazardous Waste Requirements for Metal Finishers," U.S. EPA, Washington, DC, EPA/625/4-87/018 (September 1987).

9. Additional technology and case study information can be obtained from:

   U.S. EPA
   ORD Publications
   P.O. Box 19962
   Cincinnati, OH 45219-0962

   Request "ORD Publications Announcement" (e.g., EPA/600/M-91/048) and "Technology Transfer" Newsletter (e.g., EPA/600/M-91/042).

10. "Metal Finishing Guidebook and Directory Issue," *Metal Finishing*, **89**:1A (1991).

11. "Guides to Pollution Prevention: The Printed Circuit Board Manufacturing Industry," U.S. EPA, Washington, DC, EPA/625/7-90/007 (June 1990).

12. P. Gallerani and R. McCarvill, Waste minimization and pollution prevention at Pratt and Whitney, *Plating & Surface Finishing*, **78**:3, 36 (March 1991).

13. P. Dalton, Water conservation and waste minimization efforts at Hudgins Plating, *Plating & Surface Finishing*, **78**:7, 28 (July 1991).

14. G. Horvath, In-tank filtration/solution purification: a new concept in waste minimization, *Plating & Surface Finishing*, **78**:4, 28 (April 1991).

15. F.W. Kursch and G.P. Looby, Case study: Pollution prevention in practice: applications for electroplating technology, *Pollution Prevention Review*, **2**:1, 63 (Winter 1991-92).

16. J.D. Holly, Waste treatment process for electroless copper, *Plating & Surface Finishing*, **78**:1, 24 (January 1991).

17. R.A. Katlin, Waste minimization—part 3 minimize waste at operating plants, *Chemical Engineering Progress*, **87**:7, 39 (July 1991).

18. S.F. Rudy and B. Durkin, The use of liquid cleaners in metal finishing, *Plating & Surface Finishing*, **77**:3, 26 (March 1990).

19. T.-C. Wen, Electrodeposited $PbO_2$ anode for the decomposition of low concentrations of cyanide, *Plating & Surface Finishing*, **77**:11, 54 (November 1990).

20. C.W. Walton and G.L. Poppe, Applying microcomputers to the analysis of waste treatment and recovery processes, *Plating & Surface Finishing*, **77**:6, 48 (June 1990).

21. A.C. Hillier and C.W. Walton, Modeling electroplating rinse systems using equation-solving software, *Plating & Surface Finishing*, **78**:11, 72 (November 1991).

# THE ROLE OF CATALYSIS IN INDUSTRIAL WASTE REDUCTION

David Allen

Department of Chemical Engineering
University of California, Los Angeles
Los Angeles, Ca. 90024

## INTRODUCTION

Millions of tons of hazardous waste are generated each year in the United States, and the direct costs of managing these materials are in the tens of billions of dollars. As public pressure for reduced, or even zero discharge grow, waste management costs are likely to further escalate, and conventional end-of-pipe treatment measures will become cost prohibitive. A new approach to waste management will be required, which stresses the elimination or reduction of wastes at their source, rather than capture after their formation. This new approach to waste management has been called waste reduction, pollution prevention and source reduction. The approach offers both challenges and opportunities. The challenge is to control the costs of environmental stewardship so that products produced in an environmentally benign manner can remain competitive in global markets. If this challenge is met successfully, then the clean technologies that are developed have the opportunity to dominate world markets.

Meeting the challenges of waste reduction will require key advances in the field of engineering design, particularly in catalysis and reaction engineering. The goal of this paper is to describe the role that catalysis can play in reducing waste. Our starting point will be a description of the generation and management of industrial hazardous waste. From this brief overview, it will become apparent that catalysis is currently underused as method for converting wastes into recyclable material. In addition, a more detailed examination of the processes generating hazardous wastes will show that chemical processes that rely on catalysis are responsible for a significant fraction of industrial hazardous waste generation. Clearly then, improved selectivities in catalytic processes could reduce waste generation. Our examination of the potential role of catalysis in waste reduction will conclude with the presentation of a few case studies.

## HAZARDOUS WASTE GENERATION AND MANAGEMENT: The role of catalysis

According to recent estimates (Baker and Warren, 1992), more than 750 million tons of waste, classified as hazardous under the provisions of the Resource Conservation and Recovery Act, are generated annually in the United States. Over 90% of these wastes are in the form of wastewaters, and more than 90% of the material is managed at the site where the waste was generated. Common management methods include various forms of

*Industrial Environmental Chemistry*, Edited by D.T. Sawyer
and A.E. Martell, Plenum Press, New York, 1992

wastewater treatment, incineration, reuse as fuel, land disposal technologies and recycling. The flow of material to each of these technologies is given in Figure 1. A detailed description of the data presented in Figure 1 is beyond the scope of this paper, but is provided elsewhere (Behmanesh et al., 1992; Baker, et al., 1992). The industrial sectors generating these wastes are shown in Figure 2. Chemical manufacturing operations clearly dominate hazardous waste generation, producing more than 50% of the total waste mass.

While it is clear that the chemical process industries dominate industrial hazardous waste generation, it is not clear which chemical processes bear primary responsibility. Data on waste generation and management, of the type shown in Figures 1 and 2, are generally reported for entire facilities, rather than for individual processes. So it is very difficult to determine, strictly from waste reporting, which catalytic processes might be most significant in waste generation. A sense of the relative waste generation of different chemical processes can be obtained based on industry wide yield data, however. Table 1 reports annual production rates and industry wide average yield data for a number of commodity chemicals. Within Table 1, the data listed for acrylonitrile, 1,3 butadiene and vinyl chloride illustrate a number of interesting points. For acrylonitrile, high production rates and low selectivities indicate potential for high impact improvements. In contrast, although the data for 1,3 butadiene indicate high production rates and relatively low conversions, the by-products and unreacted feed are not generally wastes in this process, but rather have direct fuel or chemical value. Finally, the data for vinyl chloride indicate moderate production rates and high selectivities, but, as we shall see in a later section, catalytic technologies can play a role in reducing wastes from this process.

Thus, while waste stream data, production data and yield data can be interesting starting points in looking for opportunities for catalysis to contribute to waste reduction, they do not tell the entire story. Until waste generation data are available on a process by process basis, it will be difficult for external observers to identify specific opportunities for waste reduction. Still, the data can be used to reveal some generic opportunities for catalysis. These include:

1. Catalysis could be used more extensively in waste treatment, either as a stand alone treatment unit or as a waste pretreatment method. The data presented in Figure 1 clearly show that catalysis currently plays a very minor role in the treatment of industrial hazardous wastes. There are a number of legal and technological reasons for this. Among the technological limitations are the wide variations in composition frequently encountered in waste treatment. Designing robust catalysts capable of resisting the wide range of deactivating agents found in waste streams can be difficult. The technological barriers are not, however, impossible to overcome and a number of treatment technologies employing oxidative and reductive catalysis have been developed (for example, Aguado et al., 1991; Kim et al., 1990; Hasan et al., 1990). These new technologies can be difficult to implement, however, due to the administrative and legal hurdles involved in certifying a new waste treatment technology (NETAC, 1990).

2. Catalysis could be used to convert waste streams into recyclable material. A number of technologies have been developed which use catalysis to convert waste streams into recyclable material. Several examples will be presented as case studies in the next section. All of these case studies will focus on hydrodechlorination chemistry, but, many other reduction and partial oxidation chemistries have promise.

3. Catalysis can play a significant role in waste reduction through selectivity improvement. Maximizing selectivity, i.e. the efficient use of raw materials, has always been a major factor in the design of chemical processes, and so it would be

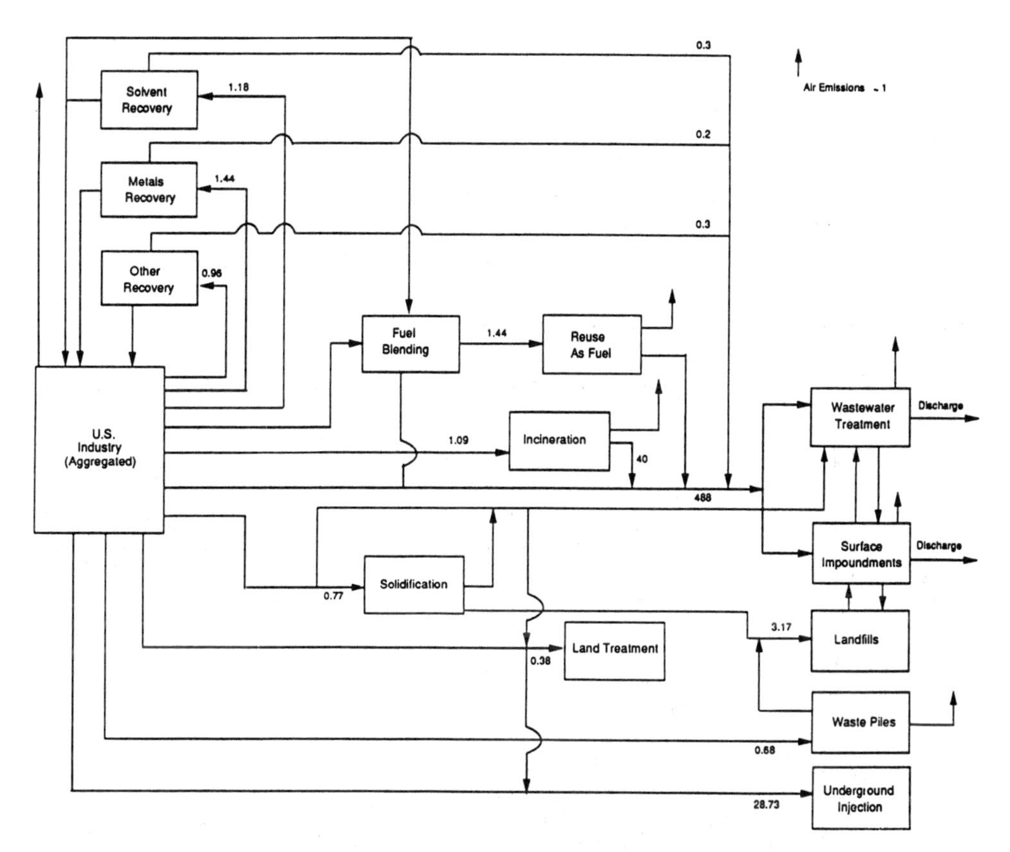


Figure 1. Management of industrial hazardous wastes in the United States; flow rates are in millions of tons per year; data are for 1986. Reproduced by permission of Mary Ann Liebert Publishers © 1992; R. D. Baker, J. L. Warren, N. Behmanesh and D. T. Allen, *Hazardous Waste and Hazardous Materials*, 9(1), 32-55 (1992).

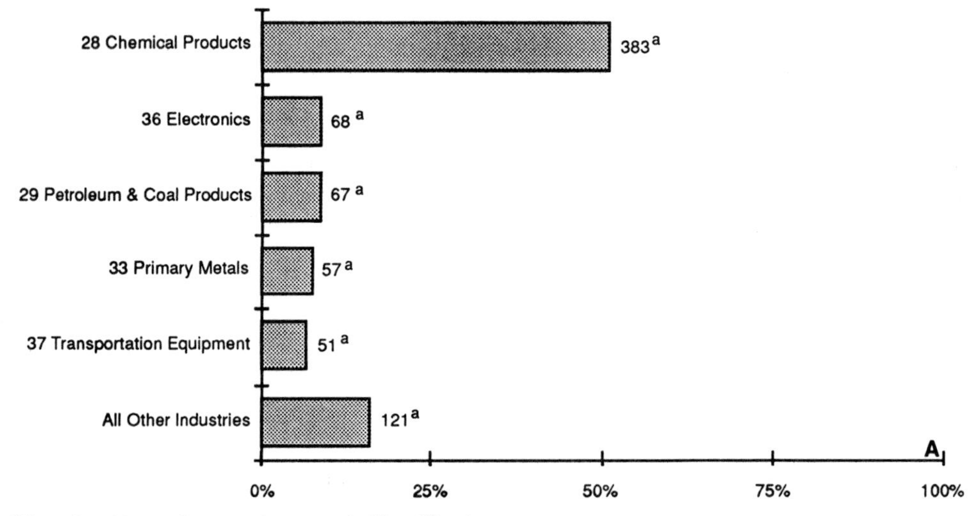


[a]Quantity of hazardous waste generated in million tons.

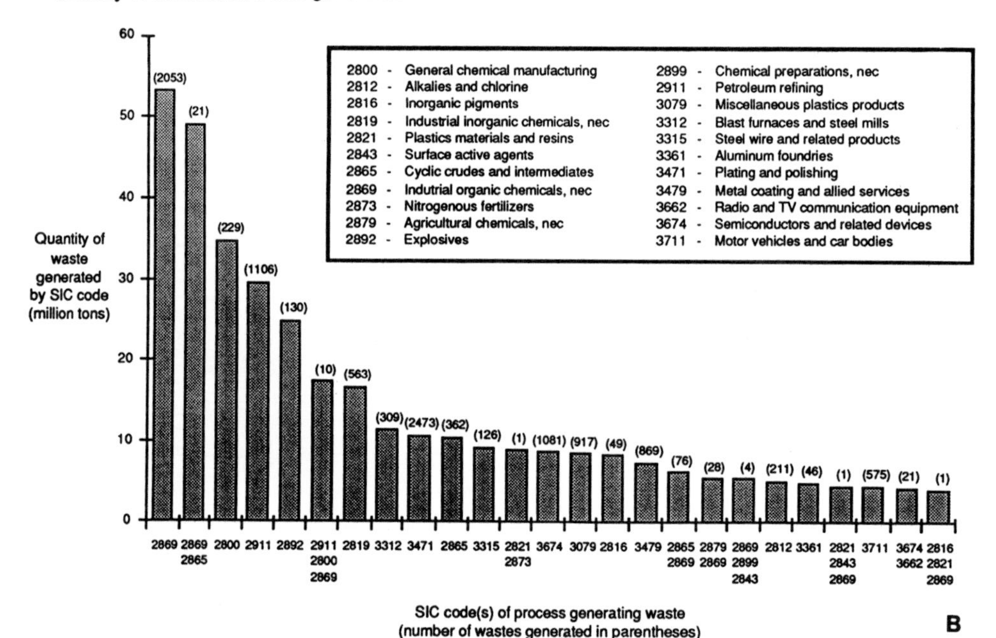


Figure 2. a) Generation of industrial hazardous waste in the United States; generation is reported by two digit Standard Industrial Classification (SIC) codes. b) Generation of industrial hazardous waste in the United States; generation, in units of millions of tons per year are reported by four digit SIC codes. Reproduced by permission of Mary Ann Liebert Publishers © 1992; R. D. Baker and J. L. Warren *Hazardous Waste and Hazardous Materials*, **9**(1), 13-31 (1992).

Table 1
Representative Production and Yield Data for
Commodity Chemical Manufacturing

| Chemical & Production Process | Annual Production, 1000 tons | Yield (%) calculated* | Yield (%)** |
|---|---|---|---|
| **Acrylonitrile** | **2,182** | | |
| 1. Ammoxidation of propylene | | 65 | 53 |
| 2. Cyanation/oxidation of ethylene | | 76 | 85 |
| **1,3-butadiene** | **2,546** | | |
| 1. Dehydrogenation of n-butylenes | | 64 | 70 |
| 2. Oxidative dehydrogenation of n-butylenes | | 87 | 74-90 |
| 3. Dehydrogenation of n-butane | | 42 | 57-63 |
| **Vinyl chloride** | **8,439** | | |
| 1. Chlorination & oxychlorination of ethylene | | 96 | 86 |
| 2. Dehydrochlorination of ethylene dichloride | | 98 | 95 |

*Calculated using data reported in "Petrochemical Technology Assessment" by, D. F. Rudd, S. Fathi-Afshar, A. A. Trevino and M. A. Stadtherr. John Wiley & Sons, 1981.

**Erskine, M. G., "Chemical Conversion Factors and Yields", Chemical Information Services, Stanford Research Institute, 1969.

very naive to assume that a sudden focus on waste reduction will magically improve selectivity. What will occur is a re-examination of existing process alternatives. A highly selective process that was not cost effective a decade ago may , due to escalating waste disposal costs, be the most cost effective alternative today.

4. Catalysis can play a role in waste reduction by reducing selectivities for highly toxic waste products. Often, what makes a waste difficult or expensive to dispose of are highly toxic trace contaminants. For example, if a waste stream contains only a few parts per billion of 2,3,7,8 tetrachlorodibenzodioxin, it will be much more difficult and expensive to manage than the same waste without the trace level of dioxin. Designing catalysts that avoid the formation of highly toxic byproducts is a selectivity problem, but it is a different selectivity problem than those faced in the past. Indeed, one may be willing to reduce overall selectivity in a chemical reaction if the generation of highly toxic contaminants in the waste streams are reduced.

## CASE STUDIES OF CATALYSIS FOR IN-PROCESS RECYCLING

The previous section described four roles that catalysis might play in waste reduction. Of the four, the only role that currently sees any significant use is catalytic conversion of wastes into recyclable materials. In this section, several case studies of catalysis for converting wastes into recyclable materials will be presented. These examples will draw heavily on the research interests and experience of the author, and as a consequence, will be devoted exclusively to hydrodechlorination reactions. The use of catalytic technology for recycling materials involves, of course, many other chemistries. Some are old (conversion of $H_2S$ to elemental sulfur in petroleum refineries). Some are new (catalytic conversion of chlorofluorocarbons, CFCs, to partially hydrogenated chlorofluorocarbons, HCFCs). The group presented here are merely samples from a large set of technologies.

Catalytic hydroprocessing of chlorinated organics is a relatively new technology that can be used to recycle waste streams from a number of chemical manufacturing operations. For example, Kalnes and James (1988) have reported on the use of catalytic hydrodechlorination for recycling liquid wastes from vinyl chloride manufacturing and still bottoms from epichlorohydrin production.

While the use of catalytic hydroprocessing to recycle chlorinated organics is a relatively new idea, the basic chemistry is merely an adaptation of petroleum hydroprocessing. Catalytic hydrotreatment has been used for decades in the petroleum refining industry to remove oxygen, nitrogen and sulfur functionalities from heavy fuels. The chemistry and kinetics of deoxygenation, denitrogenation and desulfurization have been studied extensively and several comprehensive reviews of the literature are available (Rollman, 1977; Gates, et al., 1979; Shah and Krishnamurthy, 1981) Much of this information is applicable in a general way to the development of catalytic hydrodechlorination, but, until recently, few data were available on the hydroprocessing chemistry of chlorinated compounds.

In the author's laboratory, initial work on hydrodechlorination chemistry focussed on identifying a robust dechlorination catalyst. The goal was to identify a catalyst that was reasonably active and that would not be deactivated by the wide range of functionalities found in waste streams. Eventually a commercial petroleum hydroprocessing catalyst was chosen for further study, and the rates of dechlorination for multiply chlorinated benzenes were examined. Representative results are shown in Figure 3 (Hagh and Allen, 1990a-c).

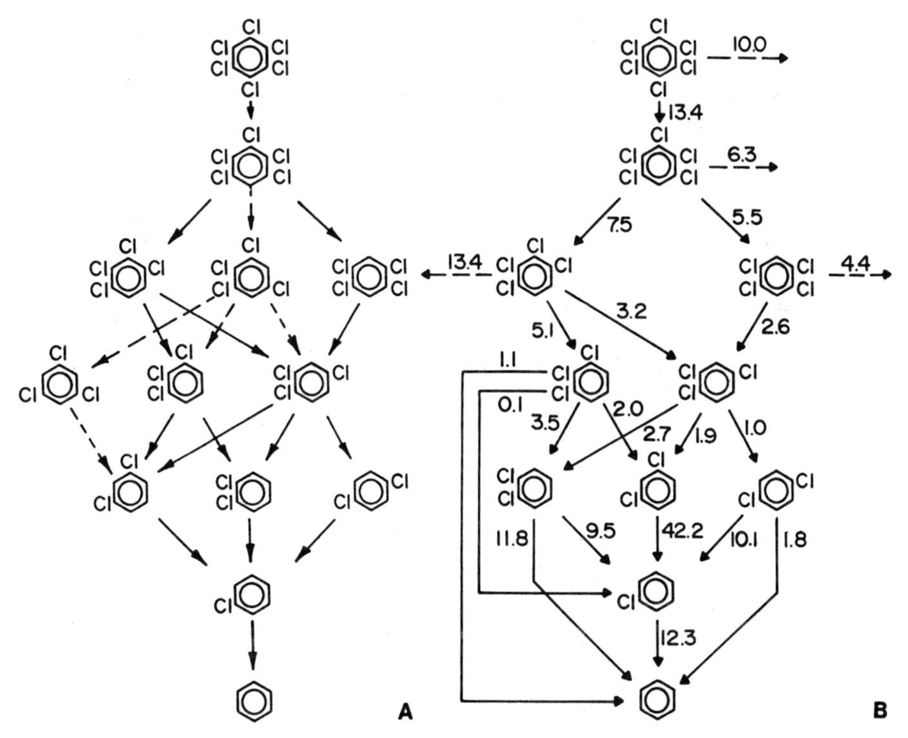


Figure 3. a) Reaction pathways for chlorinated benzenes hydroprocessed over a Ni/Mo on $Al_2O_3$ catalyst; observed pathways are shown as solid lines; pathways that were not observed are shown as dashed lines. b) Pseudo-first order kinetic rate constants for dechlorination of chlorinated benzenes in units of (liter/(g cat-min))x$10^3$; dashed lines indicate pathways in which multiple chlorine atoms are removed in a single step. Reproduced by permission of Pergamon Press © 1990; B. F. Hagh and D. T. Allen, *Chem. Eng. Sci.*, **45**, 695-702 (1990).

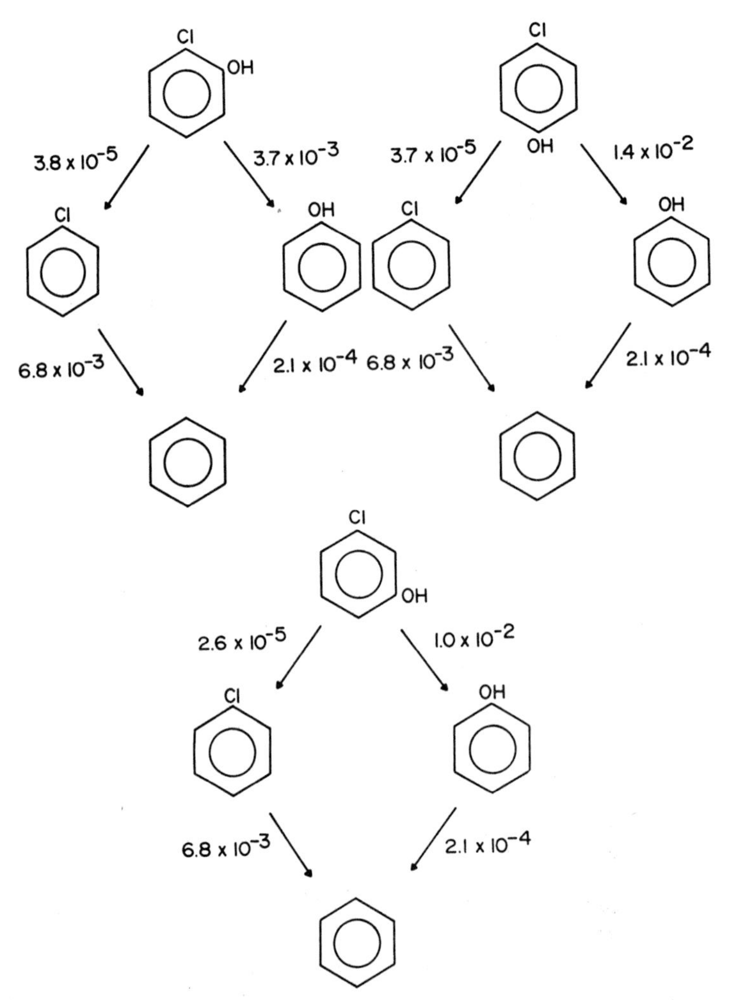


Figure 4. Hydroprocessing pathways and pseudo-first order rate constants for chlorophenols reacting over Ni/Mo on $Al_2O_3$. Reproduced by permission of the American Institute of Chemical Engineers © 1991 AIChE; S. Chon and D. T. Allen, *AIChE J.*, **37**, 1730-1732 (1991).

Dechlorination of singly and multiply chlorinated organics is only one aspect of dechlorination chemistry. Frequently wastes will contain functionalities other than chlorine that might compete for catalytic sites. For example, chlorophenols could either dechlorinate or deoxygenate under hydroprocessing conditions. Over our model catalyst, dechlorination rates for the chlorophenols were approximately two orders of magnitude greater than deoxygenation rates. Typical results are shown in Figure 4. The data show that the location of the oxygen functionality in chlorophenol influences the dechlorination rate substantially. In the case of orthochlorophenol, the dechlorination rate was reduced to roughly half that of chlorobenzene dechlorination rate at 275°C. For metachlorophenol and parachlorophenol, the presence of a hydroxyl group at the meta- and the paraposition led to approximately twofold and threefold increase in dechlorination reactivity, respectively, to that of chlorobenzene dechlorination activity. The dechlorination activities of chlorophenols are therefore para > meta > ortho, with the hydroxyl group activating the dechlorination in the para- and metaposition and deactivating dechlorination in the orthoposition. These results can be rationalized based on electron donating properties of the substituents and steric effects.

The main use of these dechlorination chemistries to date has been the complete dechlorination of waste streams to generate HCl and a hydrocarbon fuel. A number of more interesting opportunities may also exist. Referring back to the data on chlorinated benzenes, it may, for example, be possible to selectively dechlorinate meta-dichlorobenzene in a mixture of meta and para-dichlorobenzene. This might allow for increased recycle of by-products in the production of para-dichlorobenzene, a household fumigant.

## CONCLUSION

The fact that catalysis can play a major role in waste reduction should come as no surprise. As the data on industrial waste generation presented in this paper indicate, many of the materials that become hazardous wastes are initially generated in catalytic reactions. Since catalysis has played a role in promoting the reactions to form these materials, it can also play a significant role in reactions to convert these potential wastes back into useful raw materials. This paper has presented information about one class of reactions, hydrodechlorination, that can be used to convert waste streams back into raw materials. Many other chemistries exist. More still are possible. The challenge for catalysis researchers and practitioners will be to bring these chemistries into widespread use.

## REFERENCES


Aguado, M. A., M. A. Anderson and C. G. Hill, 1991 "Ceramic membrane photocatalysts for the remediation of impaired waters", AIChE Annual Meeting, Paper 97c.

Baker, R. D., and J. L. Warren, 1992 "Generation of hazardous waste in the United States", *Hazardous Waste and Hazardous Materials*, **9**(1), 13.

Baker, R. D., J. L. Warren, N. Behmanesh and D. T. Allen, 1992 "Management of hazardous waste in the United States", *Hazardous Waste and Hazardous Materials*, **9**(1), 32.

Behmanesh, N., D. T. Allen and J. L. Warren, 1992 "Flow rates and compositions of incinerated waste streams in the United States", *J. Air and Waste Management Assoc.*, **42**(4), xxx.

Chon, S., and D. T. Allen, 1991 "Catalytic hydroprocessing of chlorophenols", *AIChE J.*, **37**, 1730.

Erskine, M. G., 1969 "Chemical Conversion Factors and Yields", Stanford Research Institute, Palo Alto, CA.

Gates, B.C., J. R. Katzer, and G. C. A. Schuit, 1979 "Kinetics and reaction chemistry of catalytic hydrodechlorination of chlorinated benzenes on sulfided NiMo/$\gamma Al_2O_3$", *Chemistry of Catalytic Processes*, McGraw-Hill, New York, p. 390.

Hagh, B. F., and D. T. Allen, 1990a "Catalytic hydrodechlorination", *Innovative Hazardous Waste Treatment Technology*, Vol. 1, H. M. Freeman, ed., Technomic, Lancaster, PA.

Hagh, B. F., and D. T. Allen, 1990b "Catalytic hydroprocessing of chlorobenzene and 1,2-dichlorobenzene", *AIChE J.*, **36**, 773.

Hagh, B. F., and D. T. Allen, 1990c "Catalytic hydroprocessing of chlorinated benzenes", *Chem. Eng. Sci.*, **45**, 2695.

Hasan, S., P. Cho, K. L. Sublette, D. Pak and A. Maule, 1990 "Porphyrin catalyzed degradation of hazardous waste", AIChE Annual Meeting, Paper 49g.

Kalnes, T. N., and R. B. James, 1988 "Hydrogenation and recycle of organic waste streams", *Environ. Prog.*, **7**, 185.

Kim, S., M. A. Abraham, J. Papap, R. L. Cerro and Y. T. Shah, 1990 "Three phase catalytic oxidation of phenol in a monolith foam reactor", AIChE Annual Meeting, Paper 49b.

National Environmental Technology Applications Corporation (NETAC), 1990, University of Pittsburgh Applied Research Center, "Incentives and barriers to commercializing environmental technologies", Pittsburgh, PA.

Rollman, R. D., 1977 "Catalytic hydrogenation of model N-, S-, and O-compounds", *J. Catal.*, **46**, 243.

Shah, Y. T., and S. Krishnamurthy, 1981 "Heteroatom removal during liquefaction", *Reaction Engineering in Direct Coal Liquefaction*, Y. T. Shah, ed., Addison-Wesley, Reading, MA.

Rudd, D. F., Fathi-Afshar, S., Trevino, A. A., and Stadtherrr, M. A., 1981 "Petrochemical Technology Assessment", Wiley, New York.

# CATALYSIS, THE ATOM UTILIZATION CONCEPT AND WASTE MINIMIZATION

R.A. Sheldon

Delft University of Technology
The Netherlands

## ABSTRACT

Following the advent of the petrochemicals industry in the 1920's, catalysis was widely applied in the manufacture of bulk chemicals. Traditional, environmentally unacceptable processes have largely been replaced by cleaner catalytic technologies. Fine chemicals, in contrast, have remained largely the domain of the synthetic organic chemist who has generally clung to the use of stoichiometric methods.

But times are rapidly changing. Increasingly stringent environmental constraints are making the use of classical stoichiometric methods prohibitive. Consequently there is a general trend towards substitution of such antiquated technologies by cleaner, catalytic methods that do not generate large amounts of inorganic salts as waste.

A useful concept for evaluating the environmental acceptability of various processes for producing a particular substance is atom utilization. The latter is defined as the ratio of the molecular weight of the desired product to the sum total of all the materials (excluding solvents) used. Examples of the application of clean, low-salt technologies are illustrated by reference to alternative routes for the synthesis of the industrial monomer, methyl methacrylate and the analegesic drug, ibuprofen.

## INTRODUCTION

After an induction period of a few decades we now appear to be in the age of 'environmentality'. This is reflected both in the general trends in society as a whole and in the chemical industry in particular (see Figure 1).
A particular concern is the amount of waste produced in chemical processes, largely in the form of inorganic salts. Indeed, intergrated waste management and zero emission plants are the catch-words in the chemical corridors of power these

*Industrial Environmental Chemistry*, Edited by D.T. Sawyer
and A.E. Martell, Plenum Press, New York, 1992

days. What goes in must come out, somewhere. Preferably what comes out should be the desired product. Everything else should be considered as undesirable and its formation avoided or be kept in the system.

Obviously the key to waste minimization is selectivity in chemical processing. In describing the efficiency of chemical transformations organic chemists generally distinguish between different types of selectivity. The simplest is standard one: the yield of desired product divided by the conversion of the starting material. In reactions of complex organic molecules the concepts of chemoselectivity (competing reactions at two different functional groups in the same molecule), regioselectivity (e.g. ortho vs para attack in aromatic substitution) and stereoselectivity (enantio- or diastereoselectivity) are frequently used. However, there is another category of selectivity that is generally overlooked by organic chemists: what we shall call the atom selectivity or atom utilization.

- Cleaner, more environmentally acceptable products, e.g. 'green gasoline' (lead- and aromatics-free)
- Products that are more effective, more targeted in their action and environmentally friendly, i.e. readily recycled or biodegraded
- Zero emission plants/integrated waste management
- Cleaner technologies with negligible inorganic salt formation (i.e. catalytic processes with optimal atom utilization)
- Replacement of toxic and/or hazardous reagents, e.g. $COCl_2$, $(CH_3)_2SO_4$, $H_2CO/HCl$, heavy metal salts. Transport and storage of hazardous chemicals, e.g. halogens, becoming increasingly difficult
- Alternatives for chlorinated hydrocarbon solvents, e.g. solvent-free processes, chemistry in water
- Shorter routes, in some cases via alternate feedstocks, e.g. substitution of alkanes for alkenes and aromatics
- Utilization, where feasible, of renewable raw-materials, e.g. carbohydrates, or waste materials from other processes
- Higher chemo-, regio- and enantioselectivities

**Figure 1.** General trends in the chemical industry.

## THE ATOM UTILIZATION CONCEPT

The atom utilization of a particular process is calculated by dividing the molecular weight of the desired product by the sum total of all the materials (excluding solvents) used. It provides a simple way of assessing how 'clean' a particular process is with regard to the amount of waste that is formed, in the ideal case of 100% selective conversion of starting material.
An example, to illustrate the point, is the replacement of the traditional chlorohydrin technology for ethylene epoxidation by catalytic epoxidation:

$$H_2C{=}CH_2 + Cl_2 + H_2O \longrightarrow ClCH_2CH_2OH + HCl \qquad (1)$$

$$ClCH_2CH_2OH + HCl + Ca(OH)_2 \longrightarrow C_2H_4O + CaCl_2 + H_2O \qquad (2)$$

| atom utilization = 44/173 =25% |
|---|

$$H_2C{=}CH_2 + 1/2\ O_2 \xrightarrow{\text{catalyst}} C_2H_4O \qquad (3)$$

| atom utilization = 100% |
|---|

Although the concept of atom utilization is on the one hand trivial, it has largely been ignored by organic chemists. This is reflected in the enormous amounts of inorganic salts formed in most organic reactions. Indeed organic chemistry in general tends to involve far too many neutralization steps, all producing significant amount of salts. Consequently, in classical organic reactions inorganic salts often represent a large proportion of the total weight of material formed. Increasingly stringent environmental measures are, however, stimulating the replacement of such traditional salt-forming technologies with cleaner catalytic alternatives (i.e. low-salt technologies). One could say that the chemical industry is being subjected to a 'low-salt diet'.

Similarly, catalytic oxidation of $\beta$-phenoxy-ethanols provides a potentially interesting salt-free alternative to the classical process for the production of $\alpha$-phenoxyacetic acid herbicides (see Figure 2).

Classical process:

$$CH_3CO_2H + Cl_2 \longrightarrow ClCH_2CO_2H + HCl$$

$$ClCH_2CO_2H + ArOH \xrightarrow[\text{2. HCl}]{\text{1. 2NaOH}} ArOCH_2CO_2H + 2NaCl + 2H_2O$$

Overall stoichiometry:

$$ArOH + CH_3CO_2H + Cl_2 + 2NaOH \longrightarrow ArOCH_2CO_2H + 2NaCl + 2H_2O$$

| atom utilization* = 200/353 =57% |
|---|

* assuming Ar = 125

## METHYL METHACRYLATE MANUFACTURE

Methyl methacrylate (MMA) is an important polymer raw material the annual production of which is in excess of one million tons worldwide. Developments in manufacturing processes for this relatively simple molecule illustrate perfectly the modern trends in chemical processing. Traditionally, MMA is produced by reaction of acetone (a byproduct of phenol production) with hydrogen cyanide (a byproduct of acrylonitrile production), followed by methanolysis (Figure 3). As

Low-salt alternative:

$$\overset{O}{CH_2\text{-}CH_2} + ArOH \xrightarrow{H^+} ArOCH_2CH_2OH$$

$$ArOCH_2CH_2OH + O_2 \xrightarrow{\text{catalyst}} ArOCH_2CO_2H + H_2O$$

atom utilization = 200/218 = 92%

Figure 2. Two routes to phenoxyacetic acids.

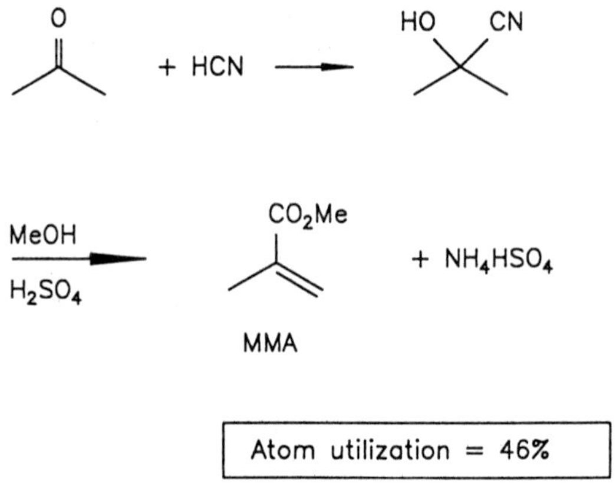

$$\text{acetone} + HCN \longrightarrow \text{acetone cyanohydrin (HO, CN)}$$

$$\xrightarrow[H_2SO_4]{MeOH} \text{MMA} (CO_2Me) + NH_4HSO_4$$

Atom utilization = 46%

Figure 3. Classical process for MMA manufacture.

such it provides a good example of the use of byproducts from one chemical process as raw materials in another.

However, the process has one serious flaw: it produces a substantial amount of ammonium bisulfate as a coproduct. Although the theoretical amount is 1.15 kg of $NH_4HSO_4$ per kg MMA, corresponding to an atom utilization of 46%, in practice a substantial excess of $H_2SO_4$ is required resulting in the formation of ca. 2.5 kg salt per kg MMA. Consequently, much effort has been devoted to the development of alternative, low-salt routes to MMA. These generally involve catalytic oxidation or carbonylation of readily available lower olefins (see Figure 4).

Interestingly, the traditional technology has recently been upgraded by the development, by Mitsubishi Gas[1] of an elegant method to circumvent the salt formation (see Figure 5).

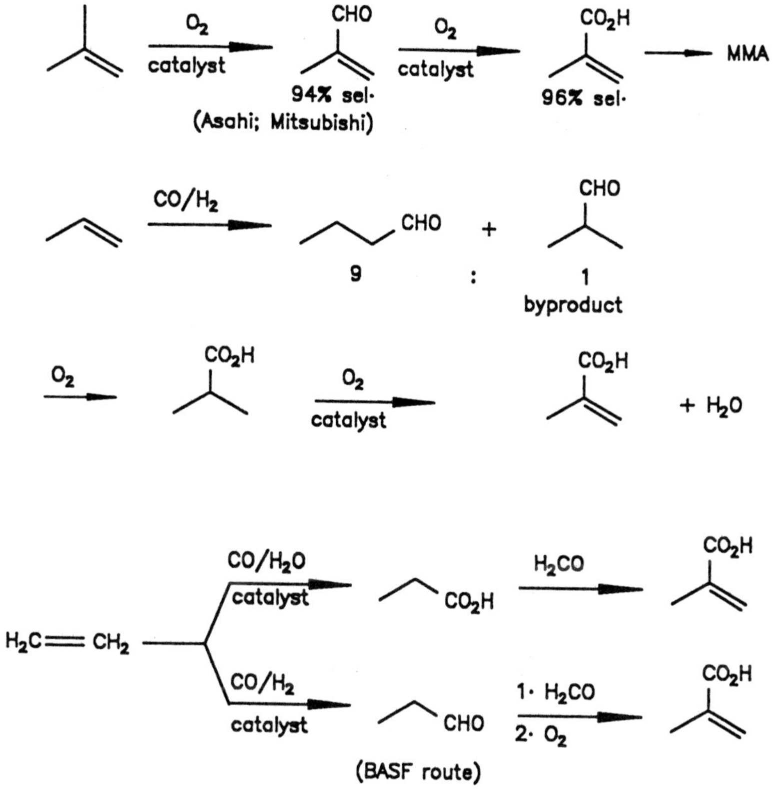

$O_2$ / catalyst → CHO (94% sel.) (Asahi; Mitsubishi) → $O_2$ / catalyst → $CO_2H$ (96% sel.) → MMA

$CO/H_2$ → CHO + CHO
9 : 1
byproduct

$O_2$ → $CO_2H$ → $O_2$ / catalyst → $CO_2H$ + $H_2O$

$H_2C{=}CH_2$ → $CO/H_2O$ / catalyst → $CO_2H$ → $H_2CO$ → $CO_2H$

$H_2C{=}CH_2$ → $CO/H_2$ / catalyst → CHO → 1. $H_2CO$ 2. $O_2$ → $CO_2H$

(BASF route)

**Figure 4.** Alternative routes to methylmethacrylate (MMA).

It should be noted, however, that this leads to a rather circuitous route. In this context, a devastatingly simple route has recently been reported by Shell workers.[2] MMA is formed in one step, in essentially quantitative yield, by palladium-catalyzed methoxycarbonylation of methylacetylene (Figure 6). The reaction is carried out under mild conditions and exhibits a remarkably high catalyst turnover frequency (up to 100.000 mol methylacetylene per g catalyst per hour). Methylacetylene is a byproduct of naphtha cracking and is produced in sufficient amounts to make it an interesting raw material for MMA. In short, the route illustrated in Figure 6 is the epitomy of modern trends in chemical manufacture: a one-step, extremely efficient catalytic process with 100% atom utilization and based on a waste stream as raw material.

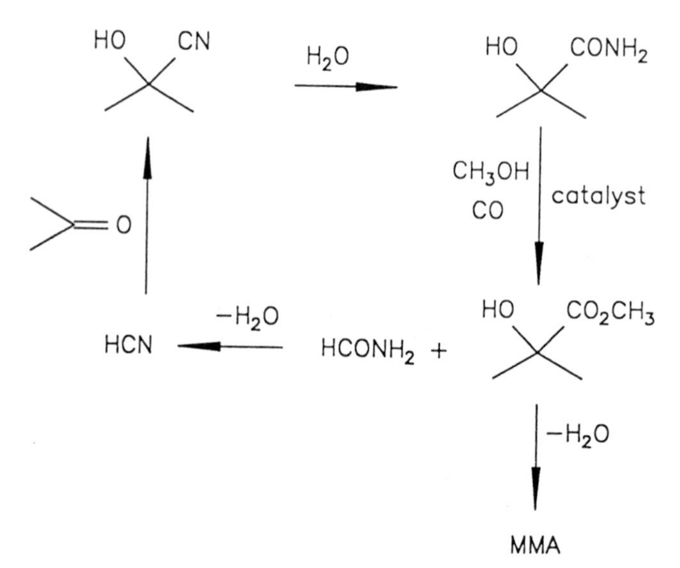


**Figure 5.** Mitsubishi Gas process.

## CATALYSIS IN BULK AND FINE CHEMICALS MANUFACTURE

In the drive towards waste minimization in chemical manufacturing catalysis will play a pivotal role. In the bulk chemicals industry classical environmentally unacceptable processes have largely (but not completely) been supplanted by cleaner catalytic alternatives. In particular, catalytic oxidation[3] and carbonylation[4-6] are widely used for the conversion of petrochemical feedstocks to industrial chemicals (see Table 1).

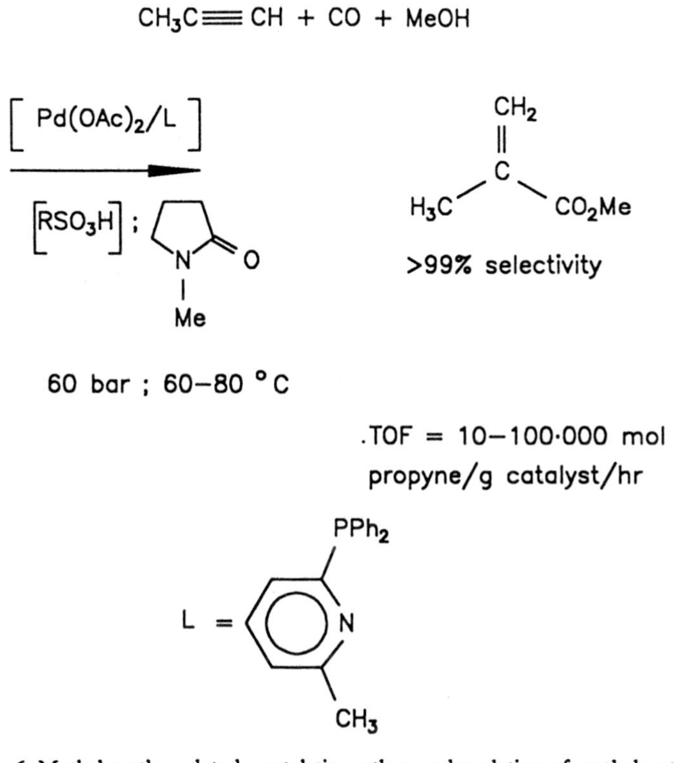

$$CH_3C{\equiv}CH + CO + MeOH \xrightarrow[\text{[RSO}_3\text{H] ; N-methylpyrrolidone}]{[Pd(OAc)_2/L]} CH_2{=}C(CH_3)CO_2Me$$

>99% selectivity

60 bar ; 60–80 °C

.TOF = 10–100·000 mol propyne/g catalyst/hr

L = 2-(diphenylphosphino)-6-methylpyridine ($PPh_2$, N, $CH_3$)

**Figure 6.** Methyl methacrylate by catalytic methoxycarbonylation of methylacetylene.

**Table 1.** Bulk chemicals manufacture USA (1990).

| CHEMICAL | VOLUME ($10^6$ tons) | REACTION | CATALYST |
|---|---|---|---|
| Terephthalic acid | 4,0 | Oxidation | Homogeneous |
| Styrene | 4,0 | Dehydrogenation | Heterogeneous |
| Methanol | 4,0 | $CO + H_2$ | Heterogeneous |
| Formaldehyde | 3,2 | Oxidation | Heterogeneous |
| Ethelyne oxide | 2,8 | Oxidation | Heterogeneous |
| Acetic acid | 1,9 | Carbonylation | Homogeneous |
| Phenol | 1,8 | Oxidation | Homogeneous |
| Propylene oxide | 1,6 | Oxidation | Homogeneous |
| Acrylonitrile | 1,5 | Ammoxidation | Heterogeneous |
| Vinyl acetate | 1,3 | Oxidation | Homogeneous |

Thus, catalytic oxidation and carbonylation are good examples of high atom utilization, low salt technolgies, e.g.

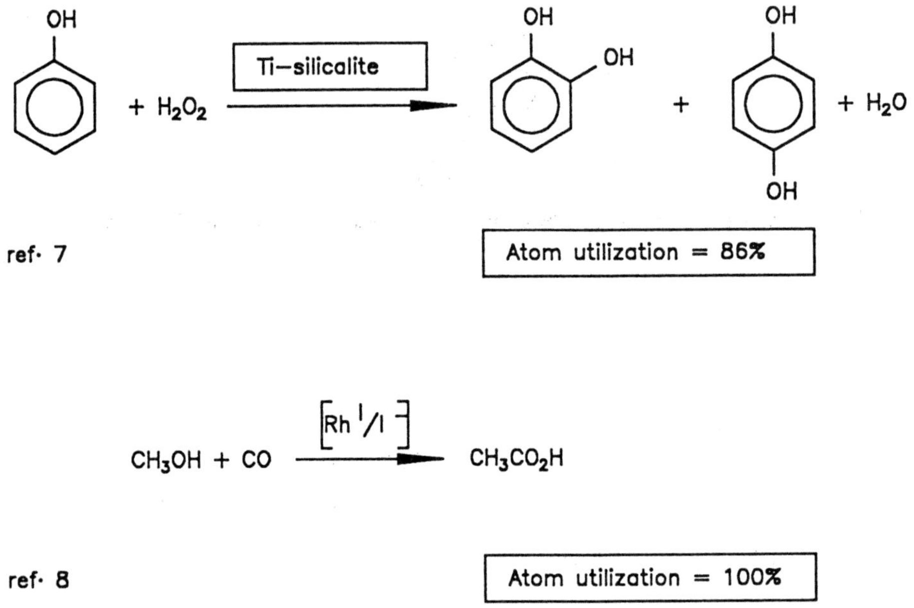

OH + $H_2O_2$ —Ti–silicalite→ OH OH + OH OH + $H_2O$

ref· 7

Atom utilization = 86%

$$CH_3OH + CO \xrightarrow{[Rh^{I}/I]} CH_3CO_2H$$

ref· 8

Atom utilization = 100%

**Figure 7.** Catalytic oxidation and carbonylation as low-salt technologies.

In the fine chemicals industry, in contrast, catalytic technologies have been only sporadically applied.[9] One reason for this is the much lower production volumes involved. Nevertheless, the number of kilos of byproducts (largely inorganic salts) per kilo of product are much higher in fine chemicals and specialties as indicated in Table 2.

**Table 2.** Byproduct formation in chemicals production.

| INDUSTRY SEGMENT | PRODUCT TONNAGE | KG BYPRODUCTS/ KG PRODUCT |
|---|---|---|
| Oil Refining | $10^6$-$10^8$ | ca. 0.1 |
| Bulk Chemicals | $10^4$-$10^6$ | <1 - 5 |
| Fine Chemicals | $10^2$-$10^4$ | 5 -> 50 |
| Pharmaceuticals | 10-$10^3$ | 25 -> 100 |

The figures in Table 2 are based on the author's experience and are intended to give an indication only. The substantial increase on going downstream is partly due to the use of stoichiometric vs catalytic processes and partly to the fact that fine chemicals and specialties are often produced in multi-step syntheses.
Of course comparing processes solely on the basis of the amount of waste is an oversimplification. Obviously the nature of the waste is also important; it certainly makes a difference if it is sodium chloride or, for example, chromium or manganese salts. A more sophisticated assessment should, therefore, take into account both the amount and the nature of waste, e.g.

| Environmental = | E | x | Q |
|---|---|---|---|
| acceptability | Environmental factor(kg waste/ kg product) | x | Unfriendliness quotient |

For example, if innocuous salts such as NaCl or $Na_2SO_4$ were arbitrarily given an unfriendliness factor of 1, then chromium salts could be assigned a factor of say 100 and toxic metals e.g. Pb, Cd a 1000. Obviously these figures are debatable and will vary from one company or production site to another, being partly dependent on the ability to recycle a particular waste stream.

## DEVELOPMENT OF ORGANIC SYNTHESIS AND CATALYSIS

Another reason why catalysis has not been widely applied in the fine chemicals industry is the more or less separate development of organic chemistry and catalysis (see Figure 8) since the time of Berzelius, who coined both terms, in 1807 and 1835, respectively.

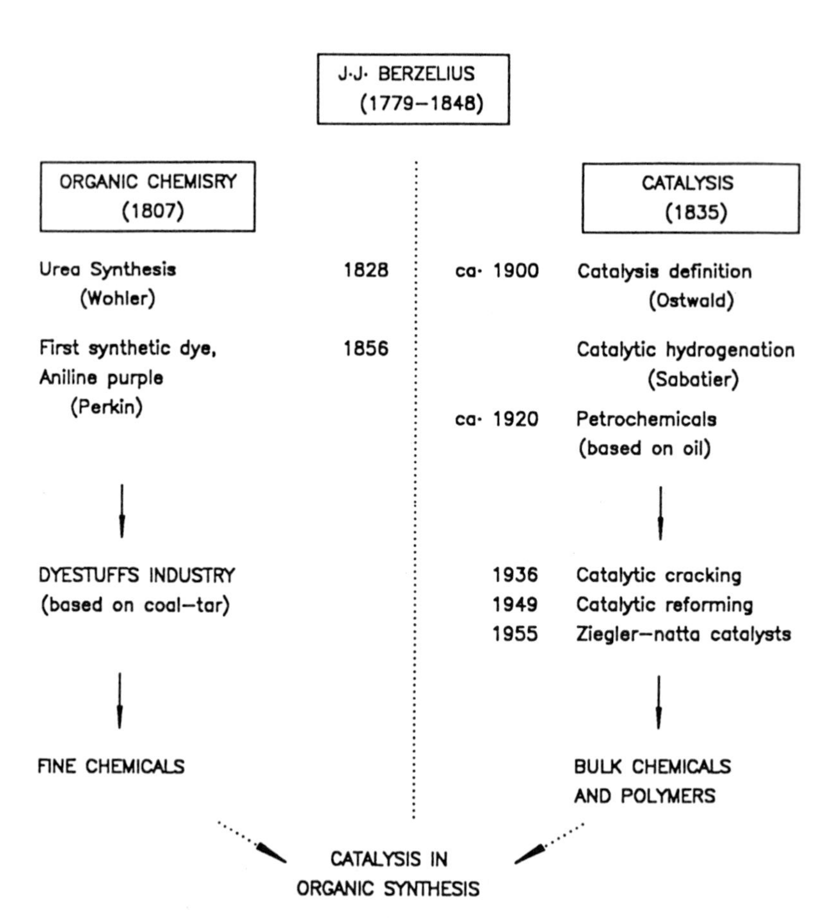


**Figure 8.** Development of catalysis and organic synthesis.

In the late nineteenth and early twentieth century catalysis developed largely as a subdiscipline of physical chemistry. Following the advent of the petrochemicals industry, catalysis was widely applied in oil refining and bulk chemicals manufacture. Industrial organic synthesis, on the other hand, really began with Perkin's serendipitous synthesis of aniline purple (mauveine) in 1856. This marked the beginning of the synthetic dyestuffs industry based on coal tar as the raw material. The present-day fine chemicals and pharmaceuticals industries developed largely as spinoffs of this activity.

Interestingly, Perkin was attempting to synthesize the pharmaceutical product, quinine, by oxidation of allyl toluidine with potassium dichromate, according to the following concept:

This, in hindsight, naive attempt was doomed to fail but one may sympathize with Perkin when one realizes that the structure of the benzene ring was not even known in 1856.

## CLASSICAL vs NON-CLASSICAL AROMATIC CHEMISTRY

Fine chemicals manufacture has, to this day, remained primarily the domain of the synthetic organic chemist who, generally speaking, clung to the use of

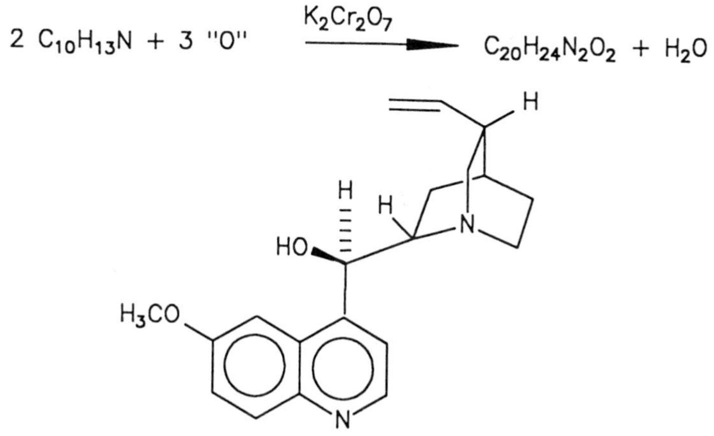

$$2\ C_{10}H_{13}N + 3\ \text{"O"} \xrightarrow{K_2Cr_2O_7} C_{20}H_{24}N_2O_2 + H_2O$$

Quinine $C_{20}H_{24}\ N_2O_2$

**Figure 9.** Perkin's attempted synthesis of quinine (1856).

stoichiometric methods, e.g. oxidations with potassium dichromate, the very same oxidant used by Perkin more than 130 years ago. Indeed, the fine chemicals industry, with its roots in coal-tar chemistry, is rampant with classical stoichiometric technologies that generate large quantities of inorganic salts. Examples include sulfonation, nitration, halogenation, diazotization, Friedel-Crafts acylations and stoichiometric oxidations and reductions (see Figure 10).
Many of these antiquated technologies are ripe for substitution by catalytic low-salt technologies (see Figure 11).

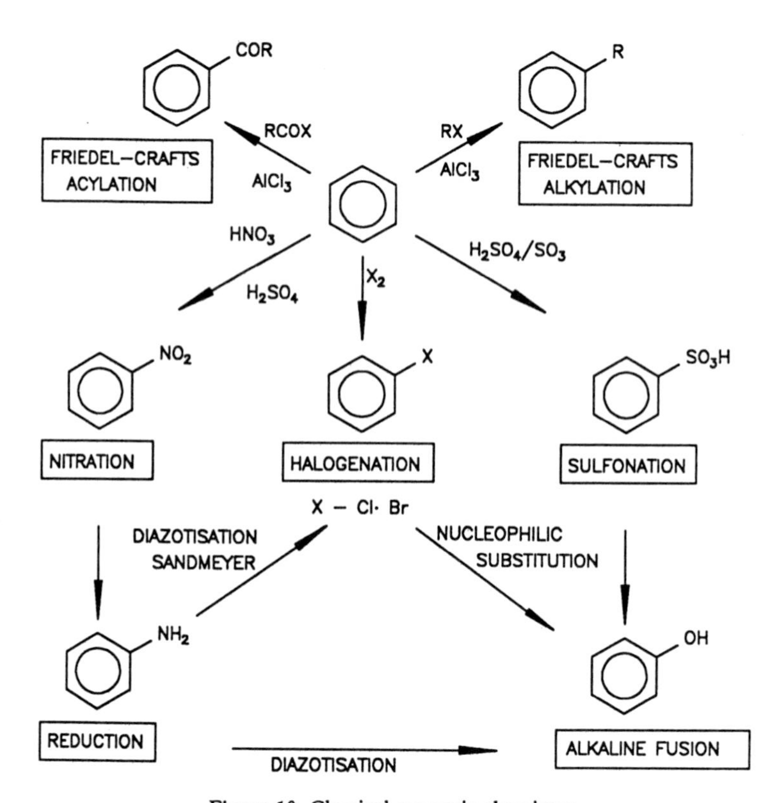


**Figure 10.** Classical aromatic chemistry.

## TWO ROUTES TO HYDROQUINONE

An instructive example is provided by hydroquinone manufacture (Figure 12). Traditionally it was produced by oxidation of aniline with stoichiometric amounts of manganese dioxide to give benzoquinone, followed by reduction with iron and hydrochloric acid. The aniline was derived from benzene via nitration and reduction and the overall process generated more than 10 kg of inorganic salts $MnSO_4$, $FeCl_2$, NaCl, $Na_2SO_4$) per kg of hydroquinone. In contrast, a more modern route to hydroquinone involves the (catalytic) autoxidation of p-diisopopylbenzene (produced by alkylation of benzene), followed by acid-catalyzed rearrangement of the bis-hydroperoxide, and produces < 1 kg of inorganic salts per kg of hydroquinone.

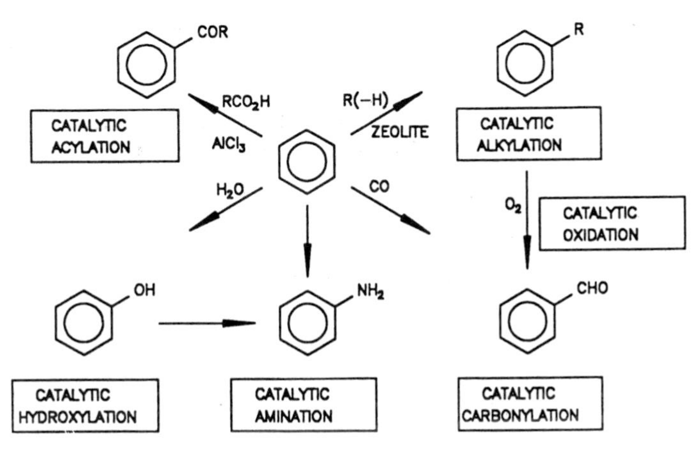


**Figure 11.** Non-classical aromatic chemistry.

## MENADIONE SYNTHESIS

Another pertinent example is the production of menadione (vitamin $K_3$), which traditionally involved the stoichiometric oxidation of 2-methylnaphthalene with chromium trioxide and afforded 18 kg of inorganic salts per kg of product. Tightening of environmental regulations has made it virtually impossible to carry out this process in developed countries, which means that less demanding production locations have to be found or alternative, cleaner catalytic technologies developed.

In the latter context the reported[10] oxidation of 2-methylnaphthalene with $H_2O_2$ in the presence of a $Pd^{II}$ catalyst immobilized on an ion exchange resin would appear to offer possibilities.

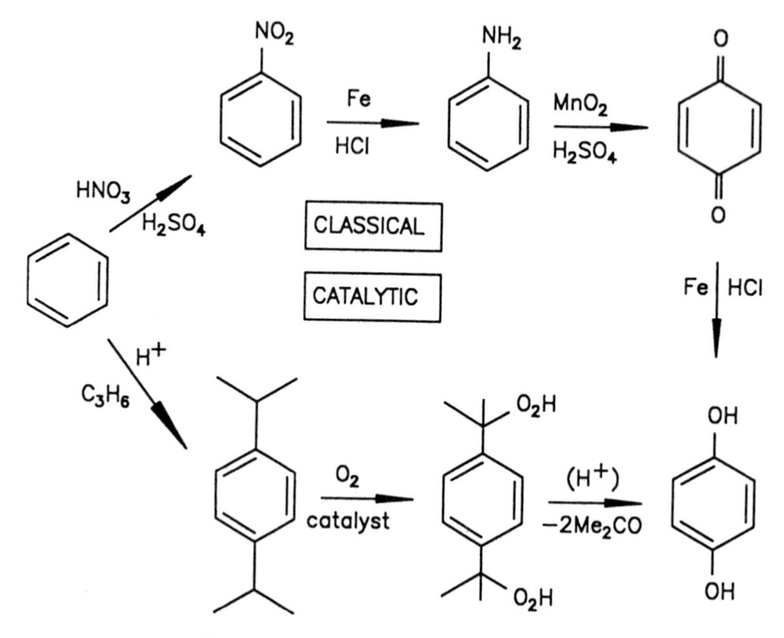


Figure 12. Two routes to hydroquinone.

| Process | Oxidant | Yield |
|---|---|---|
| Classical (Stoichiometric) | $CrO_3$ in HoAc | 50–60% |
| Alternative (Catalytic) | $H_2O_2$/[$Pd^{II}$] | 50–55% |

Figure 13. Menadione synthesis.

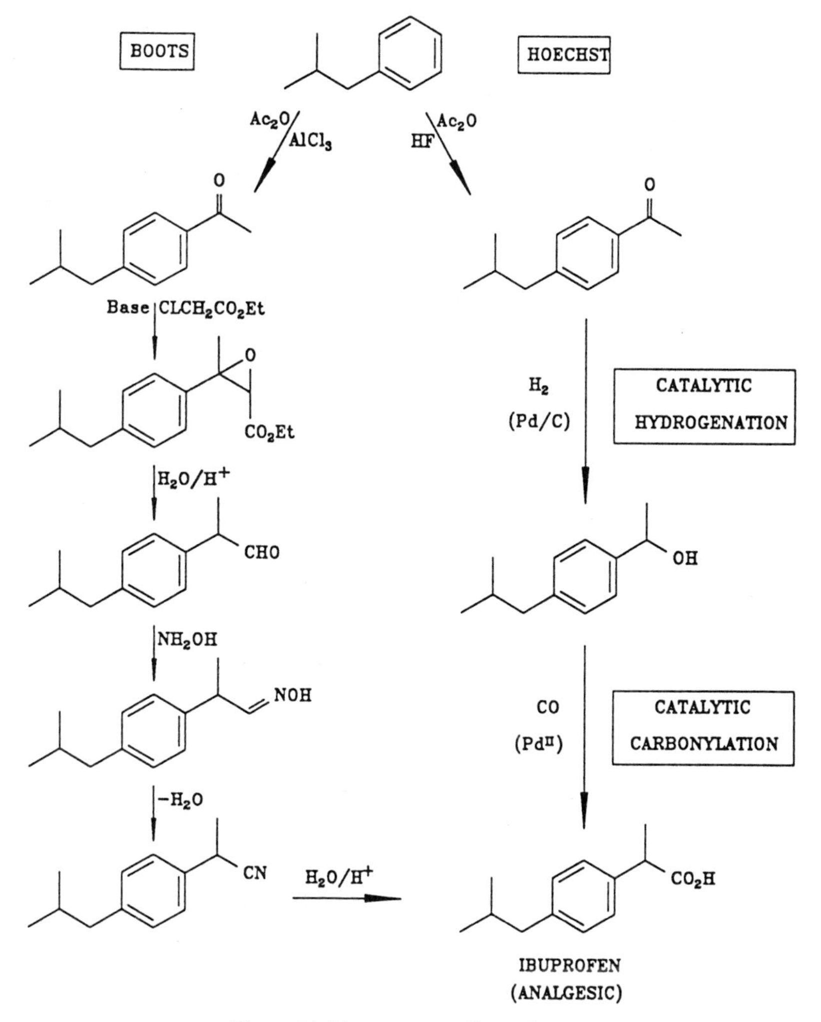

Figure 14. Two routes to ibuprofen.

## IBUPROFEN MANUFACTURE

Another case-in-point is the manufacture of the drug ibuprofen.[11] The latter is an over-the-counter analgesic with annual sales and production volume of ca. $ 1400 million and ca. 8000 tons, respectively, in 1990. Two routes for the production of ibuprofen are compared in Figure 14. Both routes proceed via a common intermediate, p-isobutylacetophenone. The classical route, used by the Boots company (the discoverer of ibuprofen) involves a further five steps, relatively low atom utilization and substantial inorganic salt formation.

The elegant alternative, developed by Hoechst Celanese[12] involves only two (catalytic) steps from the common intermediate, 100% atom utilization and negligible salt formation.

In both of the routes illustrated in Figure 14 the first step is a Friedel-Crafts acylation. The latter is also a good example of a 'classical technology' involving stoichiometric quantities of inorganic reagents, usually aluminium chloride (in the Hoechst process HF is used). Obviously this is another technology that is ripe for substitution by a cleaner catalytic method. In this context the recently reported[13] use of zeolites as heterogeneous, regenerable catalysts for Friedel-Crafts acylations (see Figure 15) would seem to offer possibilities. Interestingly, the catalytic method utilizes the carboxylic acid as the acylating agent, thus avoiding the formation of HCl.

CLASSICAL:

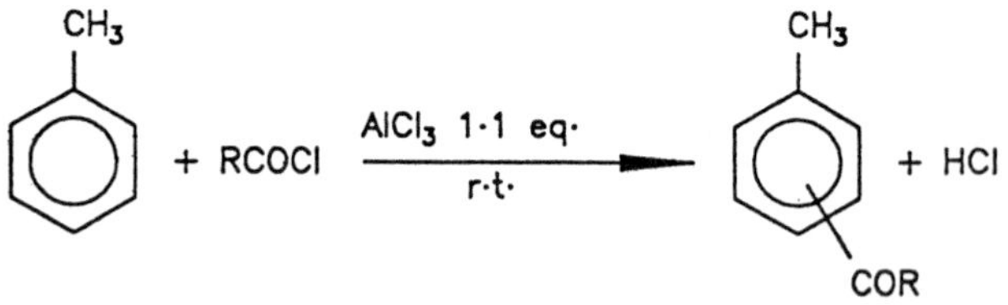

Toluene ($CH_3$) + RCOCl $\xrightarrow[\text{r.t.}]{AlCl_3\ 1.1\ eq.}$ $CH_3$–C$_6$H$_4$–COR + HCl

Stoichiometric $AlCl_3$, not regenerable

ZEOLITE CATALYSIS:

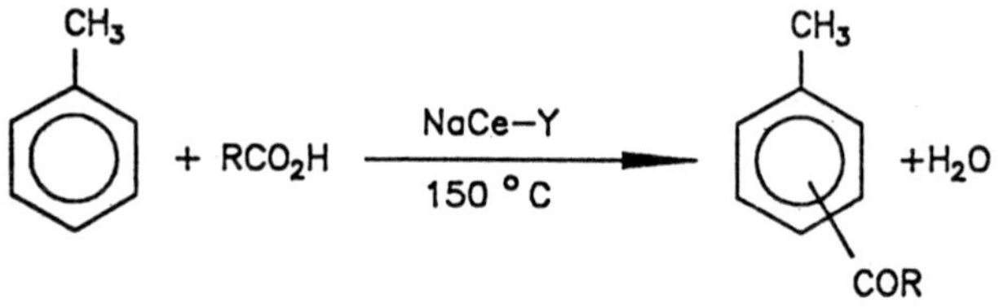

Toluene ($CH_3$) + $RCO_2H$ $\xrightarrow[150\ ^\circ C]{NaCe-Y}$ $CH_3$–C$_6$H$_4$–COR + $H_2O$

Heterogeneous, regenerable catalyst

REGIOSELECTIVITY:

$R = C_7H_{15}$

| | o | m | p |
|---|---|---|---|
| NaCe–Y | 3 | 3 | 94 |
| $AlCl_3$ | 4 | 1.6 | 80 |

**Figure 15.** Friedel-Crafts acylations.

## CATALYTIC RETROSYNTHESIS

The example of ibuprofen perfectly illustrates the benefits to be gained by paying attention to the atom utilization in different routes and for being catalysis-minded. Indeed, organic chemists should be urged to integrate these aspects into their retrosynthetic thinking. Thus, in planning an organic synthesis a 'catalytic retrosynthesis' could be constructed, identifying catalytic pathways to the desired product.

Such a catalytic retrosynthesis for ibuprofen is shown in Figure 16. We suggest that catalytic thinking would have easily identified the two routes shown, even without the benefit of hindsight.

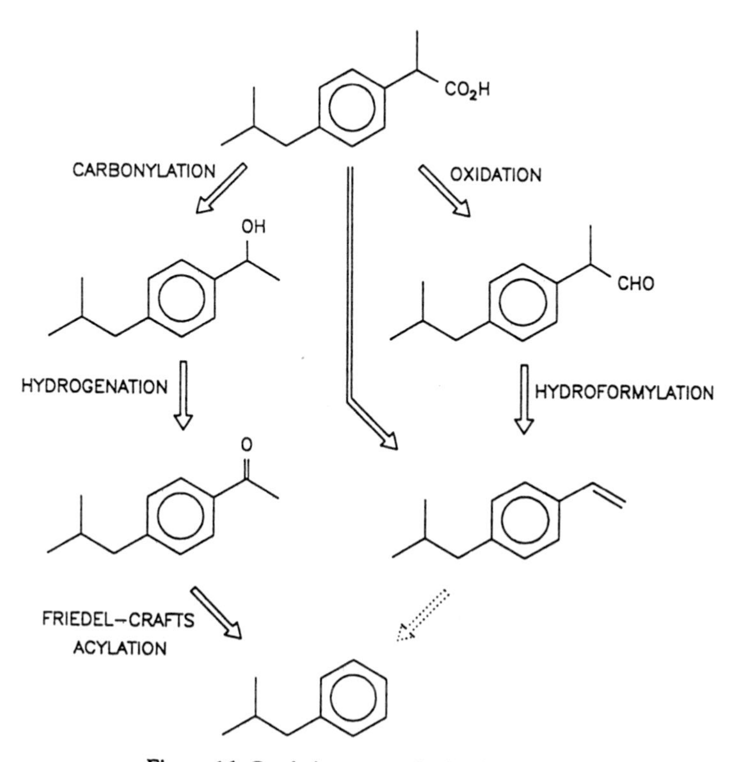


**Figure 16.** Catalytic retrosynthesis of ibuprofen.

The second catalytic route to ibuprofen, identified via retrosynthetic analysis (see Figure 16) has also been described in the literature. Thus, Neibecker and coworkers[14] have described the regioselective hydroformylation of p-isobutylstyrene in the presence of a rhodium catalyst and Riley and coworkers[15] the catalytic autoxidation of the aldehyde product to ibuprofen (see Figure 17).

The synthesis of the starting material, p-isobutylstyrene, from isobutylbenzene (see Figure 18) has also been reported.[16] Interestingly, the first step may also be amenable to replacement by one involving a zeolite as a regenerable, hetergeneous catalyst.

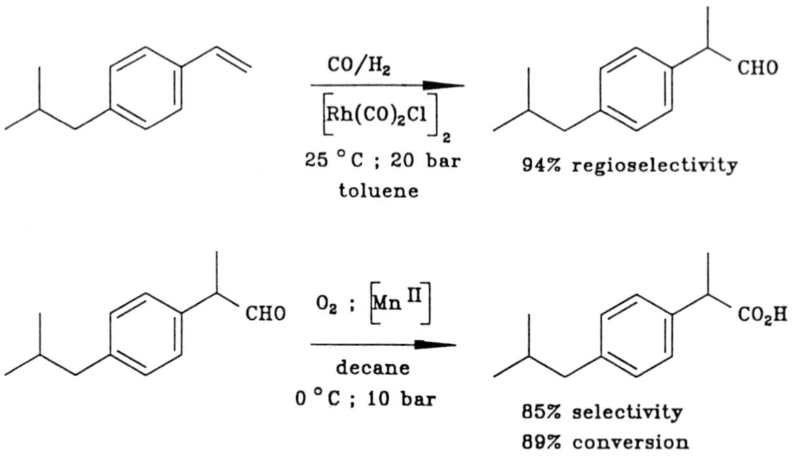

$CO/H_2$

$[Rh(CO)_2Cl]_2$

25 °C ; 20 bar

toluene

CHO

94% regioselectivity

CHO

$O_2$ ; [Mn II]

decane

0 °C ; 10 bar

$CO_2H$

85% selectivity

89% conversion

Figure 17. Synthesis of ibuprofen via hydroformylation and oxidation.

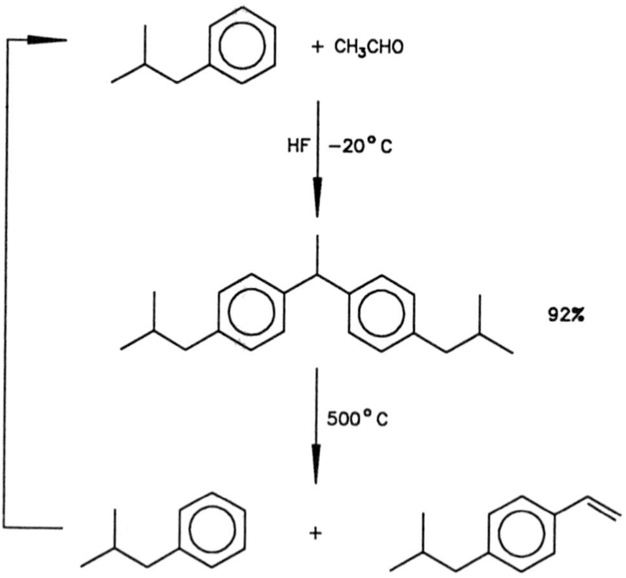

+ $CH_3CHO$

HF −20° C

92%

500° C

+

Figure 18. Synthesis of p-isobutylstyrene.

## CATALYSIS, CHIRALITY AND BIOLOGICAL ACTIVITY

Ibuprofen is an interesting synthetic target from yet another point of view: that of enantioselectivity. Many biologically active substances, e.g. pharmaceuticals, pesticides, contain one or more asymmetric centres and consist, therefore, of two or more optical isomers which should be considered as different chemical entities. In general, only one of the enantiomers (the so-called eutomer) of a racemic mixture is responsible for the desired biological activity towards the target organism. At best the other isomer (the distomer) constitutes unnecessary ballast, the removal of which would significantly reduce the chemical burden on the environment (be it the internal environment of the human body or nature at large).

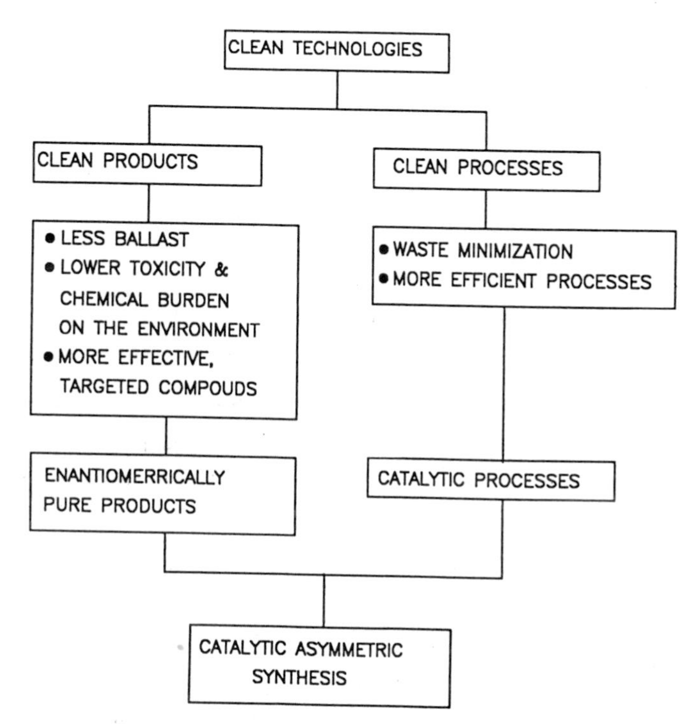


**Figure 19.** Clean technologies.

More often than not the distomer actually inhibits the desired effect of the eutomer and/or exhibits undesirable side effects. Consequently, there is an increasing trend to market such products as single enantiomers.[17,18] Thus, parallel with the development of cleaner, catalytic processes is a trend towards enantiomerically pure, i.e. 'cleaner' products. These two trends converge with the application of catalytic asymmetric synthesis (see Figure 19).

Ibuprofen is no exception in this respect. It is the S-enantiomer of ibuprofen that is solely responsible for the desired therapeutic effects.[11,19] The situation is complicated, however, by the fact that the R-enantiomer undergoes in vivo inversion to the S-enantiomer.[19] This has led many observers to conclude that it is of no consequence whether ibuprofen is administered as the racemate (as is presently the case) or as the pure S-eutomer. However, it has been shown that administering ibuprofen as the racemate (or as the R-enantiomer) leads to accumulation of ibuprofen in adipose tissue whereas this does not occur with the S-enantiomer. This could be an argument in favour of administering the pure eutomer, thus creating a need for a (catalytic) enantioselective synthesis. In principle, this could be achieved by carrying out the catalytic synthesis described above in an asymmetric mode, i.e. with the aid of chiral ligands. The reported[20] catalytic asymmetric carbonylation of p-isobutylstyrene (Figure 20) is certainly interesting in this context.

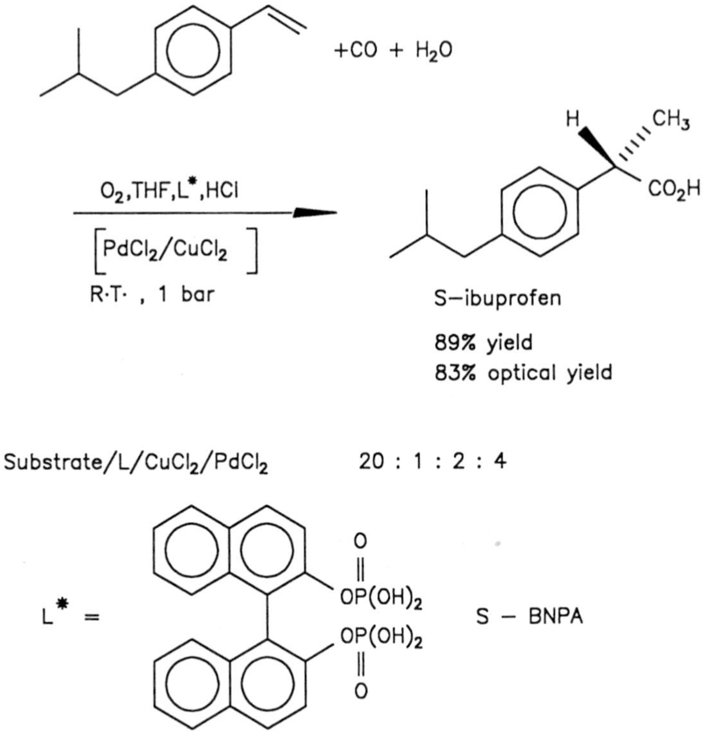

+CO + $H_2O$

$O_2$,THF,L*,HCl

[$PdCl_2/CuCl_2$]

R·T· , 1 bar

H $CH_3$ $CO_2H$

S–ibuprofen

89% yield

83% optical yield

Substrate/L/$CuCl_2$/$PdCl_2$ 20 : 1 : 2 : 4

L* = O OP(OH)$_2$ OP(OH)$_2$ O S – BNPA

**Figure 20.** Synthesis of S-ibuprofen via asymmetric carbonylation.

## SUMMARY AND CONCLUSIONS

Summarizing, in the drive towards waste minimization in chemicals manufacturing, more and more classical, 'high-salt' organic chemistry will be replaced by cleaner, catalytic methods. Companies that ignore this trend may not survive to see the 21st century. In order to provide for a smooth transition organic chemists at large should become more catalysis-minded and pay more attention to the atom utilization of their reactions. This also applies to the teaching of organic chemistry in academic institutions, which are, after all, the source of future generations of industrial organic chemists.

An appreciation of the need for greener catalytic processes is often sadly lacking in these institutions. In short, as we have noted elsewere,[9] after more than 150 years in splendid isolation, organic synthesis and catalysis are being brought together again, a development that certainly would have appealed to Berzelius.

## REFERENCES


1. *Japan Chem. Week* 30:1517 (1989); *Eur. Chem. News* 52:1375 (1989).
2. E. Drent, P.H.M. Budzelaar and W.W. Jager, *Eur. Pat. Appl.* 0386833 (1990) to Shell.
3. R.A. Sheldon and J. Kochi, "Metal-Catalyzed Oxidations of Organic Compounds", Academic Press, New York (1981).
4. R.A. Sheldon, "Chemicals from Synthesis Gas", Reidel, Dordrecht (1983).
5. J. Falbe, Ed., "New Syntheses with Carbon Monoxide", Springer-Verlag, Heidelberg (1980).
6. H.M. Colquhoun, D.J. Thompson and M.V. Twigg, "Carbonylation: Direct Synthesis of Carbonyl Compounds", Plenum Press, London (1991).
7. U. Romano et al., *Chim. Ind. (Milan)* 72:610 (1990) and references cited therein.
8. D. Forster and T.W. Dekleva, *J. Chem. Educ.* 63:204-206 (1986).
9. R.A. Sheldon, *CHEMTECH* 566-576 (1991).
10. S. Yamaguchi, M. Inoue and S. Enomoto, *Chem. Lett.* 827-828 (1985).
11. K.M. Williams, in "Problems and Wonders of Chiral Molecules", M. Simonyi, Ed., Akademiai Kiado, Budapest (1990), pp. 181-204.
12. V. Elango, M.A. Murphy, B.L. Smith, K.G. Davenport, G.N. Mott, and G.L. Moss, Eur. Pat. Appl. 0284310 (1988) to Hoechst Celanese.
13. B. Chiche, A. Finiels, C. Gauthier, P. Geneste, J. Graille and D. Pioch, *J. Org. Chem.* 51:2128 (1986); idem, *J. Mol. Catal.* 42:229 (1987).
14. D. Neibecker, R. Reau and S. Lecolier, *J. Org. Chem.* 54:5208-5210 (1989).
15. D.P. Riley, D.P. Getman, G.R. Beck and R.M. Heintz, *J. Org. Chem.* 52: 287-290 (1987).
16. I. Shimizu and Y. Tokumoto, Eur. Pat. Appl., 300,498 (1989) to Nippon Petrochemicals; CA 111:57241u.
17. R.A. Sheldon, *Chem. Ind. (London)* 212-219 (1990) and references cited therein.
18. R.A. Sheldon, in "Problems and Wonders of Chiral Molecules", M. Simonyi, Ed., Akademiai Kiado, Budapest (1990), pp. 349-386.
19. K. Williams, R. Day, R. Knihnicki and A. Duffield, *Biochem. Pharmacol.* 35: 3403-3406 (1986).
20. H. Alper and N. Hamel, *J. Am. Chem. Soc.* 112:2803-2804 (1990).

# INDUSTRIAL SCALE MEMBRANE PROCESSES

James R. Fair

Chemical Engineering Department
The University of Texas at Austin
Austin, TX 78712

The use of perm-selective membranes for separating mixtures is far from new, and in a sense is as old as the human race, since the kidney purification process involves selective passage of materials through membrane walls. For many decades there has been some commercial use of membranes for liquid mixture separation in dialysis and reverse osmosis. More recently, membrane separations have become big actors on the stage of separation technology, especially in the area of gas recovery and purification. The field has grown quite large and the literature is voluminous. There is now a North American Membrane Society and there has been for some years the Journal of Membrane Science. At several academic institutions throughout the world, including The University of Texas at Austin, there are major programs in membrane research.

The theme of this symposium is industrial waste minimization, and it is within this context that the present paper has been developed. As a further limitation on the scope of the paper, gas separations will have some emphasis, although liquid separation applications will be mentioned. Actually, many of the principles discussed will apply to either phase.

## WASTE MINIMIZATION SEPARATIONS

It is rare that a process does not generate one or more by-products that must be discharged, sold, or recycled to the processing operation. If the by-product has little intrinsic value, and is completely compatible with the environment, then discharge is the course of action likely to be followed. If it can be recovered and sold at a net profit, and the capital investment is within the capability of the manufacturer, then good economic sense dictates that this avenue be taken. If the by-product can be recycled to the process at an economic advantage, and its recycling is within the technological capability of the manufacturer, then prudence calls for doing so. These ideas are far from new, and will apply to any manufacturing operation. The problem is that many by-products do not fall into this neat categorization, or do not meet the premises just stated. Thus, there is a continuing general

*Industrial Environmental Chemistry*, Edited by D.T. Sawyer
and A.E. Martell, Plenum Press, New York, 1992

problem with process effluents that remain to be dealt with, and which contain components that are not environmentally acceptable.

For chemical manufacturing, which represents one of the greatest opportunities for waste minimization efforts, the problem by-products are often generated at a reaction step. It is commonly observed that by taking a lower conversion per pass in the reactor, quantities of by-products are decreased. A typical chemical process is shown schematically in Figure 1, where it is clear that separation steps can be important in recovering by-products for recycling, discharge or (hopefully) for sale. It has been stated many times that separations are a key ingredient of any waste reduction effort.

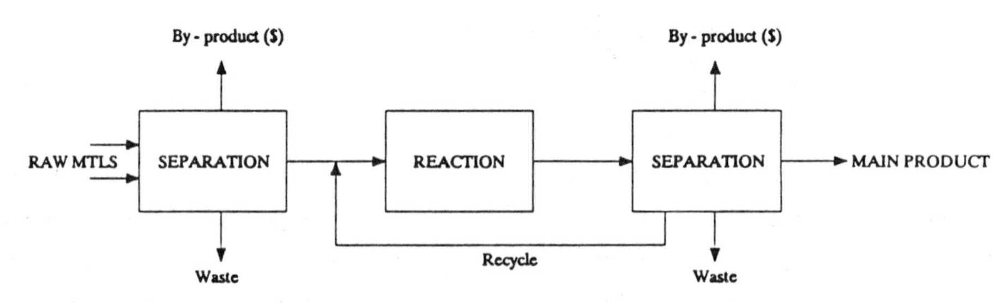


**Figure 1.** Typical chemical process flows

It is of interest here to consider the typical separation problem that is encountered in reducing amounts of contaminants in streams leaving the process boundary. The contaminants are often at low concentration and may be present in discharge gases or in aqueous outflows. This does not suggest that more concentrated mixtures are not important, but rather that conventional technologies such as distillation or extraction are likely to have been applied already to their separation requirements. This leads to the question of whether membranes can be useful for such separation needs.

## MEMBRANE SYSTEMS

When applied to chemical processing, these systems comprise one or more modules ("permeators") that contain the membrane surface in a compact, high surface/volume ratio arrangement. It may be necessary to filter the feed stream in order to avoid plugging problems. Equipment may be needed to provide the pressure driving force for flow through the membrane walls. Pre-heating of gases may be necessary to avoid condensed materials that might damage the membranes. Finally, there may be piping arrangements for recycle of streams back to the feed or to intermediate points in the process. A schematic flow arrangement is shown in Figure 2, taken from the paper by Spillman (1989). The application shown in the figure is for the removal of carbon dioxide from methane gas.

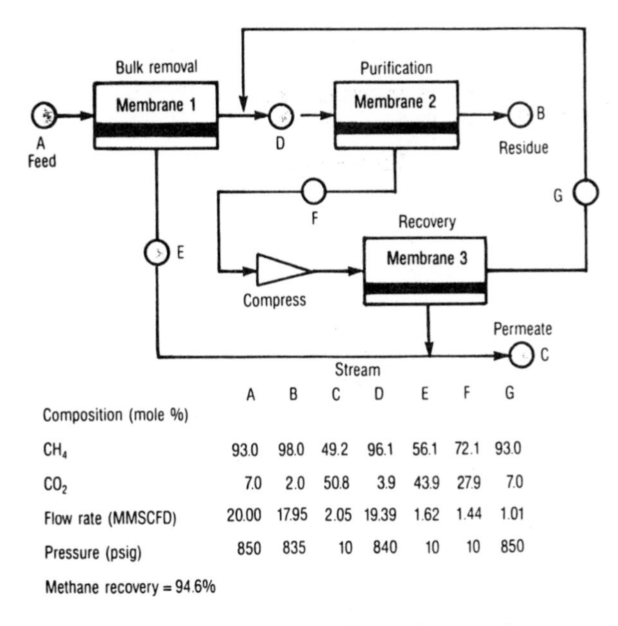


| Stream | A | B | C | D | E | F | G |
|---|---|---|---|---|---|---|---|
| Composition (mole %) | | | | | | | |
| $CH_4$ | 93.0 | 98.0 | 49.2 | 96.1 | 56.1 | 72.1 | 93.0 |
| $CO_2$ | 7.0 | 2.0 | 50.8 | 3.9 | 43.9 | 27.9 | 7.0 |
| Flow rate (MMSCFD) | 20.00 | 17.95 | 2.05 | 19.39 | 1.62 | 1.44 | 1.01 |
| Pressure (psig) | 850 | 835 | 10 | 840 | 10 | 10 | 850 |

Methane recovery = 94.6%

**Figure 2.** Membrane system for treating natural gas.

The permeators used today are of two types. One type embodies a shell and tube type affair with many hollow fiber membranes bundled into two tube sheets as shown in Figure 3. Inside diameters of the fibers are usually in the range of 300 to 600 microns. The other type of permeator is the spirally-wrapped variety as shown in Figure 4. Both types exhibit large surface-volume ratios and can be used in either gas or liquid service. Commonly, the membrane surfaces are of a polymeric nature; this suggests possible problems of compatibility between the fluid mixture and the membrane material.

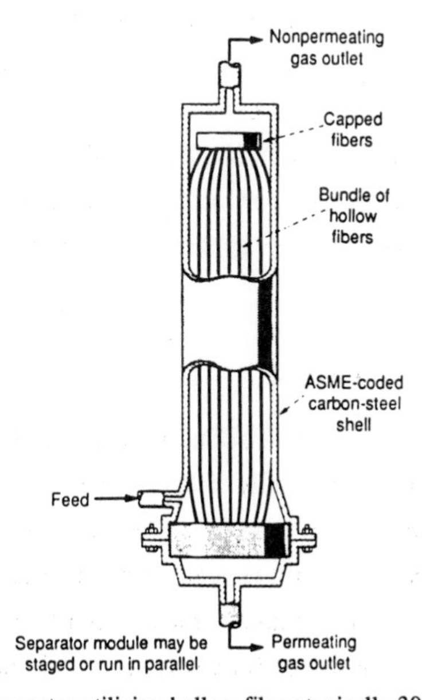


**Figure 3.** Shell and tube type permeator utilizing hollow fibers typically 300-600 microns inside diameter.

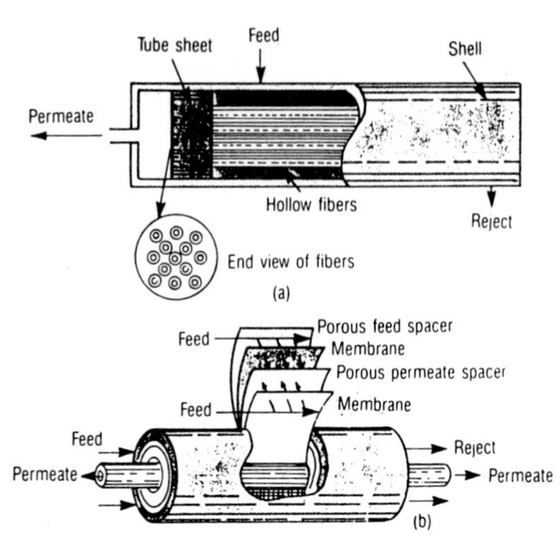


**Figure 4.** Arrangements of membranes for commercial permeators: upper, hollow fibers; lower, spirally-wrapped membrane sheets.

## THE MEMBRANE SEPARATION BUSINESS

For liquid mixtures, reverse osmosis applications predominate, and these for the purpose of water purification. An osmotic pressure must be applied to force pure water through a microporous membrane wall, leaving behind the salts or other impurities entering with the water. Fouling of the wall by these impurities ("polarization") is a major problem with reverse osmosis installations and depends, quite naturally, on the nature and concentration of the impurities.

For gas mixtures there are two large applications of membranes. One is for the production of inerting nitrogen based on an air feed. Membranes are available for this oxygen-argon-nitrogen separation, and membrane systems compete with pressure swing adsorption systems for the inert nitrogen business. The other large application, and the one that launched the modern gas permeation business, is the recovery of hydrogen from vent streams or recycle streams in plants such as those producing synthetic ammonia. In addition to these major uses there are hosts of other applications warranting the attention of membrane researchers.

## MEMBRANE RESEARCH AREAS

The research activity in membrane permeation can be classified as shown in Table 1. Much of the research could be more broadly categorized as materials research: work directed toward finding the right membrane material for a specific separation problem. For example, of the acid gases, carbon dioxide is easily separated from natural gas by polymer membranes; in fact, one of the "standard" test mixtures is methane/carbon dioxide. Another acid gas, hydrogen sulfide, has not been as successfully separated with membranes, although work is in progress and there has been developed a rubbery polymer that that can effect a good

**Table 1.** Research Areas, Membrane Permeation

- Transport through membranes
  - Rate
  - Selectivity
  - Pressure loss
  - Solubility
  - Diffusion
- Interaction of solutes with membranes
  - Enhancement of selectivity
  - Durability of membrane material
- Preparation of membranes
  - Casting
  - Spinning
  - Conditioning
- Design and fabrication of permeators
- Configuration of membrane systems
  - Staging
  - Counterflow
  - Integration with overall process

separation without serious deterioration of the membrane material.

For each material under consideration there is a tradeoff between rate of diffusion and selectivity of the separation, as shown in Figure 5. The rate of permeation is inversely proportional to the thickness of the membrane wall, and to this end asymmetric membranes have been developed that have a thin (1000 Angstroms) "skin" to make the separation and

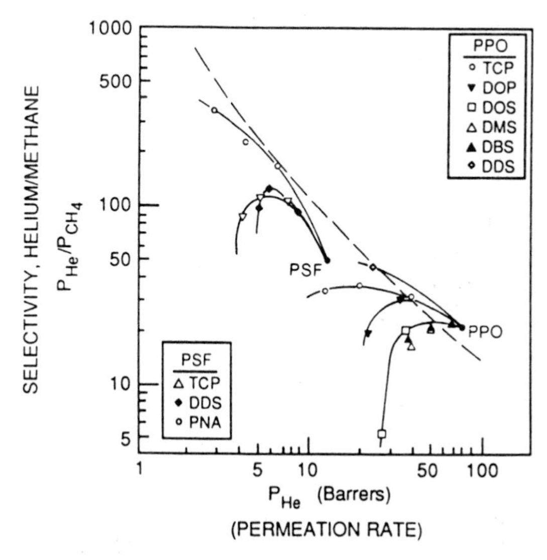


**Figure 5.** Variation of selectivity with rate of permeation for the helium-methane system. Membrane materials: PSF = polysulfone, PPO = polypropylene oxide. Symbols represent various plasticizers used in the polymer membranes.

a more substantial support (with high porosity) below the skin. Still other studies of membrane materials involve modeling of the movement through micropores and the solubility/diffusion characteristics of molecules moving through nonporous membranes.

The interaction of solutes with membrane module materials has been mentioned; some organics, for example, dissolve the membranes and/or the tube sheet materials that support hollow fibers or spiral wrapping materials. Fouling of membranes is always a potentially serious problem, and like the analogous case of heat exchanges, is not easily tractable. For bioseparation applications, the crossflow filter is used, and the fouling with filter cake is avoided by using very high stream velocities across the surface of the membrane material.

Preparation of membranes has become a major area of research. Conventional spinning or casting have their limitations, especially when the asymmetric "skin", mentioned earlier, is to be used. Should the membrane be formed and then treated to provide the skin? Or should the skin migrate to the surface during the formation process? Here there are large possibilities, particularly when co-polymers are being used for the separation.

Design and fabrication of permeators is a significant component of overall commercial membrane separation technology, yet it is one that involves mechanical know-how that is closely guarded by the vendors of permeators. One has the problem of arranging hollow fibers such that there can be good penetration of the shell side fluid into the bundle. There is the problem of pretreating the fibers or the wrappings such that early failures do not occur. As alluded to earlier, there is the problem of selecting a potting material for the fiber tube sheet that can resist such materials as ketones, aromatic solvents and alcohols.

The final entry in Table 1 deals with configuration of membrane systems. Such work is invariably based on process simulators that can handle a variety of recycle streams and membrane selectivity/rate characteristics. The fact that this is a *mass separating agent* type process means that it is difficult to generalize on the rate/selectivity criterion. At any rate, research in the area of configuration is carried out primarily by the vendors or, to some extent, by processors with a need to optimize operating equipment; very little of such work is being carried out in academia.

## WASTE MINIMIZATION APPLICATIONS

The question arises as to how membrane systems can be used in the waste minimization thrust. It should be clear from the foregoing that membranes must often be tailor-made to fit an application; this makes it difficult to generalize for waste management purposes. However, it has often been observed that membranes are much better for *concentration* than for *purification*. And when a relatively pure product such as inerting nitrogen emerges from a permeator, it is likely that the *recovery* of this material has not been high. Thus, for a membrane unit producing high purity nitrogen a good deal of the nitrogen entering with the air is discarded because of a low recovery, but considering this special case of a "no-cost" raw material, and the fact that the by-product (enriched air) is not deleterious to the environment (and might actually be saleable) makes the level of recovery only a concern for the energy expended in compressing the feed stream. Incidentally, when viewed from the oxygen balance, recovery is very high - but the recovered oxygen is not in high concentration.

For waste minimization applications, however, *recovery must be high,* although "minimization" does not necessarily equate to "elimination". This suggests that if membranes are to be used, they might best be used in conjunction with one or more other separation methods; such a combination has come to be known as a *hybrid separation system.* Some examples of such systems will be discussed in the following paragraphs.

A representative problem of waste minimization deals with the removal of acid gases from discharge gas streams. There are several methods available for this removal operation: condensation, low-temperature absorption, higher temperature physical absorption and chemical absorption. Figure 6 shows approximate regions of application of these methods (Blytas, 1986). Thus, it might be economical to make a membrane separation first, and then follow with a chemical absorption step. Studies sponsored by the Gas Research Institute have delineated operating conditions where acid gas removal is cheaper using membrane-absorption hybrids than using straight absorption. (Kellogg, 1991). A flow diagram for such an operation is shown in Figure 7.

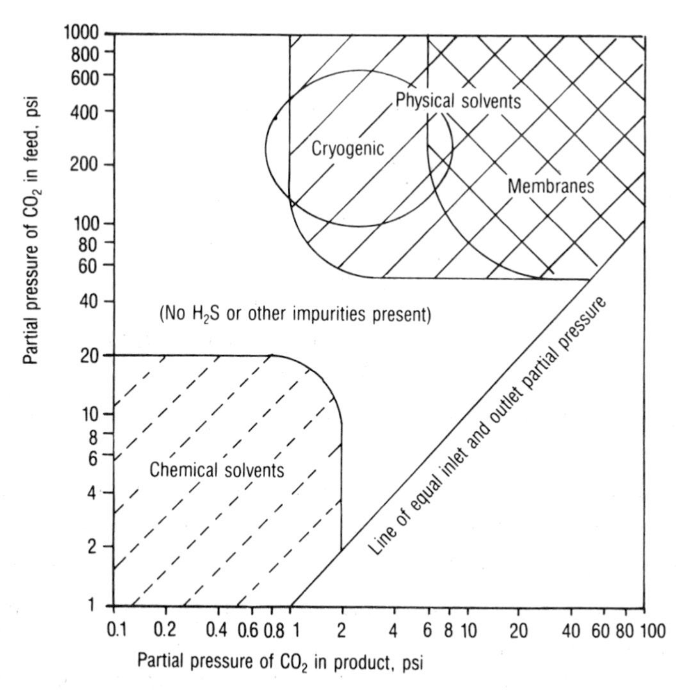


**Figure 6.** Areas of general applicability of separation methods for the removal of carbon dioxide from gas streams.

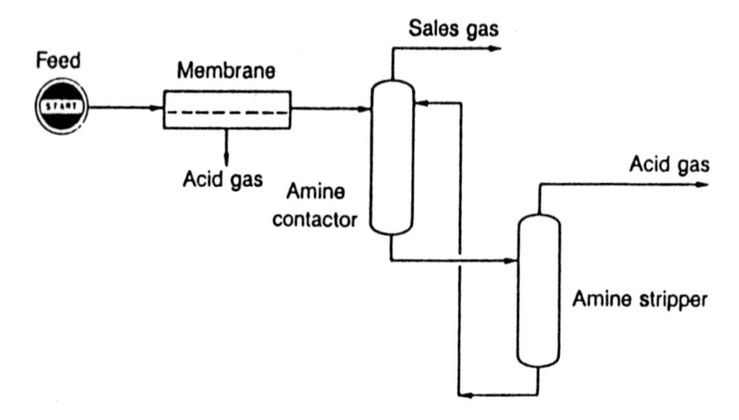


**Figure 7.** Hybrid system for the removal of acid gases from natural gas.

Another application of hybrid systems deals with hydrogen recovery. As discussed earlier, this separation may be made by membrane permeation, but classically it has also been made by cryogenic absorption. Researchers at Air Products, Inc. have found applications where a combination of these methods can be optimal (Agrawal *et al.*, 1990), as indicated in Figure 8. As shown, the membrane unit is upstream of the cold box; alternate schemes have this sequence reversed.

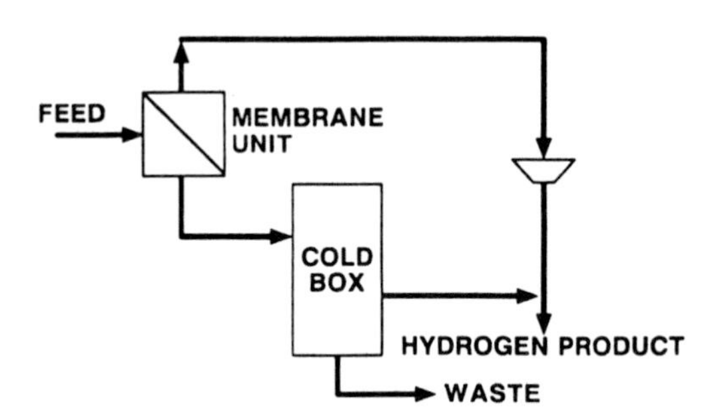


**Figure 8.** Sequential membrane/cryogenic hybrid process for hydrogen recovery.

The use of membrane permeation for removing (and recovering) organics from discharge air has been explored by Peinemann *et al.* (1986). Their proposed industrial flow diagram is shown in Figure 9. The flows appear to be similar to those of an adsorptive type recovery system. The problem is that the membranes may not be as selective as activated carbon for separating the organics from air. Permeabilities measured by Peinemann *et al.* for several organics in nitrogen, using a neoprene rubber membrane, are shown in Figure 10, and clearly are quite high. If selectivity is defined as a ratio of permeabilities, then removal of acetone, for example, from nitrogen would appear to be quite easy. Here one needs to consider the basic flux equation for membrane permeation:

$$N_i = P_i[\Delta p_i/z] \tag{1}$$

where $N_i$ is the molar flux of component i per unit area, $P_i$ is the permeability of component i through a thickness z of the membrane, with a driving force of $\Delta p_i$ partial pressure. The ratio of permeabilities of two components i and j in the feed mixture may be called a selectivity ratio:

$$\alpha_{ij} = P_i/P_j \tag{2}$$

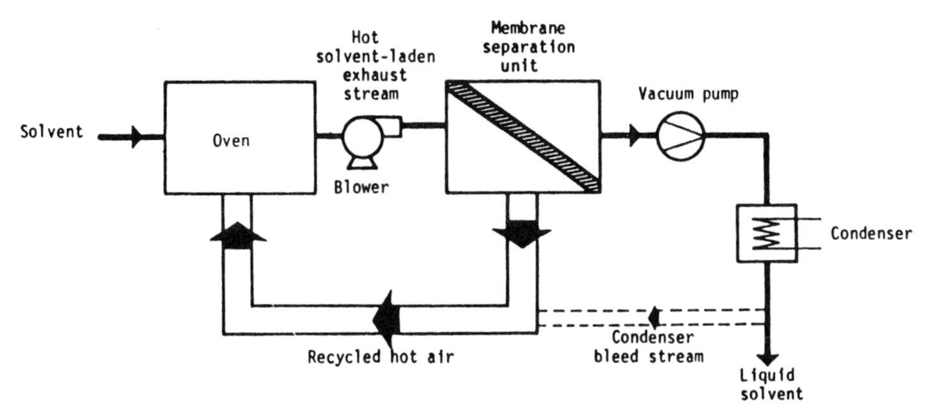


**Figure 9.** Membrane system for removing and recovering organic solvents from discharge air.

In Figure 10, the indicated selectivity ratio, acetone/nitrogen, is in the range of 1000. A convenient relationship for making use of this selectivity has been developed by Peinemann *et al.* and is shown in Figure 11. The curves are based on an entering feed composition of 0.5 vol-% organic (remainder air or nitrogen). The abscissa scale is the ratio of total pressures of permeate and feed streams. Thus, for the case of acetone removal from atmospheric air, and a vacuum of 50 mm Hg absolute, the permeate (material passing through the membrane) would contain about 20 vol-% acetone, remainder air or nitrogen. This might place a burden on the vacuum-producing equipment for large scale operation, and would require refrigeration for condensing the acetone from the permeate gas. It is unlikely that such a process could compete economically with a conventional carbon bed adsorption process. However, a combination of membrane permeation followed by an activated carbon system might represent an economical operation.

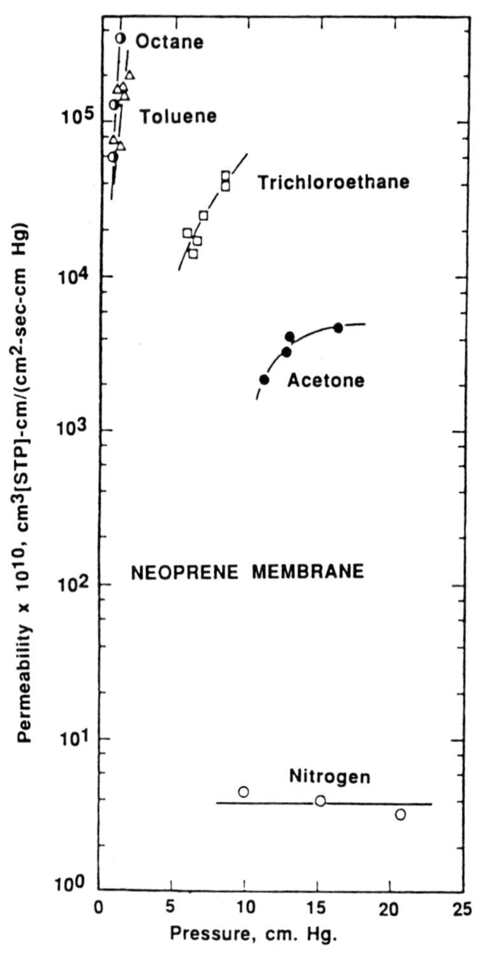


**Figure 10.** Permeabilities of several organics compared with the permeability of nitrogen at 40°C.

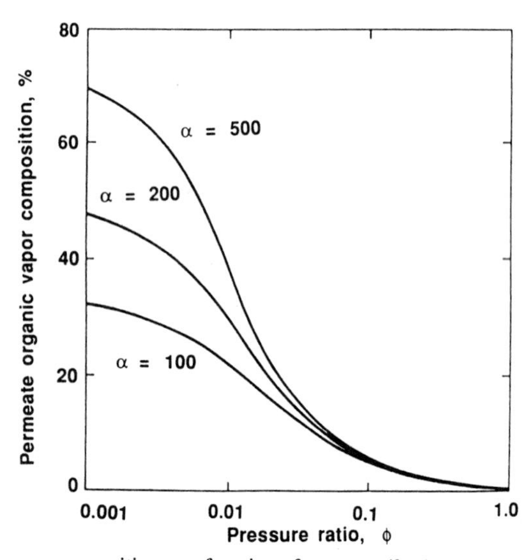


**Figure 11.** Permeate vapor composition as a function of permeate/feed pressure ratio and selectivity.

The use of membranes for the removal of small amounts of organics from water is a potential liquid-phase application with waste minimization implications. Here one would need to use a polymeric membrane in which the organics would selectively absorb and diffuse to the permeate side of the membrane wall. Microporous membranes with a hydrophobic treatment would also be a possibility. Work in this area is known to be in progress, utilizing the process of *pervaporation,* in which the permeate is taken off as a vapor. Pervaporation is often used to break organics-water azeotropes, with the water selectively transferring to the vapor side. Thus, as is the usual case, a materials development problem emerges: one must find (or design) the proper membrane material that can resist attack by the organics, yet allow those organics selectively to move through the membrane material. It would be completely infeasible to use the reverse osmosis approach of transferring the water through the membrane.

## FORWARD RESEARCH IN MEMBRANE TECHNOLOGY

Continuing research in the membrane area will concentrate on finding and developing better membrane materials. Although there may be some indication in this paper that all membranes are made from polymers, this is not the case. Membranes of ceramics and carbon are very much under investigation, especially for higher temperature applications. The ability to tailor-make membranes for specific separation needs is a key research goal. There are a number of academic institutions with research concentrations in membranes, and one of these is at the author's university, as part of the Separations Research Program. Some of the more recent accomplishments in this particular program are indicative of research accomplishments in the general field:

- The ability to make ceramic-supported polymeric composite thin film membranes.
- A new process for making ultrathin (less than 1000 Angstroms), defect-free gas separation membranes.
- Methodology for pre-conditioning membranes to improve rate without loss of selectivity.
- Development of fully aromatic polymers that have high oxidation and thermal stability.
- Improved understanding of how thin structure affects transport as well as physical properties of membranes.
- Ability to fabrication composite membranes.
- Ability to form defect-free asymmetric membranes directly as opposed to using coating techniques.

## SUMMARY AND CONCLUSIONS

This paper has been directed toward waste minimization applications of membranes, with the chemical process as the example application. The mixtures involved typically have small concentrations of contaminants which must be removed almost quantitatively if exit waste restrictions are to be recognized. Characteristically, membranes serve as concentrators, which could mean that the majority materials (e.g., water or air) must be removed from the mixture with minimum amounts of contaminants taken along. Thus, at first glance one would suspect that membranes might have a minor role to play in waste minimization efforts. At least this has been the reputation of membrane separations in the past.

Through continued and intensive research, this picture is changing. Coupling membranes with other separation techniques to provide hybrid systems is proving to have economic advantages in some cases; the concentration capability of the membranes provides a more cheaply-separated mixture for the associated separation technique located downstream. Special polymeric materials are being developed that can remove selectively organic contaminants in waste waters. In general, there is a strong thrust to find or design special membrane materials that can be used for separating particular molecules from mixtures. While most of the effort seems to be devoted to gas mixtures, backgrounds in reverse osmosis separations are leading researchers toward the large potential of removing low-concentration level contaminants from liquid streams.

The research in membrane technology is taking place both in industry and in academia. Naturally, more is heard from the academicians and reviews such as the present one must rely largely on published works or in-house research. There are large areas of research and development that are the special domain of industry: how the membrane separator is integrated with the rest of the plant; how the membrane devices are constructed and installed; how the membrane surfaces are formed most economically. Partnerships between industry and academia are flourishing, and the end result can only mean one thing: more and better ideas for minimizing wastes, and definite progress toward the ultimate zero-discharge plant.

## ACKNOWLEDGMENT

The author wishes to thank Professor W. J. Koros, Separations Research Program, The University of Texas at Austin, for his many helpful suggestions.

## REFERENCES

Agrawal, R., Auvil, S.R., DiMartino, S.P., Choe, J.S., Hopkins, J.A., 1988, Membrane/cryogenic hybrid process for hydrogen purification, *Gas Sepn. Purif.*, 2: 9.

Blytas, G.C., 1986, Separation challenges and opportunities in the oil and petrochemical industry, *in* "Chemical Separations," Vol. 2, C.J. King and J.D. Navratil, eds., Litarvan Literature, Denver.

McKee, R.L., Changela, M.K., Reading, G.J., 1991, $CO_2$ removal: membrane plus amine, *Hydrocarb. Proc.* 70: (4) 63.

Peinemann, K.-V., Mohr, J.M., Baker, R.W., 1986, The separation of organic vapors from air, *AIChE Symp. Ser. No. 250*, 82:19.
Spillman, R.W., 1989, Economics of gas separation membranes, *Chem. Eng. Progr.*, 85: (1) 41.

# NOVEL MICROPOROUS MEMBRANE-BASED SEPARATION PROCESSES FOR POLLUTION CONTROL AND WASTE MINIMIZATION

Chang H. Yun, Asim K. Guha,[1] Ravi Prasad,[2] and Kamalesh K. Sirkar[1*]

Department of Chemistry and Chemical Engineering
Stevens Institute of Technology
Hoboken, NJ 07030
[1]Department of Chemical Engineering, Chemistry and Environmental Science
New Jersey Institute of Technology
Newark, NJ 07102
[2]Hoechst Celanese SPD
Charlotte, NC 28273
[*]Corresponding Author

## INTRODUCTION

Multiphase equilibrium-based processes for separation and purification generally utilize dispersed systems in which one phase is dispersed in the other as bubbles or drops or thin films. Using microporous membranes, novel techniques have been developed such that multiphase processes can now be carried out in a nondispersive fashion for gas-liquid (Sirkar, 1992) and liquid-liquid (Prasad and Sirkar, 1992) contacting processes. Among such processes, only nondispersive solvent extraction of pollutants using microporous membranes will be of concern here. These processes employ immobilized immiscible phase interfaces at the pore mouths in a microporous membrane. Through such interfaces, solutes are extracted into the solvent as two immiscible phases flow on two sides of a microporous membrane. Many advantages of such a technique over conventional extractors have been summarized (Prasad and Sirkar, 1992).

Solvent extraction selectively extracts the solute/pollutant into the solvent from the feed aqueous solution. If it is desired to further have the pollutant selectively transferred into another aqueous phase in a concentrated form, a membrane may be employed between the two aqueous phases. Liquid membranes are particularly useful for such purification-concentration.

We have recently developed the concept of the hollow fiber contained liquid membrane (HFCLM) which, unlike the supported liquid membrane (SLM) (Danesi et al., 1987), is highly stable (Sengupta et al., 1992). Unlike membrane-based solvent extraction which requires only one immobilized phase interface in a membrane, there are two immobilized

*Industrial Environmental Chemistry*, Edited by D.T. Sawyer
and A.E. Martell, Plenum Press, New York, 1992

phase interfaces in two different sets of microporous hollow fiber membranes in a module used in HFCLM technology. Feed aqueous solution containing the pollutant flows through the bores of one set of fibers while a strip aqueous solution flows through the bores of the other set of fibers. Any loss of the liquid membrane, contained between the two fiber sets on the shell side, is automatically replenished from an external membrane liquid reservoir. The pollutant is first selectively extracted into the liquid membrane from the feed fiber set, and then it is back extracted and concentrated into the strip aqueous solution flowing through the other fiber set. Application of this technique to the removal of heavy metals from wastewater and its simultaneous concentration in the strip aqueous solution is the second novel process illustrated here.

The hierarchy of strategies utilized for environmental protection from pollutants may be roughly identified in order of desirability as pollution prevention, waste minimization, and end-of-pipe control. Although some chemical processes are being modified to approach pollution prevention, many more are not; they are ideal candidates for waste minimization, recycling and/or resource recovery. We will illustrate in this paper three types of examples to that end. The first example will focus on membrane-based solvent extraction for removal and recovery of toluene from a real aqueous waste stream whose toluene concentration is reduced to 150-160 *ppb* without introducing any other pollutant. Membrane-based solvent extraction is the basis in the second example for treating a synthetic aqueous waste stream containing five priority pollutants at high concentrations to a level less than 25 *ppm* of each pollutant. HFCLM technology is utilized next to remove heavy metals like copper or chromium from an aqueous waste stream to levels around or less than 1 *ppm* while concentrating the heavy metal simultaneously in the strip stream to a level 10-15 times the feed stream level for recycling.

## EXAMPLE I: TOLUENE REMOVAL TO SUB-*PPM* LEVEL BY MEMBRANE-BASED SOLVENT EXTRACTION

In late 1989, we were approached by B.F.Goodrich & Co.(Brecksville, OH) to study how the toluene level of a waste stream could be reduced from around 30 *ppm* to 30-40 *ppb*. The stream contained in addition heptane around saturation level besides polymeric or particulate matter. If one can reduce the toluene level to 140-160 *ppb*, then dilution with other plant streams will allow achievement of the targeted discharge level.

Recognize now that the other organic present in the waste stream, heptane, is not a priority pollutant and its saturation level is low (~50 *ppm*). Further it is an excellent solvent for toluene. We therefore decided to carry out membrane-based solvent extraction removal/recovery of toluene using heptane as the extracting solvent. The extract may then be separated by distillation or other processes to recover toluene and recycle heptane to the extractor. The extracting solvent is no longer introduced fresh into the treated stream.

A hollow-fiber module, made by Hoechst Celanese SPD, Charlotte, N.C. was selected. Fifteen *cm* in length, 1.9 *cm* in diameter, this module had 1,800 Celgard X-20 hydrophobic microporous polypropylene fibers of 240 $\mu m$ I.D. and 290 $\mu m$ O.D.. The fiber wall had a porosity of 0.4 and a pore size of 0.03 $\mu m$. The fiber packing fraction was 0.43 and the membrane surface area per unit volume was 47.5 $cm^{-1}$. A schematic of this type of module is provided in Prasad and Sirkar(1989, 1990). General procedures for membrane-based solvent extraction is available in the above references.

The experimental set up is shown in Figure 1. The extracting solvent heptane kept in pressure vessel C was driven by nitrogen pressure in cylinder 2 to flow through the shell side of the microporous hollow fiber(MHF) module. Feed aqueous waste stream from the pressure vessel B was allowed to flow through a capsule filter (pore size: 10 $\mu m$) to the tube side of the MHF module countercurrent to the flow of the solvent heptane. The mode of operation was once through in a continuous fashion.

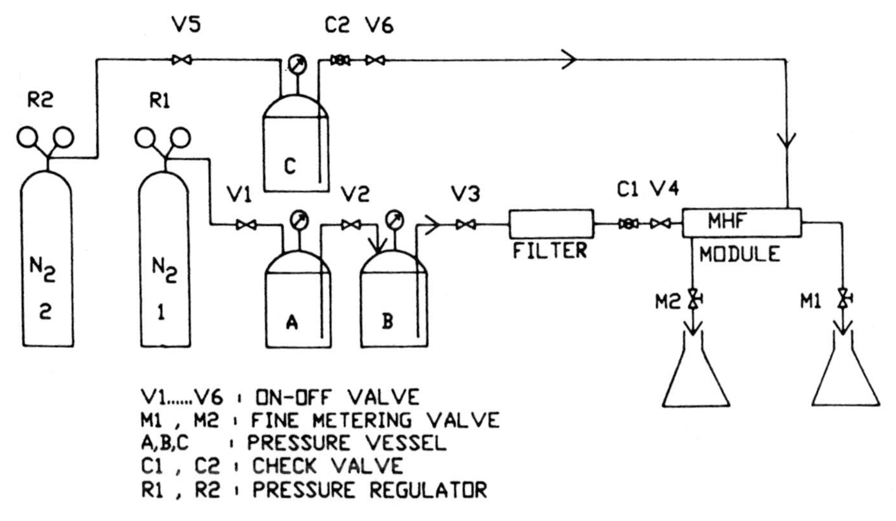


**Figure 1.** Experimental set up for membrane solvent extraction removal of toluene for an actual plant waste stream sample using hollow fiber module.

Feed aqueous waste stream was initially introduced quickly and simultaneously into pressure vessels A and B. The pressure vessel B was filled completely while the pressure vessel A had a little space on top to drive the aqueous waste solution by $N_2$ pressure from cylinder 1. This led to a certain amount of toluene stripping in pressure vessel A but none in pressure vessel B where a little dilution would occur due to mixing with the solution from vessel A. Samples of the aqueous waste were withdrawn carefully and quickly before and after the filter, capped in sample bottles filled to the top and sent to the plant for toluene analysis. Similarly, samples of the treated aqueous waste, the raffinate, were collected after M1 (Figure 1) with great care in covered sample bottles till they were full, capped and sent for analysis.

The flow rates of the aqueous waste stream and the organic solvent (heptane) during the experiment were 3.0 *ml/min.* and 7.7 *ml/min.* respectively. The pressure of the $N_2$ cylinder 1 on the aqueous side was maintained at 9.5 *psig* while that for the solvent side was 5.5 *psig*. This ensured that the solvent heptane was not dispersed in the aqueous phase.

**Experimental Results and Discussion**

The experimental results are summarized in Table 1. The original waste stream analysis indicated toluene levels between 30-35 *ppm*. The toluene concentration in the feed stream before the filter was between 13-15 *ppm*. This reduction indicates a certain amount of loss in transferring from the waste drum to the feed reservoirs and any stripping in vessel A. The toluene concentration in the feed stream after filtering varied between 9 and 11 *ppm* due to any adsorption in the filter. The treated waste stream (raffinate) concentration of toluene was in the range of 157-200 *ppb*.

These results demonstrate toluene level reduction by two orders of magnitude to sub-*ppm* levels (0.1-0.2 *ppm*). The raffinate samples were taken only after 45 minutes of operation; further samples were taken at 20 minute intervals. Because of the steady behavior of the raffinate, any effect of adsorption-based separation is ruled out.

Reduction of priority pollutant concentration by two orders of magnitude via continuous countercurrent membrane-based solvent extraction employing a microporous hollow fiber module has already been achieved (Yun et al., 1992). However, as we will see in the next section, the wastewater pollutant concentration levels in that study were far higher to start with. Thus raffinate pollutant concentrations were around ~25 *ppm*.

The results of Example I of the present study provide opportunities for the following generalizations on aqueous waste stream treatment/recovery:

(a) Multipollutant containing waste streams may be subjected to solvent extraction based remediation if the extracting solvent is relatively benign(e.g. heptane) or if the benign extracting solvent is already present in the waste stream. This assumes, however, that the pollutants are removable by extraction and the extracting solvent can be recovered and recycled.

(b) If the multipollutant waste stream contains, amongst others, a priority pollutant (e.g. toluene) which is also a good solvent for the other pollutants present, then remove and recover the other pollutants using this priority pollutant as an extracting solvent unless a relatively benign solvent is available. This method is especially useful if the different pollutants require different treatments as encountered in specialty chemical industry.

(c) On the basis of the present results and the principle of countercurrent extraction, we can argue that a membrane solvent extraction-based remediation can reduce pollutants to sub-*ppm* levels.

Along with many other advantages of microporous membrane-based solvent extraction, these features point to many opportunities for microporous membrane-based solvent extraction in pollution control and waste minimization.

**Table 1.** Toluene concentration in various streams[1].

| Sample identification | Toluene concentration *ppb* |
|---|---|
| Original waste stream | 30191<br>35693 |
| Feed #1 before filtering | 14499<br>13250 |
| Feed #1 after filtering | 9145<br>9460 |
| Raffinate #1 | 209<br>185<br>197 |
| Feed #2 before filtering | 13418<br>13454 |
| Feed #2 after filtering | 10739<br>10674 |
| Raffinate #2 | 189<br>188 |
| Raffinate #3 | 184<br>180 |
| Raffinate #4 | 163<br>170 |
| Raffinate #5 | 157<br>165 |
| Raffinate #6 | 157<br>161 |

[1]Samples analyzed at Research Laboratories of B.F. Goodrich, Brecksville, OH.

## EXAMPLE II: MULTIPOLLUTANT REMOVAL FROM HIGH STRENGTH WASTE WATER BY MEMBRANE-BASED SOLVENT EXTRACTION

One often encounters aqueous waste streams containing a large number of priority organic pollutants. Further the pollutant concentrations may be high. These pollutants may be high boiling, low boiling, polar, nonpolar, aromatic, aliphatic, chlorinated or otherwise. Incineration costs may be too high. Membrane-based solvent extraction may be employed here. The extract of pollutants and the extracting solvent may then be separated and the solvent recycled by distillation. Alternately, the extract can be incinerated at a cost order-of-magnitude lower than that for the wastewater since the solvent volume can be so much smaller with orders of magnitude higher concentration of pollutants. Only the extracting solvent is introduced into the treated stream and may have to be removed downstream.

The system studied was a synthetic waste water containing the following priority pollutants: phenol (500 *mg/l*), 2-chlorophenol (1241 *mg/l*), nitrobenzene (604 *mg/l*), toluene (216.5 *mg/l*) and acrylonitrile (3627 *mg/l*). The extracting solvents studied were isopropyl acetate (IPAc), methyl isobutyl ketone (MIBK) and hexane. This problem has been studied and analyzed; mass transfer characterization of the pollutants being extracted in the microporous hollow fiber module has been carried out and are available in Yun et al. (1992). Here we present some additional results for the cases where the solvent is either IPAc or hexane.

The hollow fiber module employed has already been described under Example I. The experimental apparatus employed is quite similar to that in Figure 1 except for the following: no filter was needed in the aqueous waste line before the membrane module; further only one aqueous waste stream reservoir was used instead of two shown in Figure 1. Aqueous phase pressure (10 *psig*) was maintained 5 *psig* higher than the organic phase pressure to prevent phase dispersion in the hydrophobic membrane module. Unlike that in Example I, the phase flow rates were varied over a significant range to study pollutant removal capabilities. Aqueous phase concentrations were determined by HPLC.

Figure 2 illustrates the aqueous stream concentrations of four pollutants exiting from the hollow fiber module at different aqueous phase flow rates for a given extracting solvent (IPAc) flow rate; the figure also provides the percent removal of the pollutants in this continuous hollow fiber extractor. At low aqueous phase flow rates, we observe from Figure 2(a) that the individual pollutant concentration of each pollutant has been brought down to 10-15 *mg/l*. The concentration of toluene is not being reported here since it was below the detection levels being employed. We also observe that reduction of acrylonitrile concentration is the most difficult since its inlet concentration is very high and it has a low distribution coefficient(~6) in IPAc compared to the other pollutants(phenol~53; 2-chlorophenol~250; nitrobenzene~300 etc.).

Figure 3 illustrates the pollutant concentrations in the aqueous treated stream at the hollow fiber module exit for a given flow rate of hexane as an extracting solvent. Since hexane has a very low affinity for polar pollutants like phenol, 2-chlorophenol and acrylonitrile, the extents of their removal are much lower. Considerable deliberation is, therefore, needed on the selection of the extracting solvent.

A polar solvent like IPAc or MIBK is very efficient (Yun et al., 1992); however, since their solubility in the treated water is significant, additional treatment will be needed. If, however, the pollutants can be removed by a nonpolar solvent like heptane or hexane, additional treatments may be avoided. On the other hand, the treatment plant size or membrane area required with hexane as the solvent may be quite large since the pollutant distribution coefficients in the case of heptane or hexane are usually low for polar pollutants.

An additional feature of membrane solvent extraction is that the maximum level of the extracting solvent introduced into the treated stream is governed by its solubility since there

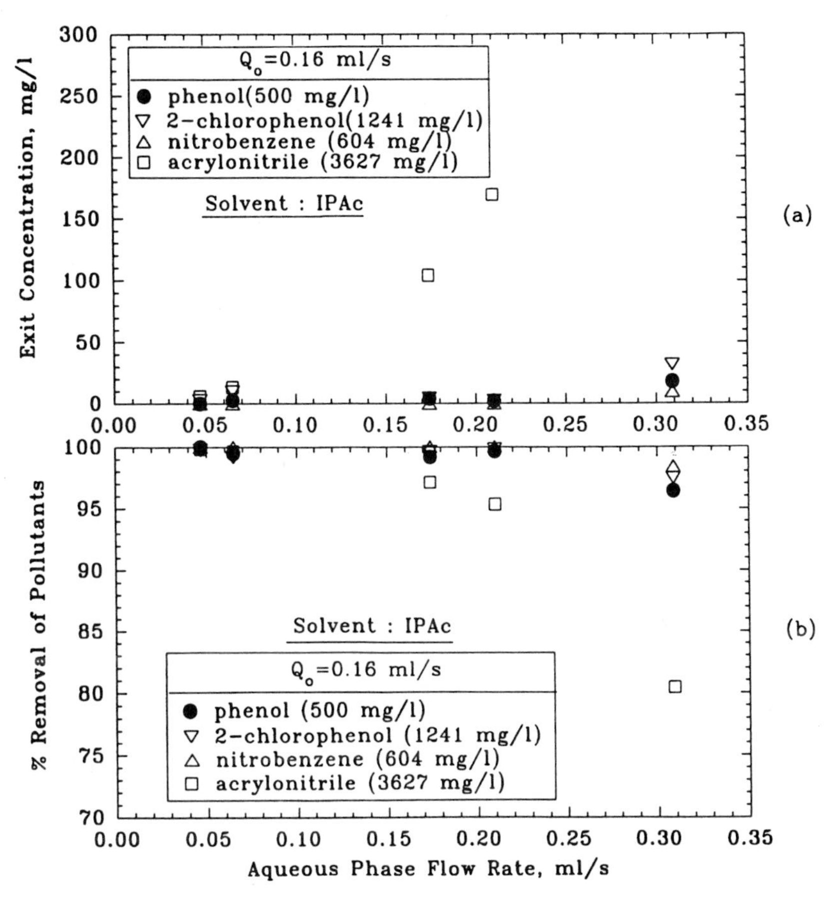


**Figure 2.** **(a)** Concentration and **(b)** percent removal of the priority organic pollutants in the exit aqueous phase from the experimental hollow fiber module (solvent: IPAc).

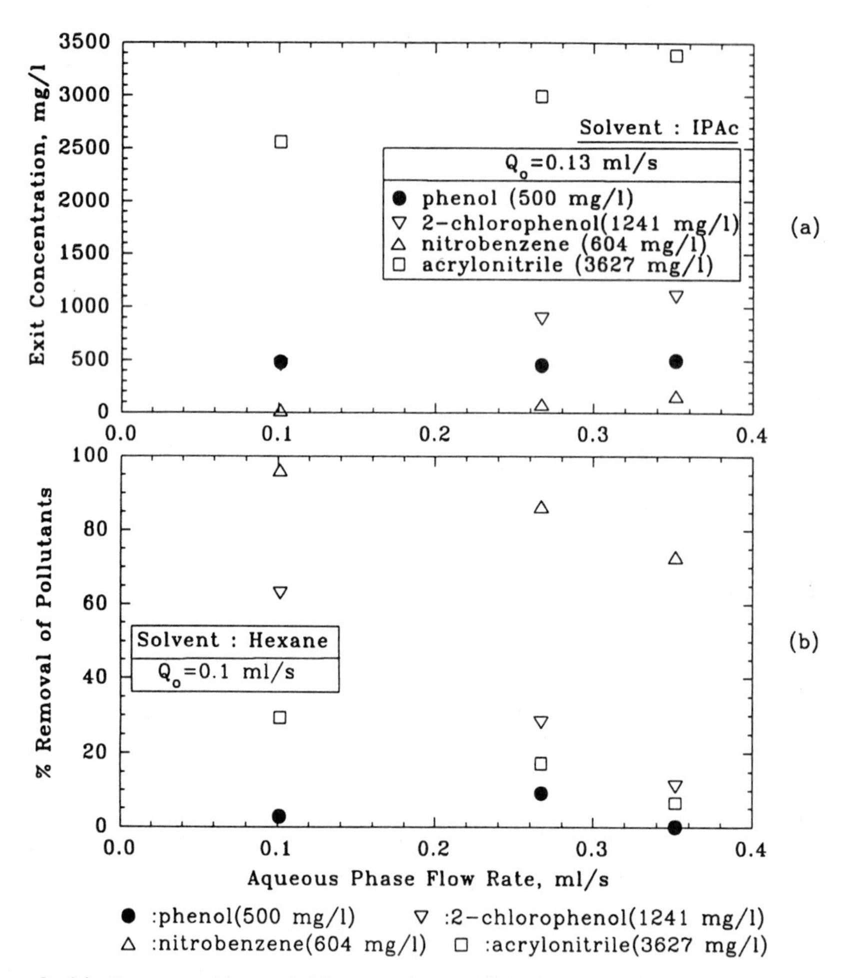


**Figure 3. (a)** Concentration and **(b)** percent removal of the priority organic pollutants in the exit aqueous phase from the experimental hollow fiber module (solvent: hexane).

is no dispersion. In fact, Yun et al.(1992) observed the level of MIBK in treated water to be lower than its solubility and decreasing with decreasing residence time. These features suggest that microporous membrane-based solvent extraction can be a highly efficient technique for waste minimization and pollution control.

## EXAMPLE III: REMOVAL OF HEAVY METAL AND ITS SIMULTANEOUS CONCENTRATION AND RECYCLE BY HOLLOW FIBER CONTAINED LIQUID MEMBRANE PERMEATOR

We illustrate now the hollow fiber contained liquid membrane (HFCLM) permeator performance in purifying an aqueous waste stream of a heavy metal like copper or chromium and simultaneously concentrating it in a strip aqueous solution for recycle. The structure of HFCLM permeators has been illustrated in Sengupta et al. (1988) and Basu and Sirkar (1991) for liquid phase separations. The permeator in this particular study had two sets of hollow fibers, each set containing 300 hydrophobic microporous polypropylene fibers (Celgard X-10, 100 $\mu m$ I.D., 150 $\mu m$ O.D.). The ends of each set of fiber at each permeator end were set apart. In the middle of the permeator, however, these two sets were well mixed and bundled into a 0.61 *cm* I.D. teflon sleeve which was inserted into a 1.05 *cm* I.D. stainless steel pipe. Each end of each fiber set was potted with epoxy. A general schematic of such a permeator in a loop with a feed solution and a strip solution in countercurrent flow is shown in Figure 4.

Figure 5 illustrates the performance of such a permeator when a synthetic aqueous waste stream containing 208 *mg/l* of copper present as cupric sulphate (feed pH~4.55) was passed through the feed fiber set while the strip aqueous solution containing 200 *g/l* of $H_2SO_4$ flowed cocurrently through the strip fiber set. The liquid membrane present on the shell-side and the fiber wall pores was 10 *v/v %* LIX84 diluted in heptane.

At the aqueous-organic interface in the feed fiber set, copper was extracted into the organic liquid membrane by the forward reaction

$$Cu^{++}(\text{aq}) + 2RH(\text{org}) \rightleftharpoons R_2Cu(\text{org}) + 2H^{+}(\text{aq}) \qquad (1)$$

whereas at the aqueous-organic interface in the strip fiber set, the reverse reaction takes place liberating copper into the aqueous strip solution. The liquid ion exchanger LIX84 exchanges $H^+$ with $Cu^{++}$ to complex copper ion and extract it into the liquid membrane from the feed solution. The strip solution pH is so low that copper is easily stripped; by maintaining a low strip solution flow rate, copper is easily concentrated.

As shown in Figure 5, at low aqueous feed flow rates, $Cu^{++}$ concentration at the outlet of feed fibers is reduced to around 1 *mg/l* while the strip outlet concentration is around 900 *mg/l*. At higher waste stream flow rates, feed outlet concentration is somewhat higher due to insufficient membrane area in this module while the strip concentration has been increased to 1250 *mg/l*. Other data not being reported here indicate strip solution concentration of copper at a level of 4000 - 6500 *mg/l* showing simultaneous cleanup of the waste stream of the toxic heavy metal copper as it is concentrated to a high level in the strip solution for recycle and recovery. Figure 6 illustrates percent $Cu^{++}$ removed for different concentrations of LIX84 in the liquid membrane.

Figure 7 illustrates a similar purification-concentration behavior in the removal of $Cr^{6+}$ (pH=2.43, obtained by dissolving $K_2Cr_2O_7$ in deionized water). The level of chromium is reduced to less than 1 *mg/l* whereas the strip stream concentration is raised to 750 *mg/l*. The liquid membrane employed was 20 *v/v %* trioctylamine diluted in xylene and the strip solution was 0.1 (N) NaOH. We have since studied this system more extensively; $Cr^{6+}$ concentration in the exiting strip stream has been raised to 4000 *mg/l* from a 250 *mg/l* feed

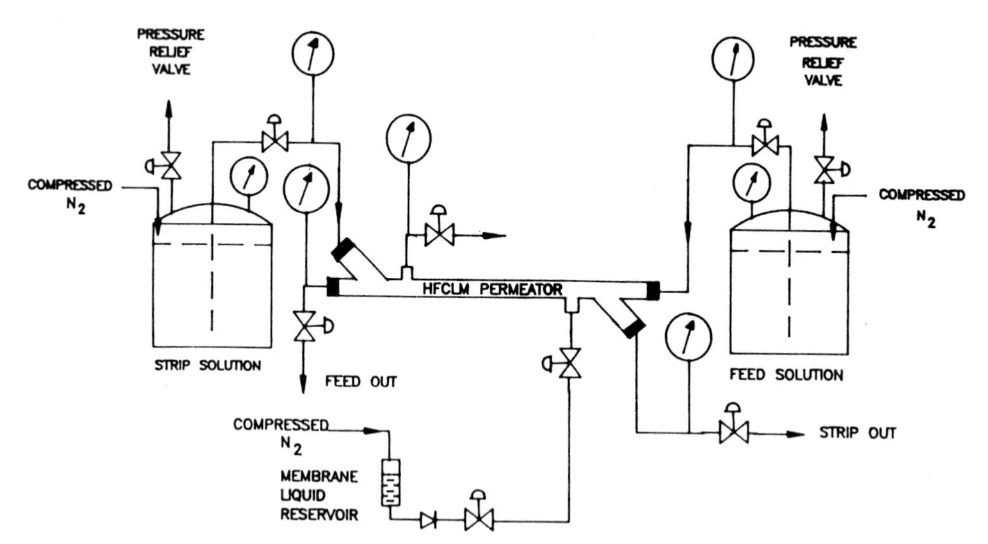


**Figure 4.** Schematic of the experimental set up using HFCLM permeator.

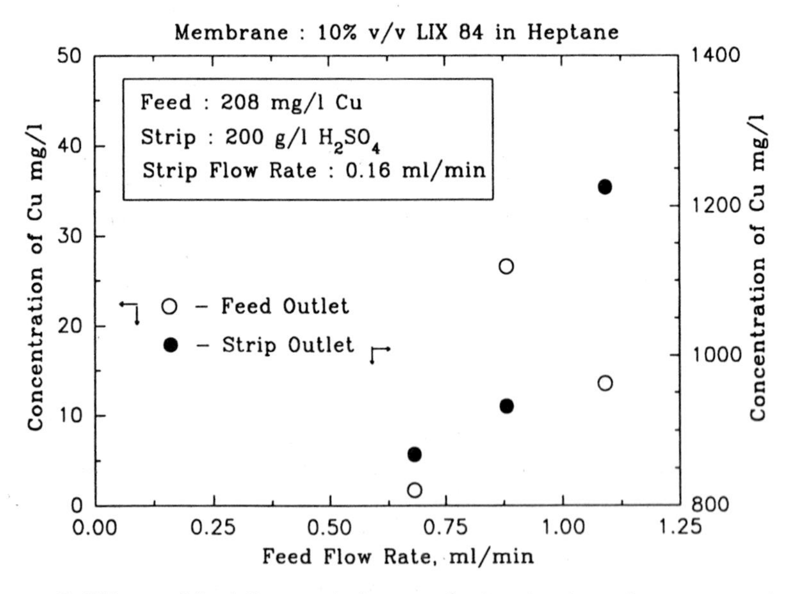


**Figure 5.** Effects of feed flow variations on feed and strip outlet concentrations.

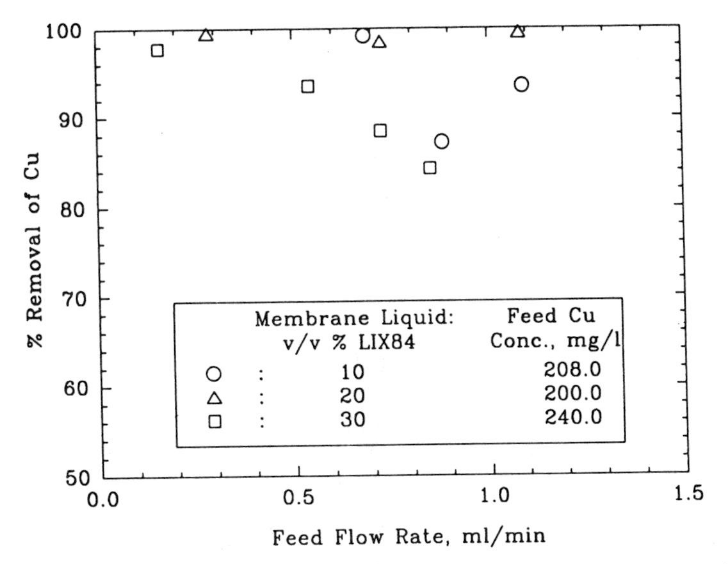


**Figure 6.** Percent Removal of copper for different concentrations of LIX84 in heptane.

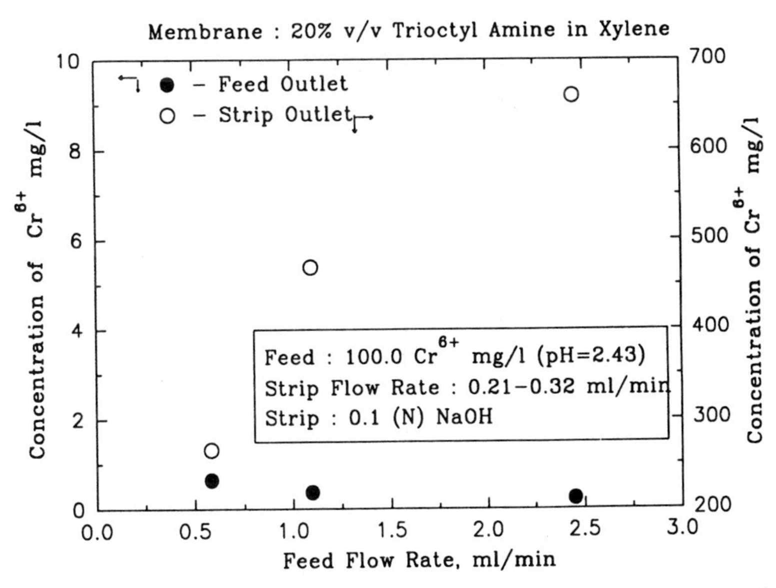


**Figure 7.** Effects of feed flow variations on feed and strip outlet concentrations.

solution. Further, on the based of long term studies, we have found such a system to be quite stable in its behavior (Basu, 1990).

These and other studies being currently investigated suggest that removal of heavy metals from waste streams and their simultaneous concentration and recycle via HFCLM permeators are feasible and should be explored further. The strip stream, concentrated in the heavy metal, may easily be the basis for recycling and resource recovery.

## REFERENCES


Basu, R., 1990, "Mass Transfer Enhancement in Phase Barrier Membrane Separators for Reactive Systems," *Ph.D. Diss.*, Stevens Institute of Technology, Hoboken, NJ.

Basu, R., and Sirkar, K.K., 1991, Hollow fiber contained liquid membrane separation of citric acid, *AIChE J.* 37:383.

Danesi, P.R., Reichley-Yinger, L., and Rickert, P.G.,1987, Lifetime of supported liquid membranes: the influence of interfacial properties, chemical composition and water transport on the long term stability of the membranes, *J. Membr. Sci.* 31:117.

Majumdar, S., Sirkar, K.K., and A. Sengupta, 1992, Hollow fiber contained liquid membrane, *in*: "Membrane Handbook," W.S. Winston Ho and K.K. Sirkar, eds., Van Nostrand Reinhold, New York.

Prasad, R., and Sirkar, K.K., 1989, Hollow fiber solvent extraction of pharmaceutical products: a case study, *J. Membr. Sci.* 47:235.

Prasad, R., and Sirkar, K.K., 1990, Hollow fiber solvent extraction: performances and design, *J. Membr. Sci.* 50:153.

Prasad, R., and Sirkar, K.K. 1992, Membrane-Based solvent extraction, *in*: "Membrane Handbook," W.S. Winston Ho and K.K. Sirkar, eds., Van Nostrand Reinhold, New York.

Sengupta, A., Basu, R., and Sirkar, K.K., 1988, Separation of solutes from aqueous solutions by contained liquid membranes, *AIChE J.* 34:1698.

Sirkar, K.K., 1992, Other new membrane processes, *in*: "Membrane Handbook," W.S. Winston Ho and K.K. Sirkar, eds., Van Nostrand Reinhold, New York.

Yun, C.H., Prasad, R., and Sirkar, K.K., 1992, Membrane solvent extraction removal of priority organic pollutants from aqueous waste streams, *I & E Chem. Res.* (in press).

# ION EXCHANGE FOR GLYPHOSATE RECOVERY

Lowell R. Smith and Jane L. Barclay

Monsanto Agricultural Company
A Unit of Monsanto Company
800 N. Lindbergh Boulevard
St. Louis, Missouri 63167

## INTRODUCTION

The first generation Monsanto glyphosate (Active ingredient in Monsanto's Roundup® herbicide, N-phosphonomethylglycine) manufacturing process, operated since 1976, produced a stream to waste treatment containing a significant amount of glyphosate. As a result of several research and engineering innovations, a new process was developed and a new plant recently commenced operations which reduced the glyphosate lost to waste treatment by a very significant amount. This result was mainly due to the use of an innovative ion exchange process for glyphosate recovery from the waste stream which also contained the process byproducts.

## THE GLYPHOSATE WASTE STREAM

The glyphosate process produces a spectrum of impurities as shown in Table 1, some of which are stronger acids than glyphosate and some of which are weaker acids. In addition, glyphosate and some of the impurities carry a basic function.

## ION EXCHANGE RESIN EVALUATION

A process was conceived which would use these differences in acidity as the basis for an ion exchange recovery method. For this purpose, ion exchange resins would have to be found which were capable of separating glyphosate from these impurities and which would concentrate it up to useful levels.

Ion exchange resins are of four general types as shown in Figure 1. In addition to acidity, selectivity in ion exchange depends on several factors among which are molecular size, charge and polarity. The

*Industrial Environmental Chemistry*, Edited by D.T. Sawyer
and A.E. Martell, Plenum Press, New York, 1992

**Table 1.** Composition of typical filtrate.

| COMPONENT | WT.% | |
|---|---|---|
| PHOSPHONOMETHYLIMINODIACETIC ACID | 0.01 | |
| N-FORMYL GLYPHOSATE | 0.01 | |
| HYDROXYMETHYLPHOSPHONIC ACID | 0.002 | STRONG ACIDS |
| PHOSPHATE | 0.02 | |
| PHOSPHITE | 0.004 | |
| IMINOBISMETHYLENE PHOSPHONATE | 0.035 | |
| GLYPHOSATE | 1.50 | |
| METHYL GLYPHOSATE | 0.06 | |
| FORMALDEHYDE | 2.85 | |
| FORMIC ACID | 1.50 | WEAK ACIDS |
| AMINOMETHYLPHOSPHONIC ACID | 0.35 | |

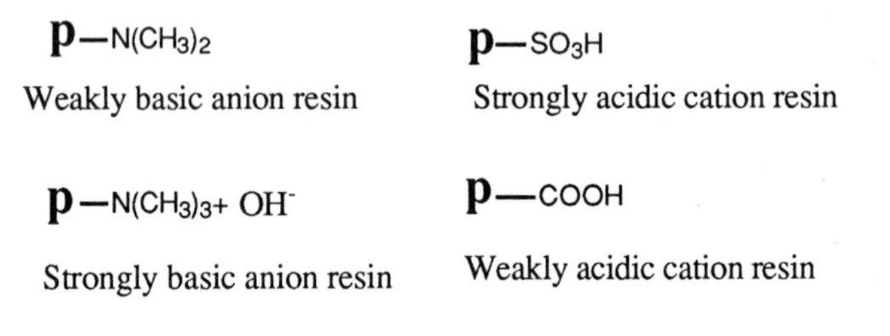


**P** = Polymer bead, polystyrene, acrylic, etc.

**Figure 1.** Types of common ion exchange resins.

ability of a molecule or ion to diffuse into the resin can be as important as the strength of the bond made with the resins functional group. There are no ways to predict resin selectivity for organic molecules but empirical tests can be used to determine selectivity of the resins. These tests generally are done in a batch or column mode. In the batch mode, the resin is simply stirred with a solution of an excess and equal amount the substances which are to be separated, the resin is filtered, extracted and the extract analysed. This gives a crude idea of the preference of the resin for the various substrates. Promising resins then may be tested in the more accurate column mode in which the solution is passed through a column of the resin and fractions of the effluent are taken and analysed.

The glyphosate waste problem mainly required the separation of

N-phosphonomethyliminodiacetic acid (PIA) and formic acid from the glyphosate. This is because formic acid will react with glyphosate downstream to give N-formylglyphosate and PIA was thought to be representative of the strong acids chromatographic behavior. Cation resins were found to have no affinity for any of the materials in the filtrate. Formaldehyde, being neutral, would not be absorbed. In general, weakly basic anion resins are preferable to strongly basic resins because of their greater ease of regeneration after use. Batch resin tests using commercially available basic anion resins were performed for the purpose of studying the preference of the resins for PIA over glyphosate and for glyphosate over formic acid. The resin selected for these separations need not be the same resin. The results are shown in Tables 2 and 3.

**Table 2.** Separation of N-phosphonomethyliminodiacetic acid (PIA) and glyphosate.

| RESIN | RATIO PIA/GLYPHOSATE ADSORBED |
|---|---|
| AMBERLITE®*1 IRA-68 | 17.5 |
| DUOLITE®*2 A-392 | 34.0 |
| AMBERLYST®*1 A-21 | no glyphosate |
| IONAC®*3 A-305 | 12.1 |
| IONAC AFP-329 | 53.8 |
| IONAC A-365 | 9.7 |
| IONAC A-380 | 17.1 |
| AMBERLITE IRA-400 | 21.3 |
| AMBERLITE IRA-900 | 23.0 |
| AMBERLITE IRA-93 | no glyphosate |
| LEWATIT®*4 MP64 | 24.1 |

*1 AMBERLITE and AMBERLYST are registered trademarks of Rohm & Haas Co.
*2 DUOLITE is a registered trademark of Diamond Shamrock Corp.
*3 IONAC is a registered trademark of Sybron Chemicals, Inc.
*4 LEWATIT is a registered trademark of Farbenfabriken Bayer

**Table 3**. Separation of glyphosate and formic acid.

| RESIN | RATIO GLYPHOSATE/ FORMIC ACID ABSORBED |
|---|---|
| AMBERLITE IRA-93 | 1.79 |
| DUOLITE A-392 | 4.66 |
| AMBERLITE IRA-68 | 11.3 |
| IONAC 305 | 10.0 |
| IONAC 380 | 6.6 |
| IONAC 365 | 5.4 |

In addition to these commercial resins several experimental anion resins from Dow were tested. Because of their superior adsorption ratios Amberlite IRA-93 and Amberlite IRA-68 were chosen for separation of glyphosate from PIA and formic acid respectively.

## COLUMN EXPERIMENTS

Experiments were performed using columns of 80 ml. (1" diameter tubes) of IRA-93 and IRA-68 in sequence. Conductivity measurements were recorded on the effluent from the columns to determine when the columns were exhausted. After washing the columns with water, the materials were recovered and the resins were regenerated by passing dilute sodium hydroxide or an amine solution through the columns. By analysing fractions of the effluent and the regenerant solution, the capacity of the columns for strong acids and glyphosate could be determined. It was found that, because of the relatively low concentration of strong acids, the column for the absorption of glyphosate (main column) was exhausted long before the strong acid column (precolumn) and that a large excess of glyphosate was needed to force the formic acid off the main column.

## THE ION EXCHANGE SYSTEM

The ion exchange recovery system which was developed based on the above study consisted of four columns of equal size, one contained IRA-93 and three contained IRA-68. The system was run as shown in Figure 2.

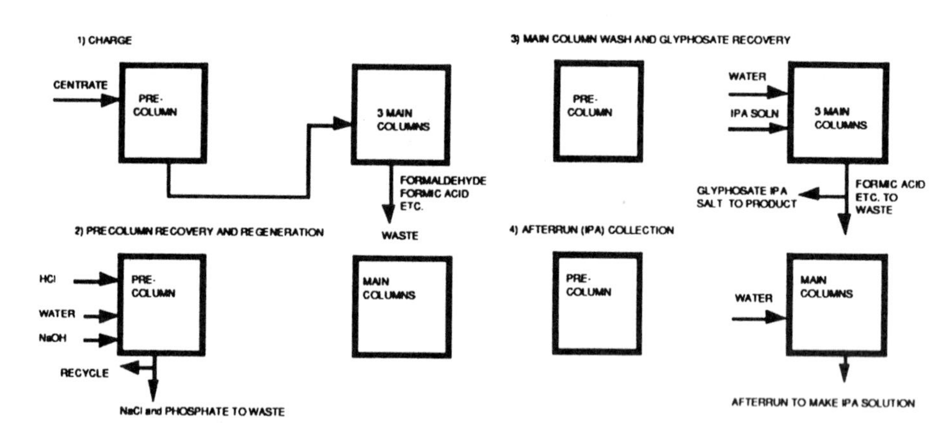


**Figure 2.** Ion exchange system.

During filtrate addition, effluent from the precolumns is sent directly to the main columns and the strong acids are retained. When the precolumn is saturated with strong acids, it is washed with water and, if recovery of the strong acids is desired, a dilute hydrochloric acid solution is added. This is followed by a water wash and dilute sodium hydroxide solution which regenerates the resin. An additional water wash is then provided. During the caustic addition and subsequent water wash the precolumn effluent is sent to waste.

The effluent from the precolumns during filtrate addition flows through three of the main columns until significant glyphosate breaks through the second column which shows that the first column is sufficiently loaded with glyphosate. The three columns are washed with water. Then an isopropylamine solution in an afterrun solution from a previous run is applied to the first column. When effluent from the first column contains glyphosate isopropylammonium salt of sufficient concentration, as indicated by conductivity, columns two and three are taken off line and the effluent from column one is sent to a product tank. Water is applied to column one to wash the last portion of the product off the column. When the effluent reaches a dilution which is unsatisfactory for product collection an afterrun is collected which is used to prepare the next isopropylamine solution. Results are shown in Table 4.

Two precolumns and four main columns are needed for continuous operation.

**Table 4.** Results of ion exchange, % of total.

| Applied Centrate(g) | Glyphosate | N-phosphonomethylimino diacetic acid | Formaldehyde | Formic acid | N-Methylglyphosate | N-formylglyphosate | Aminomethyl phosphonic | Phosphate |
|---|---|---|---|---|---|---|---|---|
| 1098.7 | 1.89 | 0.020 | 1.99 | 0.85 | 0.156 | 0.11 | 0.33 | 0.02 |
| 1078.8 | 1.41. | 0.005 | 2.00 | 0.87 | 0.074 | 0.076 | 0.36 | 0.07 |
| 592.2 | 1.49 | 0.005 | 2.12 | 0.94 | 0.081 | 0.076 | 0.37 | 0.02 |
| 542.2 | 1.49 | 0.005 | 2.12 | 0.94 | 0.081 | 0.076 | 0.37 | 0.03 |
| 266.0 | 1.35 | 0.003 | 2.01 | 0.92 | 0.069 | 0.076 | 0.35 | 0.02 |
| **Recovered Products(g)** | | | | | | | | |
| 300 | 3.66 | ND* | ND | 0.074 | 0.18 | 0.01 | 0.03 | 0.002 |
| 300 | 4.34 | ND | ND | 0.101 | 0.31 | 0.025 | 0.06 | 0.005 |

ND*= not detected

## RESULTS

As can be seen, formaldehyde and PIA have been reduced to undetectable levels and the other impurities have been reduced to managable levels. The glyphosate concentration has been increased by a factor of about 2.5. The ion exchange product can be mixed with the main product stream and sent to formulation. An accelerated resin life test in the laboratory which put the resins through 1500 recovery cycles indicated little or no deterioration of its effectiveness. This would be more than two years of plant use.

## SUMMARY

This ion exchange technology has been transferred to plant use with a scaleup factor of 90,000 and is very successful. The plant had an easy startup and has operated extremely well. There is no evidence of resin degradation.

In summary, reliable new ion exchange technology has been developed which has reduced glyphosate loss to waste treatment by a very significant amount.

# SUPERCRITICAL EXTRACTION IN ENVIRONMENTAL REMEDIATION AND RESTORATION

Aydin Akgerman

Chemical Engineering Department
Texas A&M University
College Station, TX 77843

## INTRODUCTION

A supercritical fluid (SCF) is a fluid at conditions above its critical temperature and pressure. Interest in the extraction of solid and liquid media by SCFs have increased during the last decade due to: (1) environmental problems associated with common solvents (mostly chlorinated hydrocarbons); (2) the increasing cost of energy intensive separation processes (for example distillation); and (3) the inability of conventional separation processes to provide the necessary separations needed in emerging new industries. The attractive physicochemical properties of SCFs qualify them as a viable alternative to conventional solvents used in extraction processes.

At temperatures and pressures above its critical point a pure substance exists in a state that exhibits gas-like and liquid-like properties. The fluid's density would be very close to that of a liquid, the surface tension is very close to zero, the diffusivity and viscosity have a value somewhat in between that of a liquid and a gas, and most important the solvent power of a SCF is related to its density which can be varied over a very wide range by small variations in temperature and/or pressure in the supercritical region. These properties result in several advantages in extraction such as ease of solvent recovery, elimination of residual solvent in the extracted medium, lower pressure drops, and higher mass transfer rates.

In extraction of a solute from a matrix (such as water) the choice of the solvent depends on two criteria, its immiscibility with the matrix and the solubility of the solute in the solvent. The solubility in SCFs is a strong function of density. In the vicinity of the critical point, $1<T_r<1.1$ and $1<P_r<2$, the density is a very strong function of both the temperature and the pressure. The solvent characteristics of a SCF can therefore be adjusted as desired, an important advantage compared to conventional solvents. At extraction conditions the solvent power would be high so the SCF fluid can remove the solute from the matrix and at the separation/solvent recovery stage (where the solute is removed from the solvent) the solvent power would be reduced to close to zero.

*Industrial Environmental Chemistry*, Edited by D.T. Sawyer
and A.E. Martell, Plenum Press, New York, 1992

The use of a compressed gas in a separation process was first proposed for oil deasphalting (Wilson et al., 1936). Although the process was not strictly supercritical, it did take advantage of the change in solubility associated with a pressure reduction. Later, Elgin and Weinstock (1959) reported on a phase-splitting technique for recovery of methyl-ethyl-ketone (MEK) from water using supercritical ethylene. Since then SCFs have been used as solvents to extract a variety of solid and liquid matrices such as coal (Wilhelm and Hedden, 1983), caffeine from coffee (Zosel, 1978; Vitzthum and Hubert, 1975), tobacco (Hubert and Vitzthum, 1978), fruit aromas (Schultz and Randall, 1970), and alcohols from water (McHugh and Krukonis, 1986; Kuk and Montagna, 1983). Various symposia proceedings and books that are on supercritical extraction phenomena are available in the literature.

The broad range of organic solutes that can be extracted from aqueous and solid waste (wastewater, contaminated soils, sludges etc.) by SCFs as well as the availability of inexpensive, readily available, and non-toxic SCF solvents such as $CO_2$, has directed attention to this process as a viable method for removing toxic organic compounds. In the following discussion aqueous waste and solid waste are treated under separate headings since the fundamentals of the technology as well as the application techniques are slightly different.

## SUPERCRITICAL EXTRACTION OF AQUEOUS WASTE

The advantages of employing SCF extraction to remove organic contaminants from wastewater can be realized by considering a typical extraction process (Figure 1). Contaminated water is extracted by a SCF in a countercurrent flow extraction column. The extract stream goes through a pressure reduction to separate the contaminant from the SCF which is recompressed and recirculated. The raffinate (decontaminated water) is also expanded to recover dissolved SCF solvent. The process would typically be at ambient

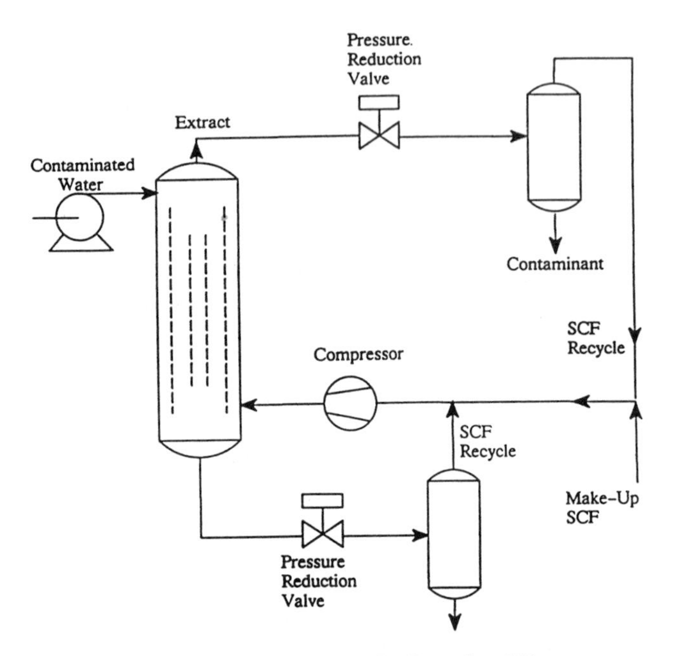


**Figure 1.** Schematic representation of a typical supercritical extraction process (from Akgerman et al., 1991).

temperature and pressures above 5 MPa. If a liquid solvent was used for the process two additional separation processes would have been necessary, separation of the solute from the solvent for solvent recovery and separation of residual solvent from the water stream.

Energy shortages of 1970s fueled studies on separation of alcohols from water. Therefore earlier work on SCF extraction of an aqueous medium concentrated on alcohol/water separations. Alcohols of interest were ethanol, n-propanol, i-propyl alcohol, and n-butanol and the thermodynamics of the ternary systems (alcohol/water/SCF) are well studied (Paulaitis et al., 1984). SCF extraction of various other organic compounds are also reported such as acetone from water (Panagiotopoulos and Reid, 1985), aroma constituents of fruit and other foods (Schultz and Randall, 1970), 23 organic compounds typically found in water for analytical purposes (Ehntholt et al., 1983).

Stringent environmental regulations demand removal of trace organics from water making SCF extraction a viable alternative to other remediation technologies such as bioremediation, adsorption, liquid extraction, air-stripping, distillation, incineration, etc. all of which have their advantages and disadvantages. Supercritical carbon dioxide (SCCO2) is the preferred SCF solvent for environmental applications since it is non-toxic, inexpensive, and readily available. In addition, it has a conveniently low critical temperature (304 K) and a moderate critical pressure (7.39 MPa).

Economic evaluation of a separation process necessitates the distribution coefficient (also called partition coefficient, equilibrium constant, and K value) of the extracted component between the phases, i.e. for SCF extraction of a solute from water the phases would be the aqueous phase and the SCF phase. Thus, studies to date concentrated on the thermodynamics of the SCF/water equilibrium. We have reported on distribution coefficients of single components such as phenol, benzene, toluene, naphthalene, and parathion (Roop and Akgerman, 1989; Roop et al. 1989; Yeo and Akgerman, 1990) as well as mixtures such as a solution containing benzene, toluene, naphthalene and parathion (Yeo and Akgerman, 1990), a phenolic mixture (Roop and Akgerman, 1990), and petroleum creosote (Akgerman et al., 1991). Similarly Ghonasgi et al. (1991) and Knopf (1991) reported on extraction of phenol, m-cresol, p-chlorophenol, and benzene both individually and as a mixture.

The thermodynamic modeling of the equilibrium starts with equating fugacities of the component distributed between the two phases

$$\hat{f}_i^{\,WP} = \hat{f}_i^{\,SCFP}$$

where the superscripts WP and SCFP refer to the water phase and the supercritical fluid phase respectively. Expressing the fugacities in terms of the mole fractions and fugacity coefficients results in the distribution coefficient

$$K_i = X_i^{\,SCFP}/X_i^{\,WP} = \hat{\phi}_i^{\,WP}/\hat{\phi}_i^{\,SCFP}$$

The fugacity coefficient $\hat{\phi}_i$ can be calculated from an equation of state using standard rigorous thermodynamic relationships. In our studies we have used the Peng-Robinson equation of state (Roop and Akgerman, 1989; Yeo and Akgerman, 1990) whereas Ghonasgi et al (1991) used the Carnahan-Starling-DeSantis- Redlich-Kwong equation of state. Figures 2 & 3 present data on benzene extraction (Yeo and Akgerman, 1990; and Ghonasgi et al., 1991) and Figure 4 on phenol extraction (Roop and Akgerman, 1989) with predictions. It is important to note the difference in the magnitude of distribution coefficients for hydrophobic and hydrophillic compounds.

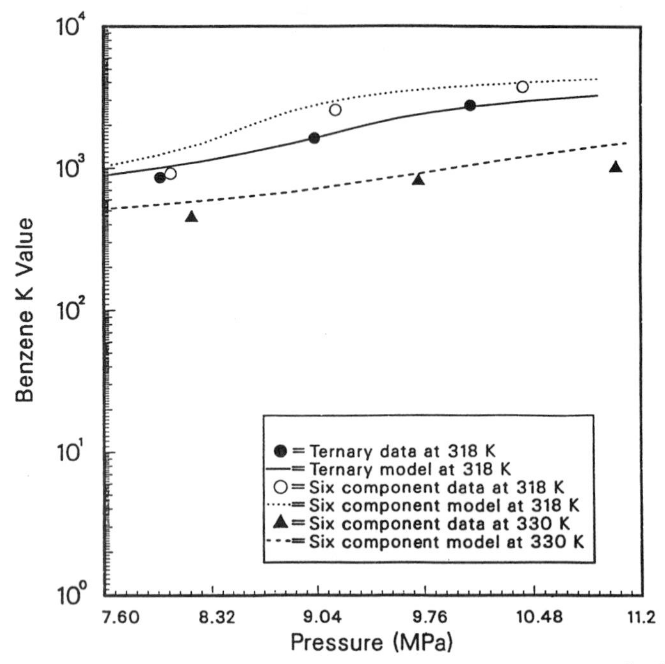


**Figure 2.** Distribution coefficients of benzene between water and supercritical carbon dioxide in ternary and six component systems (from Akgerman et al., 1991).

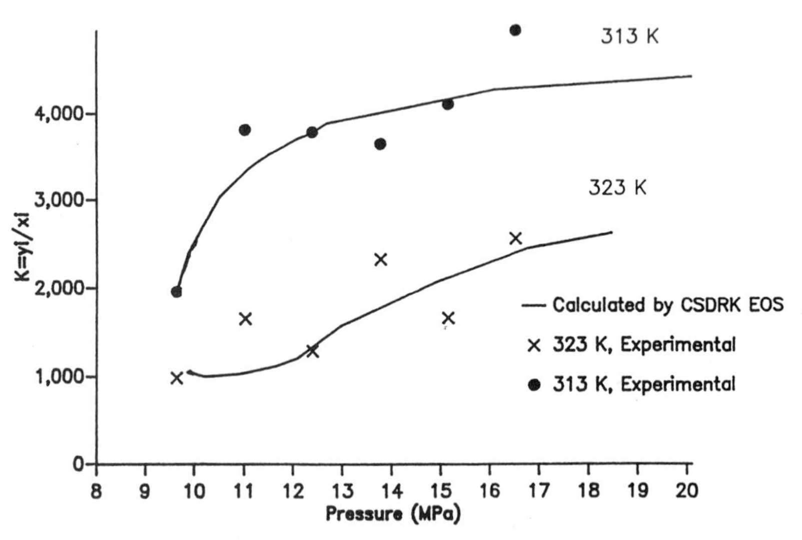


**Figure 3.** Distribution coefficients of benzene between water and supercritical carbon dioxide (from Ghonasgi et al., 1991).

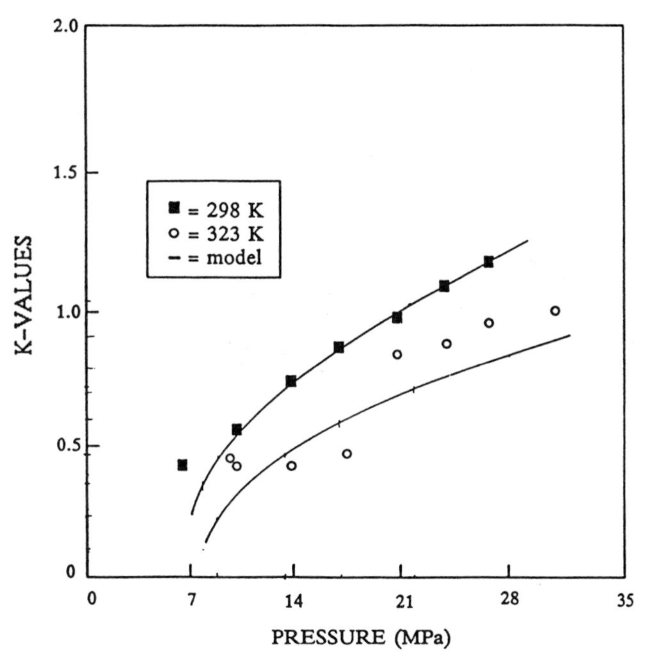


**Figure 4.** **Distribution coefficients of phenol between water and supercritical carbon dioxide** (from *Entrainer Effect for Supercritical Extraction of Phenol from Water*, Roop and Akgerman, 1989, American Chemical Society)

## EFFECT OF ENTRAINERS IN AQUEOUS SYSTEM EXTRACTIONS

Solvent power of SCFs can be increased significantly by addition of small amounts of co-solvents, called entrainers. Entrainers usually are polar compounds that are added to the SCF in small amounts (typically <5%) and they increase the solubility of organics in the SCF by orders of magnitude. Methanol and toluene are the most widely studied entrainers. In the literature there are significant amount of studies reporting on the solubility increase of solids in SCFs by entrainers. The increase in solubility is explained using solution theories. However, studies on the effect of entrainers on SCF extraction, i.e. the distribution coefficients, are more limited.

Roop and Akgerman (1989; 1990) showed that methanol has no effect on the distribution coefficients of phenol and a phenolic mixture proving that an entrainer that increases the solvent power does not necessarily affect the partitioning of a compound between phases. They have explained the entrainer effect in terms of multicomponent molecular interactions in calculation of thermodynamic equilibrium. Through a quaternary system equilibrium calculation it was shown that methanol will not have any effect of phenol distribution coefficients and that benzene would increase the distribution coefficient of phenol by 50-70% and the prediction was verified experimentally. Figures 5 and 6 indicate the entrainer effect of benzene on the distribution coefficient of phenol and a phenolic mixture. Similarly Yeo and Akgerman (1990) determined the effect of benzene, toluene, naphthalene and parathion on each other's distribution coefficient. Figure 7 indicates the increase in the distribution coefficient of each component in the mixture due to the presence of others as compared to single component extraction. A significant conclusion is that the extraction of all the components are enhanced in mixtures, i.e. extraction thermodynamics of mixtures is more favorable.

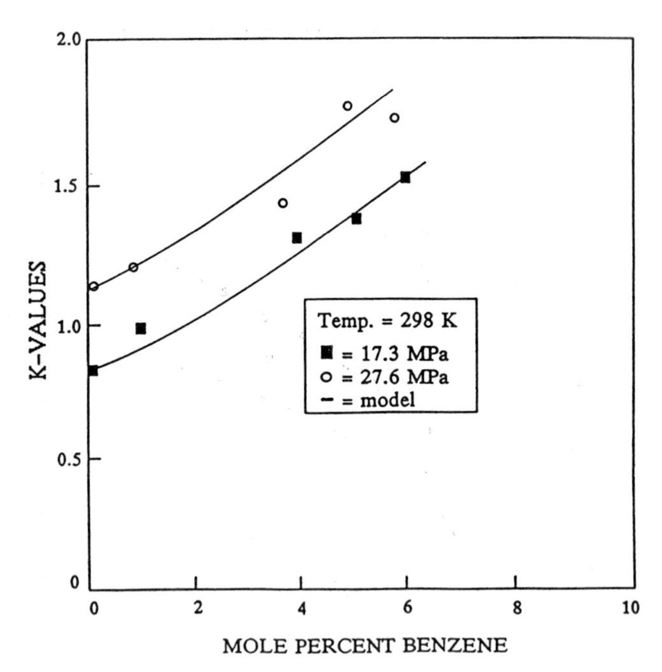


**Figure 5.** Data and predicition on the effect of benzene as an entrainer on the distribution coefficients of phenol between water and supercritical $CO_2$ (from *Entrainer Effect for Supercritical Extraction of Phenol from Water*, Roop and Akgerman, 1989, American Chemical Society)

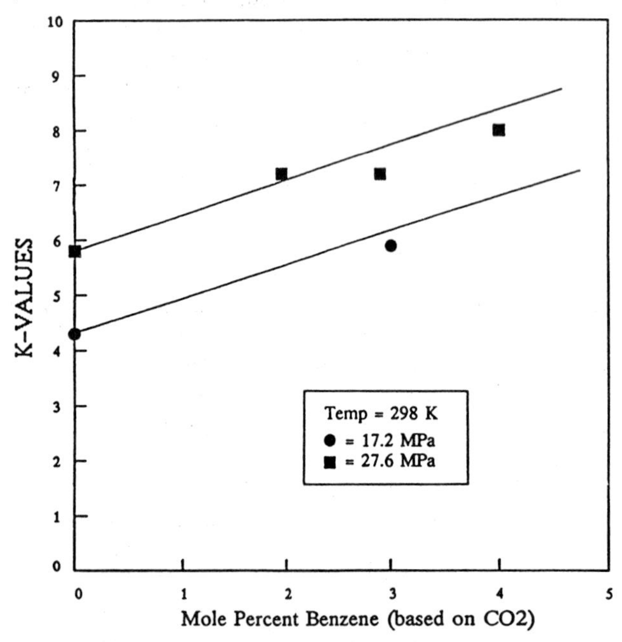


**Figure 6.** Data and prediction on the effect of benzene as an entrainer on the distribution coefficients of a phenolic mixture between water and supercritical $CO_2$ (from *Distribution of a Complex Phenolic Mixture Between Water and Supercritical $CO_2$, Roop and Agkerman,* 1990, American Chemical Society)

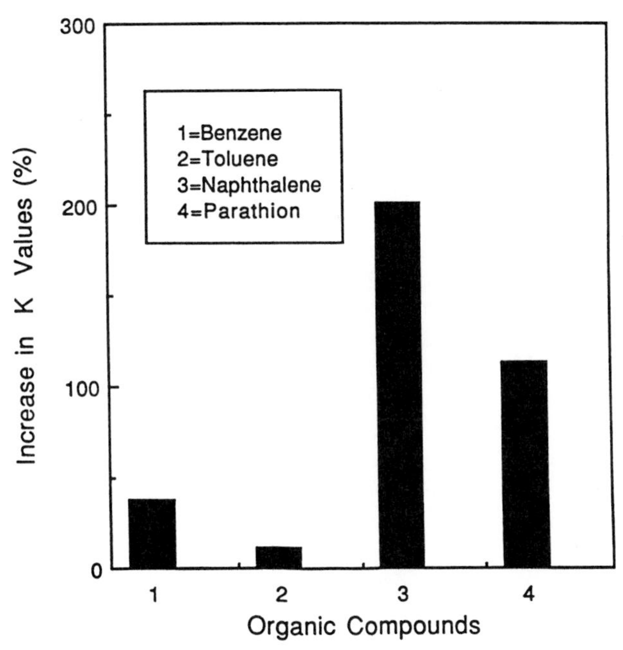


**Figure 7.** Increase in the distribution coefficients of each component in a six component system comparted to the ternary system values (from Yeo and Akgerman, 1990).

## SUPERCRITICAL EXTRACTION OF SOLID WASTE

Supercritical extraction has been demonstrated in the literature at the bench scale for extraction of organic contaminants from a variety of solid matrices. Capriel et al. (1986) used supercritical methanol to extract bound pesticide residues from soil and plant residues. Hawthrone and Miller (1986) extracted polycyclic aromatic hydrocarbons (PAH) from diesel soot and Tenax packing for gas chromatographic columns by supercritical carbon dioxide. Schantz and Chesler (1986), similarly, used supercritical $CO_2$ to extract polychlorinated biphenyls (PCB) from sediment and PAHs from urban particulate matter. Methanol/$N_2O$ mixtures are also used to extract PAHs from river sediments and urban particulate matter (Hawthrone and Miller, 1987). Supercritical $CO_2$ is used to remove hexachlorocyclohexane, parathion, PCBs, and PAHs from Tenax packing (Raymer and Pellizari, 1987) and polyimide based adsorbents (Raymer et al., 1987). Supercritical $CO_2$ is also the solvent of choice for extraction of PAHs from various adsorbents selectively by varying the operating conditions (Wright et al., 1987a; 1987b). A significant amount of work concentrated on activated carbon regeneration, such as desorption of phenol (Kander and Paulaitis, 1983), pesticides (DeFilippi et al., 1980), acetic acid and alachlor (Picht et al., 1982), and ethyl acetate (Tan and Liou, 1988).

The results of these studies have demonstrated at the bench scale that it is possible to extract compounds with molecular weight as high as 400 at mild conditions and selectively if desired. This conclusion constitutes the basis of supercritical chromatography. Recently we have reported on the application of chromatography theory to supercritical extraction from solid matrices (Erkey and Akgerman, 1990).

Concerning environmental applications, Kingsley (1985) applied subcritical and supercritical $CO_2$ for extraction of oil from metal fines (mill scale) and bleaching clay in the pilot scale. The process operated on a semi-batch mode and the results indicated that the recovery of extractable material depended on the solvent flow rate. It was also observed that, for the bleaching clay, the recovery was improved by a static soaking period before extraction. Brady et al. (1987) demonstrated the ability of SCFs to extract PCBs and DDT from contaminated topsoil and subsoils using SCCO2. The effect of entrainers on the solvent power was also determined (Dooley et al., 1987). They demonstrated that 100% DDT extraction was possible when 5% by weight methanol was used as entrainer whereas when toluene was used as the entrainer (5% by weight) only 75 % of the DDT was recovered. Eckert et al. (1986a) studied the removal from soil of chlorinated aromatics such as trichlorophenol as a model compound for PCBs and dioxins. Using supercritical ethylene virtually all trichlorophenol was removed from soil.

Supercritical extraction of organic contaminants from dry solids, such as soil, adsorbents, etc., involves two different phenomena simultaneously. The organic contaminant on the solid surface can exists in two states, adsorbed state and the deposited state. The portion of the organic that is deposited as a separate phase on the solid surface is extracted by simple dissolution in the supercritical phase. On the other hand the extraction of the portion of the organic that is adsorbed on the solid phase is controlled by the adsorption/desorption equilibrium. Entrainers play a significant role in extraction of the deposited separate organic phase since the solubility of the organic in the SCF phase is significantly increased due to the entrainer. In addition, SCFs have an advantage compared to conventional solvents since they penetrate the pore structure more efficiently due to their zero surface tension and remove the organic condensed/deposited in the pores of the matrix.

Extraction of organics from a solid matrix that is also wet (contains water such as soil moisture) is more complicated since the organic now can exist in four different states, adsorbed on dry soil, dissolved in soil moisture, adsorbed on soil which is covered by a water layer (partitioning between soil and water), and as a separate organic phase. Again, the extraction of the separate organic phase is a dissolution phenomenon. However, the extraction of the other forms involves different types of equilibrium (soil/water/SCF partitioning) combined with extraction of multiple species (water and the organic contaminant[s]).

Most important of these extraction processes is the extraction of the adsorbed species which involve solid/organic/SCF binary and ternary interactions. Thermodynamics of solids extraction by supercritical fluids is formulated in a similar manner to any adsorption process. The distribution constant of a species between the solid phase and the SCF phase is more often referred to as the adsorption equilibrium constant or the partition coefficient. The partitioning is explained in terms of an adsorption isotherm. The adsorption equilibrium constant is defined as

$K_i^{ads}$ = mass fraction in the SCF phase/mass fraction in the solid phase

which, for dilute systems (mass of contaminant compared to the solid and the SCF mass negligible) is related to the true thermodynamic distribution coefficient in terms of the mole fractions by the expression

$$K_i = x_i^{SCFP}/x_i^{SP} = K_i^{ads}(MWT_{solid}/MWT_{SCF})$$

where the superscript SP refers to the solid phase (Hess et al., 1991).

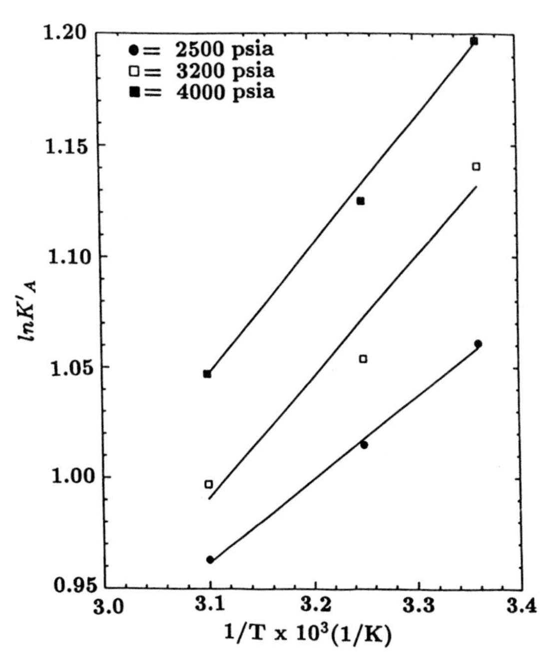


**Figure 8.** Isobaric temperature dependence of the partition coefficient of hexachlorobenzene between soil and supercritical $CO_2$ (from *Supercritical Extraction of Hexachlorobenzene from Soil*, Akgerman et al., 1992, American Chemical Society)

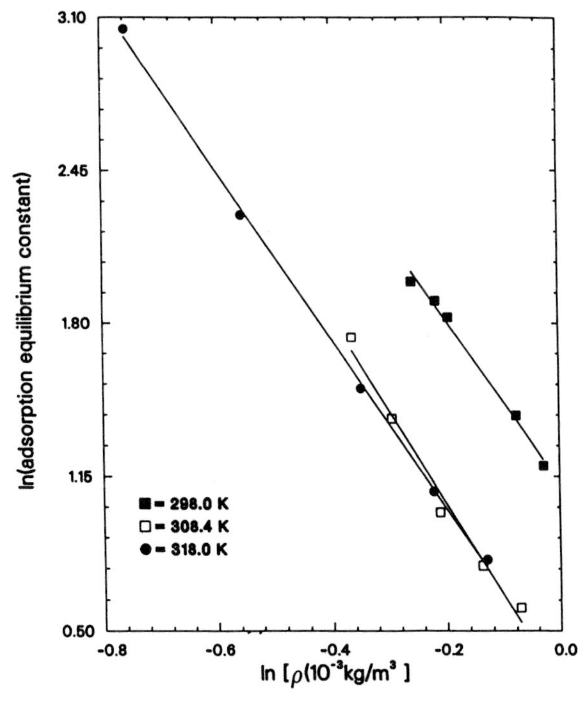


**Figure 9.** Isothermal density dependence of the partition coefficients of naphthalene between activated alumina and supercritical $CO_2$ (from Erkey and Akgerman, 1990).

The adsorption equilibrium constant has important thermodynamic properties. Through the van't Hoff equation, isobaric reciprocal temperature dependency of the logarithm of the adsorption constant yields the partial molar enthalpy of adsorption, which is a measure of the strength of adsorption (Hess et al., 1991; Akgerman et al., 1992). Similarly, the isothermal density dependence of the adsorption constant yields the partial molar volume of the solute in the supercritical phase, which is related to its solubility in the supercritical phase (Erkey and Akgerman, 1990).

Figure 8 shows the isobaric temperature dependence of the partition coefficient of hexachlorobenzene between soil and SCCO2 ($K'_A$ is reciprocal of the adsorption equilibrium constant at infinite dilution). The positive slopes indicate exothermic heats of adsorption increasing slightly with pressure (Akgerman, et al., 1992). Figure 9 shows the isothermal density dependence of the partition coefficients of naphthalene between activated alumina and SCCO2 and Figure 10 shows the partial molar volume of naphthalene in SCCO2 compared with data from Eckert et al. (1986b). Both Figures are taken from Erkey and Akgerman (1990).

A common myth in environmental circles postulates that organic contaminants partition only to the organic portion of soil (soil humic material) and correlations are developed for

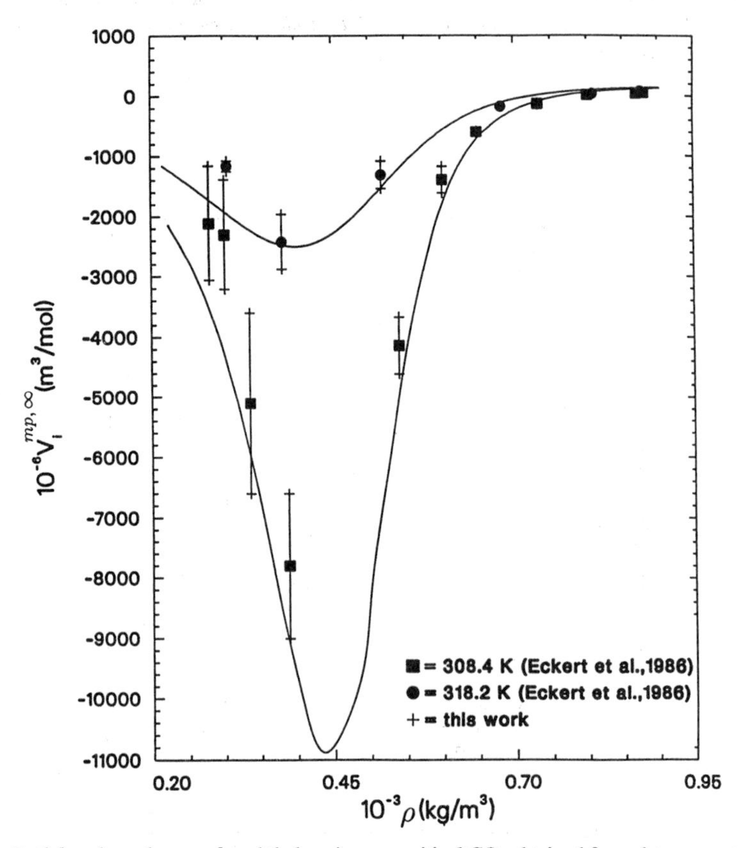


**Figure 10.** Partial molar volumes of naphthalene in supercritical $CO_2$ obtained from data presented in Figure 9, and comparison to experimental values (from Erkey and Akgerman, 1990).

partitioning in terms of soil organic content. Although these correlations are useful for practical applications they do not have a fundamental significance. Figures 11 and 12 compare the partition coefficients for phenol between SCCO2 and 8 solid matrices, six soils of varying organic content, sand and montmorillonite (Hess et al., 1991). These results indicate that solid matrix surface area correlates the partition coefficient over a wide range.

We have recently measured the adsorption isotherms of phenol from SCCO2 on to soil and activated carbon over the concentration range from zero to saturation solubility limit

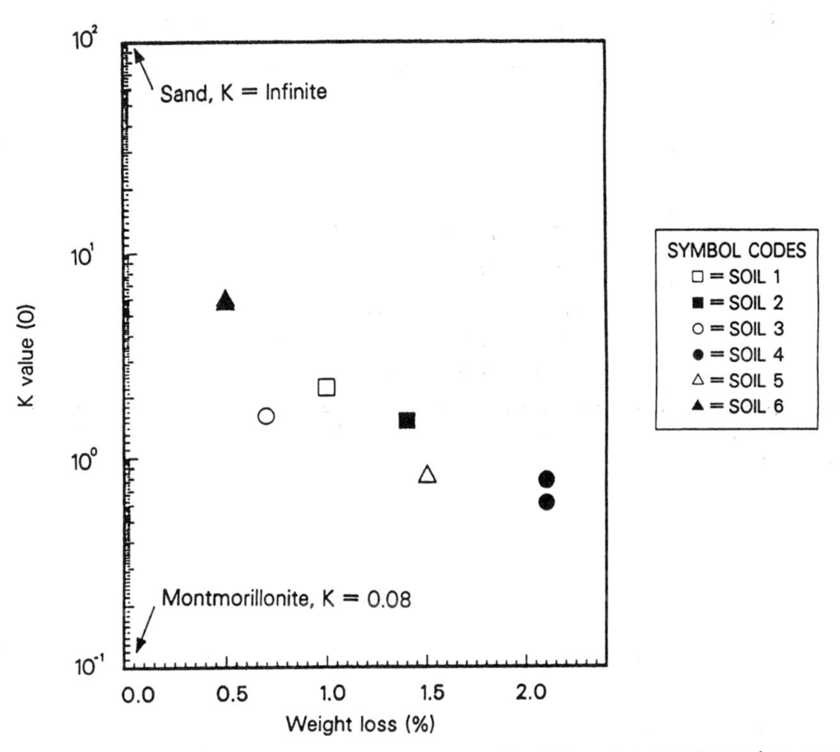


**Figure 11.** The partition coefficients of phenol between supercritical $CO_2$ and eight solid matrices, correlation with organic content (from *Supercritical Extraction of Phenol from Soil*, Hess et al., 1991, American Chemical Society)

(Figures 13 and 14). Both isotherms are Type I - Langmuir Isotherms indicating monolayer surface coverage. Similar behavior is observed for adsorption of anthracene, triphenylene, perylene, and phenanthrene on soil from SCCO2 (Andrews et al., 1990). In fact, by estimating the projected area of the adsorbed species and measuring the surface area of the adsorbent, it is possible to calculate the maximum amount that can be adsorbed on the solid. These results indicate that the total accessible surface area of the solid is the available area for uniform adsorption and the amount of solute adsorbed in excess of uniform monolayer coverage would be deposited as a separate phase.

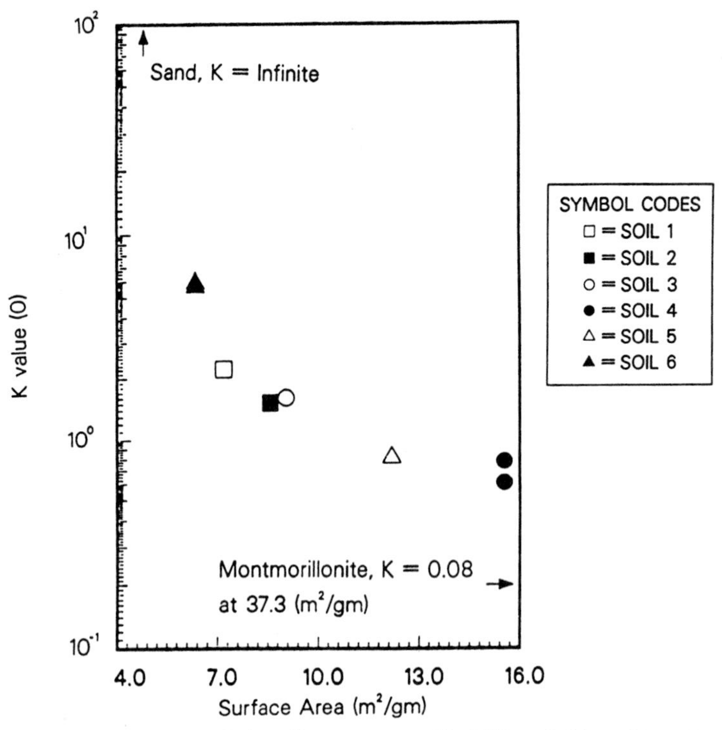


**Figure 12. The partition coefficients of phenol between supercritical $CO_2$ and eight solid matrices, correlation with solid phase surface area** (from *Supercritical Extraction of Phenol from Soil*, Hess et al., 1991, American Chemical Society)

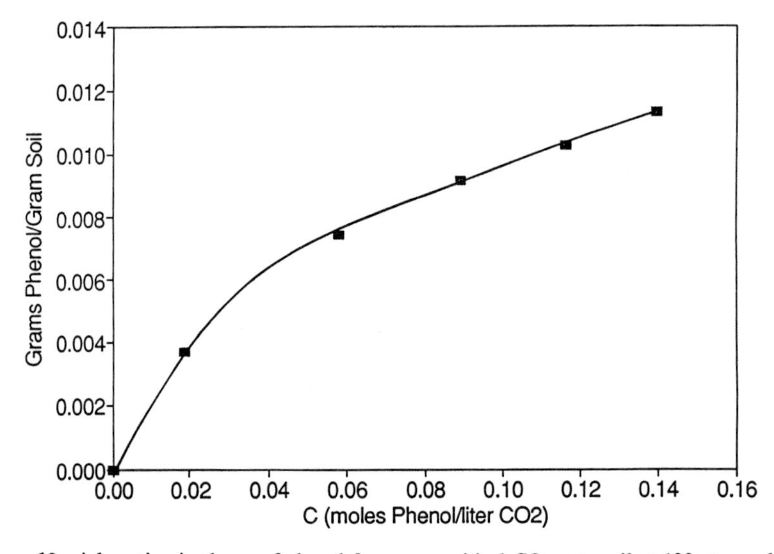


**Figure 13. Adsorption isotherm of phenol from supercritical $CO_2$ onto soil at 100 atm and 36$^{oC}$**

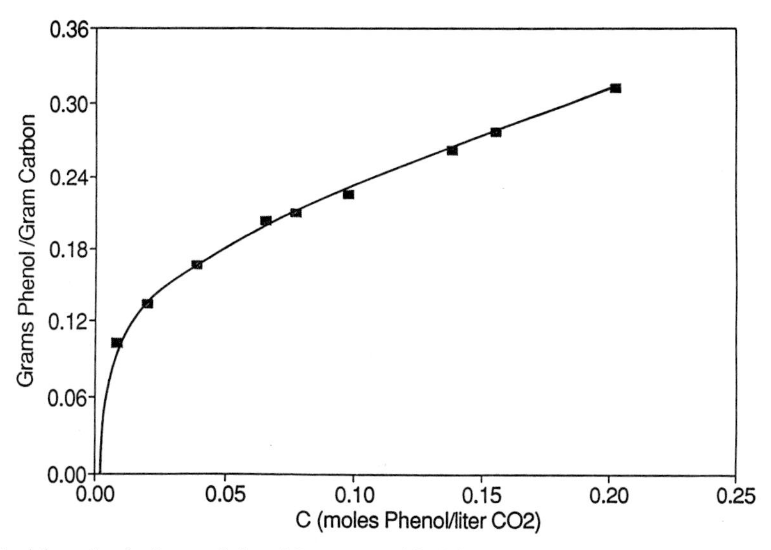


**Figure 14.** Adsorption isotherm of phenol from supercritical $CO_2$ onto activated carbon at 100 atm and 36°C.

Modeling of a solids extractor involves hydrodynamic and mass transfer parameters in addition to the adsorption equilibrium and kinetics. The model used is normally a heterogeneous model that employs a dispersed flow equation in the extractor bed combined with an equation for diffusion into solid particles which includes an accumulation term for adsorption/desorption. The two equations are coupled through the external film mass transfer boundary condition and the surface accumulation term is related to the adsorption isotherm. Solution of the set of equations yield the adsorption or desorption breakthrough profiles (Sherwood et al., 1983). The parameters involved are the axial dispersion coefficient, the film mass transfer coefficient, the effective diffusivity, and the adsorption isotherm parameters. We have shown that the effective diffusivity and the adsorption parameters are the controlling parameters at normal operating conditions (Erkey and Akgerman, 1990). Normally, at least at laboratory conditions, the breakthroughs are sharp, indicating control by adsorption equilibrium constant.

## APPLICATIONS

The CF Systems Corporation developed a SCF extraction process to separate and recover oils from refinery sludges and to extract hazardous organic compounds from wastewater, sludge, sediment and soil (Hall et al., 1990). The process uses supercritical propane on contaminated solids and SCCO2 to treat wastewater. The solid feed materials are reduced in size and are slurried so that they can be pumped to the extractor. Wastewater is used directly. The process closely resembles the flow chart presented in Figure 1. Reportedly, up to 90% of the solvent is recycled in the system; the remaining 10% retains the extracted contaminants. The CF Systems process was demonstrated at pilot scale for the U.S. EPA's SITE program and shown to be capable of removing PCBs from sediments.

Recently we proposed an alternate scheme for extraction of solids as shown in Figure 15. This scheme eliminates the expansion of the SCF for separation of the extracted

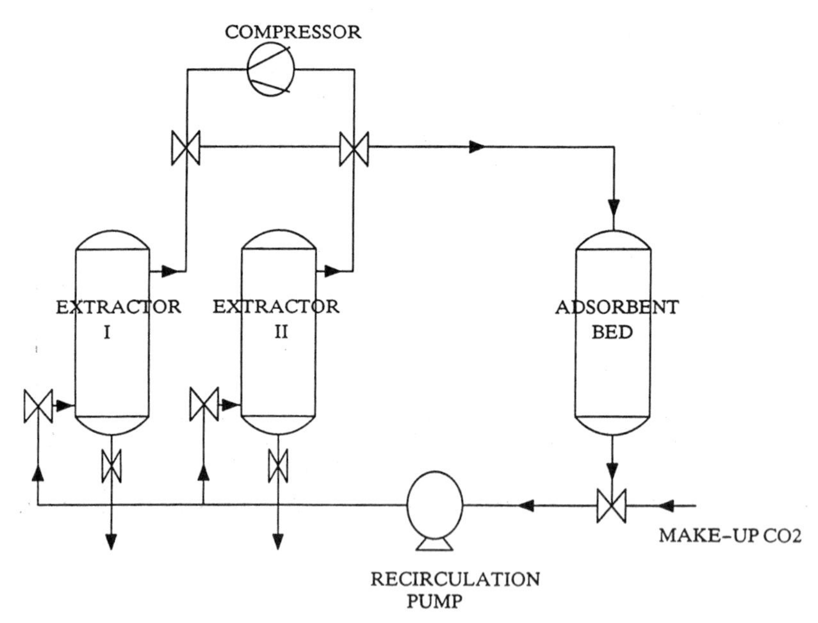


**Figure 15.** Schematic representation of a soil supercritical extraction process with subsequent adsorption of the extracted contaminants.

TABLE I. Comparison of Costs of Alternative Technologies for Soil Remediation, from Carpenter (1986)

| Technology | Cost |
|---|---|
| Supercritical Water Oxidation | \$250 - 733 per $m^3$ soil |
| KPEG | \$211 - 378 per $m^3$ soil |
| O. H. M. Methanol Extraction | \$400 - 514 per $m^3$ soil |
| Acurex Solvent Wash | \$196 - 569 per $m^3$ soil |
| Vitrification | \$255 - 548 per $m^3$ soil |
| Bio-Clean | \$191 - 370 per $m^3$ soil |
| Soilex Solvent Extraction | \$856 - 913 per $m^3$ soil |
| Chemical Waste Landfill | \$260 - 490 per $m^3$ soil |
| Supercritical $CO_2$ Extraction (Combined with Adsorption) | \$120 per $m^3$ soil |

contaminants that necessitates a costly re-compression stage. Instead, the extracted contaminants are deposited on a suitable adsorbent for destruction by incineration. Costs compare favorably with other treatment techniques as summarized in Table 1. The cost data are based on phenol and details of cost estimate are given elsewhere (Akgerman and Erkey, 1992).

## CONCLUSIONS

There is no single magic process that will be applicable to remediation and restoration of all contaminated sites, in fact, more than one technology is usually needed for remediation of the media at a given site. Thus SCF extraction is a viable alternative for remediation of wastewater contaminated with hydrophobic compounds and soils, sludges and sediments contaminated with non-volatile heavy compounds such as polycyclicaromatic hydrocarbons (PAHs), polychlorinatedbiphenyls (PCBs), pesticides, dioxins, etc.

SCF extraction has been shown to be capable of removing organics from water, soil, sediments, sludges, and a variety of solid adsorbents. Although the available data are still very limited, the behavior of SCFs and their solvent powers have only recently begun to be understood. As the ability to predict the behavior of contaminants in a SCF increases, the use of SCF extraction will find new applications in environmental remediation.

## REFERENCES

Akgerman, A., Roop, R. K., Hess, R. K., and Yeo, S.-D., 1991, Supercritical extraction in environmental control, Chapter *in*: "Supercritical Fluid Technology: Reviews in Modern Theory and Application", Bruno, T. J. and Ely, J. F., Eds., CRC Press, Boca Raton.

Akgerman, A., Erkey, C., and Ghoreishi, S. M., 1992, Supercritical Extraction of hexachlorobenzene from soil, *Ind. Eng. Chem. Res.*, 31:333.

Akgerman, A., and Erkey, C., 1992, Supercritical extraction combined with adsorption for environmental remediation of contaminated soils, paper presented at AIChE Meeting, Minneapolis.

Andrews, A. T., Ahlert, R. C., and Kosson, D. S., 1990, Supercritical extraction of aromatic contaminants from a sandy loam soil, *Environ. Progr.*, 9:204.

Brady, B. O., Kao, C. C., Dooley, K. M., Knopf, F. C., and Gambrell, R. P., 1987, Supercritical extraction of toxic organics from soil, *Ind. Eng. Chem. Res.*, 26:261.

Capriel, P., Haisch, A., and Khan, S. U., 1986, Supercritical methanol: an efficacious technique for extraction of bound pesticide residues from soil and plant samples, *J. Agric. Food Chem.*, 34:70.

Carpenter, B. H., 1986, PCB Sediment Decontamination - Technical/Economic Assessment of Selective Alternative Treatments, EPA/600/2-86/112, Environmental Protection Agency, Washington, D.C.

DeFilippi, R. P., Krukonis, V. J., Robey, R. J., and Modell, M., 1980, Supercritical fluid regeneration of activated carbon for adsorption of pesticides, EPA-600/2-80-054, Environmental Protection Agency, Washington, D. C.

Dooley, K. M., Kao, C., Gambrell, R. P., and Knopf, F. C., 1987, The use of entrainers in the supercritical extraction of soils contaminated with hazardous organics, *Ind. Eng. Chem. Res.*, 26:2058.

Eckert, C. A., Van Alsten, J. G., and Stoicos, T., 1986a, Supercritical fluid processing, *Environ. Sci. Technol.*, 20:319.

Eckert, C. A., Ziger, D. H., Johnston, K. P., and Kim, S., 1986b, Solute partial molal volumes in supercritical fluids, *J. Phys. Chem.*, 90:2738.

Ehntholt, D. J., Thrun, K., and Eppig, C., 1983, The concentration of model organic compounds present in water at parts-per-billion levels using supercritical fluid carbon dioxide, *Int. J. Environ. Anal. Chem.*, 13:219.

Elgin, J. C., and Weinstock, J. J., 1959, Phase equilibria molecular transport thermodynamics, *J. Chem. Eng. Data*, 4:3.

Erkey, C., and Akgerman, A., 1990, Chromatography theory: application to supercritical extraction, *AIChE J.*, 36:1715.

Ghonasgi, D., Gupta, S., Dooley, K. M., and Knopf, F. C., 1991, Supercritical $CO_2$ extraction of organic contaminants from aqueous streams, *AIChE J*, 37:944.

Hall, D. W., Sandrin, J. A., and McBride, R. E., 1990, An overview of solvent extraction treatment technologies, *Environ. Progr.*, 9(2):98

Hawthrone, S. B., and Miller, D. J., 1986, Extraction and recovery of organic pollutants from environmental solids and Tenax-GC using supercritical carbon dioxide, *J. Chromatogr. Sci.*, 24:258

Hawthrone, S. B., and Miller, D. J., 1987, Extraction and recovery of polycylic aromatic hydrocarbons from environmental solids using supercritical carbon dioxide, *Anal. Chem.*, 59:1705.

Hess, R. K., Erkey, C., and Akgerman, A., 1991, Supercritical extraction of phenol from soil, *J. Supercrit. Fluids*, 4:47.

Hubert, P., and Vitzthum, O. G., 1978, Fluid Extractions of hops, spices and tobacco with supercritical gases, Angew. *Chem. Int. Ed.*, Engl., 17:710.

Kander, R. G., and Paulaitis, M. E., 1983, The adsorption of phenol from dense carbon dioxide onto activated carbon, in: "Chemical Engineering at Supercritical Conditions", Paulaitis, M. E., Penninger, J. M. L., Gray, R. D., Jr., and Davidson, P., Eds., Ann Arbor Science, Ann Arbor, MI.

Kingsley, G. S., 1985, Pilot plant evaluation of critical fluid extractions for environmental applications, EPA/600/2-85/081, Environmental Protection Agency, Washington, D.C.

Knopf, F. C., 1991, private communication.

Kuk, M. S., and Montagna, J., 1983, Solubility of oxygenated hydrocarbons in supercritical carbon dioxide, in: "Chemical Engineering at Supercritical Conditions", Paulaitis, M. E., Penninger, J. M. L., Gray, R. D., Jr., and Davidson, P., Eds., Ann Arbor Science, Ann Arbor, MI.

McHugh, M. A. and Krukonis, V. J., 1986, "Supercritical Fluid Processing: Principles and Practice", Butterworth, Stoneham, MA.

Panagiotopoulos, A. Z., and R. C. Reid, 1985, High pressure phase equilibria in ternary fluid mixtures with a supercritical component, *ACS Prep. Div. Fuel Chem.*, 30:46.

Paulaitis, M. E., Kander, R. G., and DiAndreth, J. R., 1984, Phase equilibria related to supercritical fluid solvent extractions, Ber. Bunsenges. *Phys. Chem.*, 88:869.

Picht, R. D., Dillman, T. R., and Burke, D. J., 1982, *AIChE Symp. Ser.*, 78:136.

Raymer, J. H., and Pellizzari, E. D., 1987, Toxic organic compounds recoveries from 2,6-diphenyl-p-phenylene oxide porous polymer using supercritical carbon dioxide and thermal desorption methods, *Anal. Chem.*, 59:1043.

Raymer, J. H., Pellizzari, E. D., and Cooper, S. D., 1987, Desorption charactersitics of four polyimide sorbent materials using supercritical carbon dioxide and thermal desorption methods, *Anal. Chem.*, 59:2069.

Roop, R. K., and Akgerman, A., 1989, Entrainer Effect for Supercritical Extraction of Phenol from Water, *Ind. Eng. Chem. Res.*, 28:1542.

Roop, R. K., and Akgerman, A., 1990, Distribution of a complex phenolic mixture between water and supercritical carbon dioxide, *J. Chem. Eng. Data*, 35:257.

Roop, R. K., Akgerman, A., Dexter, B. J., and Irvin, T. R., 1989, Extraction of phenol from water with supercritical carbon dioxide, *J. Supercrit. Fluids*, 2:51.

Schantz, M. M., and Chesler, S. N., 1986, Supercritical fluid extraction procedure for the removal of trace organics from soil samples, *J. Chromatogr.*, 363:397.

Schultz, W. G., and Randall, J. M., 1970, Liquid carbondioxide for selective aroma extraction, *Food Technol.*, 24:94.

Sherwood, T. K., Pigford, R. L., and Wilke, C. R., 1983, "Mass Transfer", McGraw Hill, New York.

Tan, C., and Liou, D., 1988, Desorption of ethyl acetate from activated carbon by supercritical carbon dioxide, *Ind. Eng. Chem. Res.*, 27:988.

Vitzthum, O. G., and Hubert, P., 1975, U.S. Patent 3,879,569.

Wilhelm, A., and Hedden, K., 1983, Nonisothermal extraction of coal with solvents in liquid and supercritical state, *Proc. Int. Conf. Coal Sci.*, p. 6, Pittsburgh.

Wilson, R. E., Keith, P. C., and Haylett, R. E., 1936, Liquid propane use in dewaxing, deasphalting, and refining heavy oils, *Ind. Eng. Chem.*, 28:1065.

Wright, B. W., Wright, C. W., Gale, R. W., and Smith, R. D., 1987a, Analytical supercritical fluid extraction of adsorbent materials, *Anal. Chem.*, 59:38.

Wright, B. W., Frye, S. R., McMinn, D. G., and Smith, R. D., 1987b, On-line supercritical fluid extraction-capillary gas chromatography, *Anal. Chem.*, 59:640.

Yeo, S.-D., and Akgerman, A., 1990, Supercritical extraction of organic mixtures from aqueous solutions, *AIChE J.*, 36:1743.

Zosel, K., 1978, Separation with supercritical gases: practical applications, Angew. *Chem. Int. Ed.*, Engl., 17:702.

## OZONE FOR WASTE REMEDIATION

## AND WASTE WATER TREATMENT

Henry J. Ledon

Chemoxal, Air Liquide Group
75, Quai d'Orsay
75321 Paris Cedex 07, France

### INTRODUCTION

Environmental concerns expressed by more and more stringents regulations provide a strong incentive to design new technologies for the destruction of hazardous wastes. Ozone, which powerful oxidizing properties have been recognized during the second half of the XIX$^{th}$ Century, is a choice chemical reagent for the treatment of water.

The first large scale application of ozone was implemented during the 1890's for the disinfection of drinking water and is established nowadays as a standard procedure in Western Europe where more than 850 ozonation plants are in operation.

Improvement of the economics of large scale production of ozone as well as the development of new methods to enhance its chemical reactivity (usually referred as "Advanced Oxidation Processes") offer a large scope of opportunities for the chemical oxidation of water pollutants.

### PHYSICAL AND CHEMICAL PROPERTIES OF OZONE

Ozone is a metastable allotropic form of oxygen which formation is highly endothermic:

$$3O_2 \rightleftharpoons 2O_3 \qquad \Delta H° = -284.5 \text{ kJ}$$

$$\Delta S° = 69.9 \text{ J mol}^{-1} \text{ deg}^{-1}$$

Ozone is slightly soluble in water. The value of the Bunsen coefficient (STP $m^3$ of gas dissolved per STP $m^3$ of water under a partial pressure of gas of 0,1 MPa) is 0.24 at 20°C compared to 0.031 for oxygen.[1]

*Industrial Environmental Chemistry*, Edited by D.T. Sawyer
and A.E. Martell, Plenum Press, New York, 1992

The determination of ozone concentration in water solutions could easely be carried out by using the Indigo trisulfonate method [2] whereas the direct monitoring of ozone in the gas phase is possible by UV absorption measurement: molar absorptivity = 3 000 ± 30 $l.M^{-1}cm^{-1}$ at $\lambda = 254$ nm [3].

Water solutions of ozone are moderately stable, the rate of decomposition being first order with respect to ozone and hydroxide ion [4].

Ozone is a very strong oxidizing agent as shown in Table 1.

The reactivity of molecular ozone is governed by a strong electrophilic character as illustrated by the influence of substituents on the rate of ozonation of benzene derivatives, as shown in Table 2.

However in an aqueous basic medium, ozone reacts with the hydroxide ion in a very complex chain process giving the hydroxyl radical $HO^{\bullet}$ as an intermediate [4, 7-12]. As shown in Table 1 above $HO^{\bullet}$ is an even more powerful oxidant than ozone and reacts non selectively at a rate close to the diffusion limits. Comparison of the rate of oxidation of quite refractory hydrocarbons with molecular ozone or the hydroxyl radical are presented in the Table 3.

**TABLE 1.** Oxidation potentials of some usual oxidants.

| OXIDANT | | OXIDATION POTENTIAL (Volt / NHE) |
|---|---|---|
| FLUORINE | $F_2$ | 3.06 |
| HYDROXYL RADICAL | $HO^{\bullet}$ | 2.80 |
| OZONE | $O_3$ | 2.07 |
| PERSULFATE | $S_2O_8^{2-}$ | 2.0 |
| HYDROGEN PEROXIDE | $H_2O_2$ | 1.78 |
| HYDROPEROXYL RADICAL | $HO_2^{\bullet}$ | 1.70 |
| PERMANGANATE | $MnO_4^-$ | 1.68 |
| HYPOCHLOROUS ACID | HClO | 1.49 |
| CHLORINE | $Cl_2$ | 1.36 |
| CHLORINE DIOXYDE | $ClO_2$ | 1.27 |
| OXYGEN | $O_2$ | 1.23 |

**TABLE 2.** Second order rate constants for the ozonation of substituted aromatics in water [5,6].

| COMPOUND | $kO_3$ ($M^{-1}$ $S^{-1}$) |
|---|---|
| NITRO BENZENE | 0.09 ± 0.02 |
| CHLORO BENZENE | 0.75 ± 0.2 |
| BENZENE | 2 ± 0.4 |
| TOLUENE | 14 ± 3 |
| PARA XYLENE | 140 ± 30 |
| ANISOLE | 290 ± 50 |
| PHENOL | 1 300 ±200 |
| PHENATE ION | $(1.4\pm0.4).10^9$ |

**TABLE 3.** Second order rate constants for the oxidation of saturated hydrocarbons.

| COMPOUNDS | $kO_3 (M^{-1}\ sec^{-1})$ [7] | $kHO^{\bullet} (M^{-1}\ sec^{-1})$ [13] |
|---|---|---|
| TETRACHLORO ETHYLENE | < 0.1 | $1.7\ .10^9$ |
| TERBUTANOL | 0.03 | $4\ .10^8$ |
| OXALATE ION | < 0.04 | $1\ .10^7$ |
| ACETATE ION | $3.10^{-5}$ | $7\ .10^7$ |

## OZONE GENERATION

Ozone is produced on an industrial scale by passing a dry gas containing oxygen in a corona discharge.

Air or oxygen can be used as feed stock. The energy consumption for ozone production is highly dependant on purity of the feed gas, in particular a very low water containt is essential (dew point below -50°C to -80°C). Several contaminents like hydrogen, hydrocarbons, fluorinated hydrocarbons,.... have a very detrimental effect on ozone yield, even at minute concentration. Thus if air is used as feed gas, a very careful purification is required, whereas oxygen can be used as delivered.

The theoretical energy required to produce 1kg of ozone is 0,82 kwh, but in practice, using dry air as feed gas, the actual consumption for producing ozone (1 to 3% in air) is in the range of 15 to 20 kwh/kg. Use of oxygen results in a much better energy efficiency and the most advanced technology only requires 7 to 10 kwh/kg for a much higher output concentration (6 to 10% in oxygen).[14]

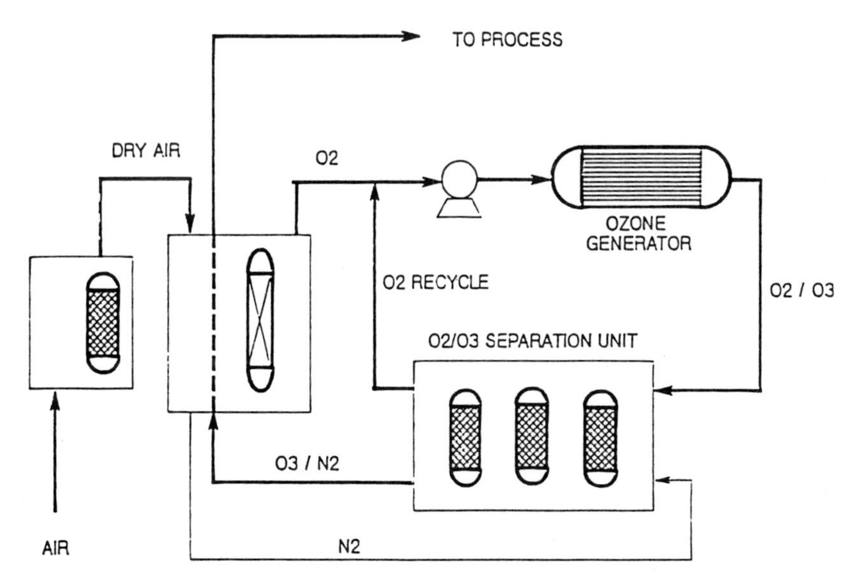


**Figure 1.** Simplified flow-sheet for ozone production with oxygen recycling.

Incentives to benefit from the numerous advantages of using oxygen as feed gas for generating ozone have led to the development of a new technology for separation of ozone from the outlet mixture gas and the recycling of oxygen to the ozone generator. Figure 1 shows a schematic flowsheet of the L'AIR LIQUIDE - DEGREMONT loop-process to produce ozone. [15]

## OZONATION OF HAZARDOUS ORGANIC COMPOUNDS

### OZONE ACTIVATION

Molecular ozone reacts extremely rapidly with electron rich unsaturated compounds (Table 2) but only slowly with saturated derivatives (Table 3). In order to extend the scope of chemical oxidation which could be used in water treatment, several authors have examined the combination of ozone whith UV-Irradiation or ozone with hydrogen peroxyde. [16-30]

Ozone dissolved in water can be photodecomposed by irradiation at $\lambda$ = 254 nm with a quantum yield of 0,62 [31] :

$$O_3 + H_2O \xrightarrow{h\nu} O_2 + H_2O_2$$

Hydrogen peroxide is a weak acid (pKa = 11,6) which reacts slowly with ozone whereas its conjugated base $HO_2^-$ leads to a fast decomposition of ozone:

$$O_3 + HO_2^- \longrightarrow O_3^{\bullet -} + HO_2^{\bullet} \qquad k = 2.2\ 10^6\ M^{-1}sec^{-1}$$

The ozonide radical ion $O_3^{\bullet -}$ decomposes to afford the hydroxyl radical $HO^{\bullet}$

$$O_3^{\bullet -} + H_2O \longrightarrow HO^{\bullet} + O_2 + HO^- \qquad k = 20 - 30\ M^{-1}sec^{-1}$$

whereas the hydroperoxide radical $HO_2^{\bullet}$ induce the chain decomposition of ozone:

$$HO_2^{\bullet} + H_2O \rightleftharpoons O_2^{\bullet -} + H_3O^+ \qquad pKa = 4.8$$

$$O_3 + O_2^{\bullet -} \longrightarrow O_3^{\bullet -} + O_2 \qquad k = 1.6\ .10^9\ M^{-1}sec^{-1}$$

Numerous other free radical pathways have been described for the reaction between species present in solution $O_3$, $HO^-$, $HO^{\bullet}$ $HO_2^{\bullet}$, $O_2^{\bullet -}$, $HO_2^-$, $H_2O_2$,... which leads to an extremely complicated mechanistic scheme. The main feature of the activation of ozone either by UV irradiation or hydrogen peroxide is the rather fast production of the hydroxyl radical $HO^{\bullet}$.

The effect of the pH and the addition of hydrogen peroxide on oxidation of organic compounds is illustrated in Figure 2 which exhibits the ozonation of methionine monitored by the chemical oxidation demand "COD" with time.

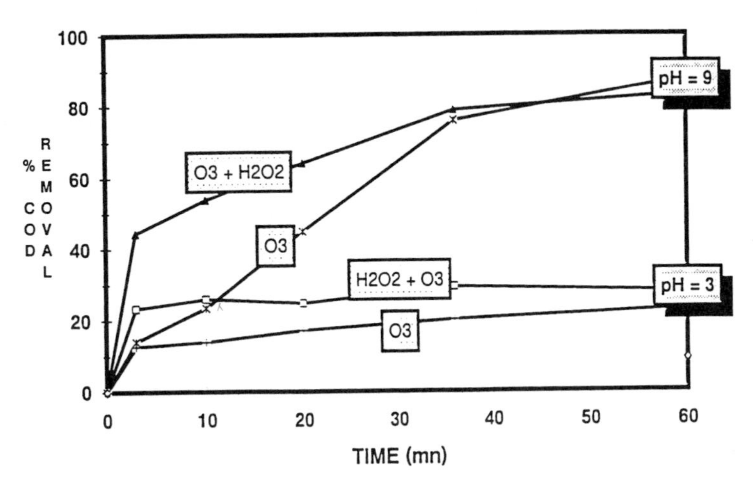


**Figure 2.** Oxidation of an aqueous solution of methionine (250mg in 250ml; $O_3$ 25mg/min; $H_2O_2$ 240mg)

Ozonation at low pH gives only a limited removal of COD whereas ozonation at basic pH lead to almost total oxidation. Higher rates of oxidation are observed in the presence of hydrogen peroxide.

## INHIBITORS

However, the rate of oxidation of the organic compounds could be dramatically reduced if inhibitors are present in the reaction medium. The most common ones in water treatment are the carbonate/bicarbonate ions which are efficient scavengers for the hydroxyl radical.[22,32]

$$HO^{\bullet} + HCO_3^{-} \longrightarrow H_2O + CO_3^{\bullet -} \qquad k = 1.5\ .10^{7}\ M^{-1}sec^{-1}$$

$$HO^{\bullet} + CO_3^{2-} \longrightarrow HO^{-} + CO_3^{\bullet -} \qquad k = 4.2\ .10^{8}\ M^{-1}sec^{-1}$$

During the course of the reaction acids are formed which lower the pH and thus reduce the rate of oxidation by depleting the medium from $HO^-$ and $HO_2^-$ ions. In order to obtain a high rate of destruction, controlled addition of a base is desirable to maintain the pH value in the basic range. However, when the total destruction of hazardous organics is required, oxidation in basic medium leads to the build up of carbonate/bicarbonate ions which inhibit the reaction.

Thus it may be advisable for full oxidation of chemicals to have a process including a first stage of ozonation in neutral or basic medium followed by a second stage which could be carried out at low pH like Fenton's type reaction( $H_2O_2$ in the presence of ion salts) or UV irradiation of hydrogen peroxide.

## OZONE - HYDROGEN PEROXIDE RATIO

The theoretical molar ratio of oxidants in ozone-hydrogen peroxide system is $H_2O_2:O_3 = 0.5$ [21]. If hydrogen peroxide is present at high concentration it could react with the $HO^{\bullet}$ radicals inducing self decomposition reactions without oxidation of the organic compounds.

A much more efficient use of chemicals is obtained with continuous addition of $H_2O_2$ as shown in Table 4 for the oxidation of methionine.

**TABLE 4** Effect of the mode of H2O2 addition for the oxidation of methionine in aqueous solution (1g/l)

| | SINGLE ADDITION [1] | STEPWISE ADDITION [2] |
|---|---|---|
| REACTION TIME FOR 85% COD REMOVAL (MN) | 60 | 36 |
| $O_3$ INTRODUCED (mg) | 1 500 | 900 |
| $O_3$ CONSUMMED (mg) | 615 | 615 |
| $H_2O_2$ INTRODUCED (mg) | 240 | 155 |
| $H_2O_2$ CONSUMMED (mg) | 240 | 110 |

[1] The total amount of $H_2O_2$ is introduced at t=0

[2] 35 mg of $H_2O_2$ are introduced at t=0 then 40 mg each 10 minutes.

## EVOLUTION OF THE TOXICITY OF SOLUTIONS DURING THE OXIDATION

The removal of toxic compounds by total oxidation to carbon dioxide and water is an expensive process and frequently only partial reactions are carried out.

However care must be taken that the intermediate products formed during the oxidation are safe for the environment Figure 3 shows the evolution of the toxicity of a $6.7 \cdot 10^{-3}$ molar solution of methionine in water with respect to the COD removal using the $O_3$ / $H_2O_2$ system.

Toxicity of the solution was estimated following the concentration which reduce by 50% (EC 50) the light emission of photobacteria using the MICROTOX® equipment.

Methionine which is an innocuous compound used as additive for animal nutrients affords toxic intermediates by oxidation which are themself destroyed by further reaction.

In conclusion ozone alone or activated by UV Irradiation or by association with hydrogen peroxide is a powerful and clean oxidizing agent as well as a strong desinfectant. However its uses at the present time are mainly limited to the treatment of drinking water.

Over the last decade extensive research has led to a much better understanding of the parameters which control these very complexes Advanced Oxidation Processes. Technical efficiency has been demonstrated in several pilot scale or field applications. Reductions of the cost of ozone production and optimization of the consumption or reagents will open new area of economically viable processes for the treatment of effluents.

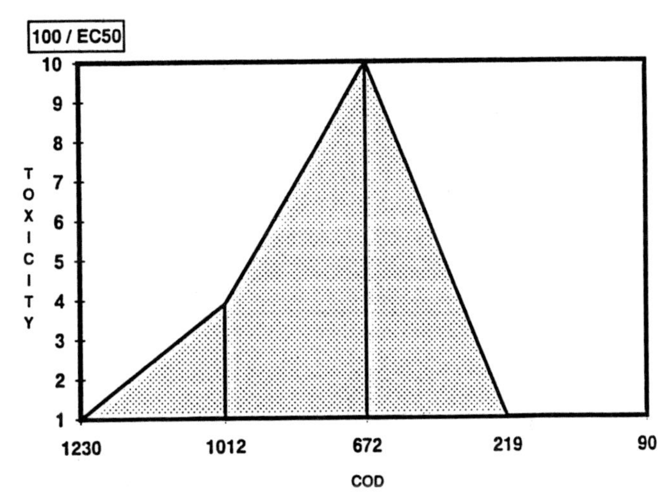


**Figure 3.** Evolution of the toxicity of an aqueous solution of methionine during its oxidation with the $O_3/H_2O_2$ system.

## ACKNOWLEDGEMENTS

The Author is grateful to "Ministère de l'Industrie et de l'Aménagement du Territoire" for support through Cooperative Agreement 90-4-90-124: Technologies Industrielles pour l'Environnement, to A. DE LADONCHAMPS, D. GOBARD and C. MESSUTA for technical assistance and to C. COSTE of DEGREMONT for sharing his expertise in ozone technology through stimulating discussions.

## REFERENCES

General readings

P.S. Bailey."Ozonation in Organic Chemistry", Academic Press, New York (1982)

R.G. Rice and A. Netzer. "Handbook of Ozone Technology and Applications". Ann Arbor Science Publishers, Ann Arbor (1982)

S.D. Razumouskii and G.E. Zaikov. "Ozone and its Reactions with Organic Compounds", Elsevier, Amsterdam (1984)

M. Dore "Chimie des Oxydants et Traitement des Eaux", Lavoisier Tech & Doc, Paris (1989)

H. Roques "Fondements Théoriques du Traitement Chimique des Eaux", Lavoisier Tech & Doc, Paris (1990)

W.J. Masschelein "Ozone et Ozonation des Eaux", I.O.A-Lavoisier Tech & Doc, Paris (1991)

B. Langlais, D.A. Reckhow and D.R. Brink. "Ozone in Water Treatment: Application and Engineering" Lewis Publishers, Chelsea (1991)

W.W. Eckenfelder, A.R. Bowers and J.A. Roth. "Chemical Oxidations, Technologies for the Nineties". Technocomic, Lancaster (1992)

1. L'AIR LIQUIDE "Gaz Encyclopedia". Elsevier, Amsterdam, (1976)

2. H. Bader and J. Hoigne, Determination of Ozone in Water by the Indigo Method: A Submitted Standard Method, *Ozone. Sci. Eng.* 4:169 (1982)

3. International Ozone Association Standardisation Committee Europe, 002/87

4. J. Staehelin and J. Hoigne, Decomposition of Ozone in Water. Rate of Initiation by Hydroxide Ions and Hydrogen Peroxide, *Envir. Sci. Technol.* 16:676 (1982)

5. J. Hoigne and H. Bader, Rate constants of Reactions of Ozone with Organic and Inorganic Compounds in Water. I. Non Dissociating Organic Compounds, *Water Res.* 17:2:173 (1983)

6. J. Hoigne and H. Bader, Rate constants of Reactions of Ozone with Organic and Inorganic Compounds in water.II. Dissociating Organic Compounds, *Water Res.* 17:2:185 (1983)

7. J. Hoigne and H. Bader, The Role of Hydroxyl Radical Reactions in Ozonation Processes in Aqueous Solutions, *Water Res.* 10:377 (1976)

8. L. Forni, D. Bahnemann and E.J. Hart, Mechanism of the hydroxide Ion Initiated Decomposition of Ozone in Aqueous Solution, *J. Phys. Chem.* 86:255 (1982)

9. R.E. Buhler, J. Staehelin and J. Hoine, Ozone Decomposition in Water Standard by Pulse Radiolysis.1. $HO_2/O_2$ and $HO_3/O_3$ as Intermediates, *J. Phys. Chem.* 88:2560 (1984)

10. H. Tomiyasu, H. Fukutomi and G. Gordon, Kinetics and Mechanisms of Ozone Decomposition in Basic Aqueous Solutions, *Inorg. Chem.* 24:2962 (1985)

11. G. Gordon, S. Nakareseisoon and G.E. Pacey, The Very Slow Decomposition of Aqueous Ozone in Highly Basic Solutions, Proceedings of the 8th Ozone World Congress. International Ozone Association, Zurich, E27 (1987)

12. C.C.D. Yao, W.R. Haag and T. Mill, Kinetics features of Advanced Oxidation Process for Treating Aqueous Chemical Mixtures, Proceedings of the Second International Symposium on Chemical Oxidation, Technology for the Nineties, Vanderbilt University, Nashville, Tennessee (1992)

13. Farhataziz and A.B. Ross, Selected Specific Rates of Reactions of transients from Water in Aqueous Solutions, National Bureau of Standards NSRDS-NBS 59, Washington, DC (1977)

14. P.E. Erni, Advanced in Ozone Generator Technology, Proceedings of the 10th Ozone World Congress, International Ozone Association, Zurich, 1:13 (1991)

15. P. Petit, P. Gastinne, P. Jan, J.M. Benas, R.S. Bes and J.C. Mora, Ozone Production with recycling of Non Transformed Oxygen, Proceedings of the 8th Ozone World Congress, International Ozone Association, Zurich, A72 (1987)

16. R.L. Garrison, H.W. Prengle, Jr and C.E. Mauk, Method of Destroying Cyanides US Patent 3,920,547 to Houston Research Inc, (1975)

17. S. Nakayama, Y. Taniguchi, K. Namba and N. Tabata, Improved Ozonation in Aqueous Systems, *Ozone Sci. Eng.* 1:119 (1979)

18. K. Namba and S. Nakayama, Hydrogen Peroxide-Catalyzed Ozonation of Refractory Organics. I. Hydroxyl Radical Formation, *Bull. Chem. Soc. Jpn.* 55:3339 (1982)

19. G.R. Peyton, F.Y. Huang, J.L. Burleson and W.H. Glaze, Destruction of Pollutants in Water with Ozone in Combination with Ultra violet Radiations 1. General Principles and Oxidation of Tetrachloroethylene *Environ. Sci. Technol.* 16:448 (1982); ibid, 2. Natural Trihalomethane Procursor, *Environ. Sci. Technol.* 16:454 (1982)

20. G.R. Peyton Modeling Advanced Oxidation Processes for Water Treatment Proceedings of the 9th Ozone World Congress, International Ozone Association Zurich, 579 (1989)

21. W.H. Glaze, J.W. Kang and D.H. Chapin, The Chemistry of Water Treatment Processes Involving Ozone, Hydrogen Peroxide and Ultra violet Radiation, *Ozone Sci. Eng.* 9:335 (1987)

22. W.H. Glaze and J.W. Kang, Evaluation of the Ozone-Hydrogen Peroxide Process in a Semi-Batch Reactor using Tetrachloro ethylene as a model Compound, proceedings of the 9th Ozone World Congress, International Ozone Association Zurich, 596 (1989)

23. R. Brunet, M.M. Bourbigot and M. Dore, Oxidation of Organic Compounds through the Combination or Ozone-Hydrogen Peroxide, *Ozone Sci. Eng.* 6:163 (1984)

24. H. Paillard, R. Brunet and M. Dore, Application of Oxidation by a combined Ozone/Ultraviolet Radiation System to the Treatment of Natural Water, *Ozone Sci. Eng.* 9:391 (1987)

25. H. Paillard, R. Brunet and M. Dore, Optimal Conditions for applying an Ozone - Hydrogen Peroxide Oxidizing System, *Wat. Res.* 2:91 (1988)

26. J.P. Duguet, E. Brodard, B. Dussert, J. Mallevialle, Improvement in the Effectiveness of Ozonation of Drinking Water Through the Use of Hydrogen Peroxide, *Ozone Sci. Eng.* 7:241 (1985)

27. P.D. Francis, Ozonation by UV and Ozone of Organic Contaminants Dissolved in Deionized and Raw Mains Water, *Ozone Sci. Eng.* 9:369 (1987)

28. J.L. Wallace, B. Vahadi, J.B. Fernandes and W.H. Glaze, Destruction of Pollutants in Water with Ozone in Combination with UltraViolet Radiation 1. General Principles and Oxidation of Tetrachloro ethylene Environ. Sci. Technol. 16:448 (1982); ibid, 2. Natural Trihalomethane Precursors, *Environ. Sci. Technol.* 16:454 (1982)

29. N. Nakanishi, Ozonation of several Organic Compounds Having Low Molecular Weight under Ultra Violet Irriadiation, *Ozone Sci. Eng.* 2:1 (1990)

30. J.D. Zeff and E. Leitis, Oxidation of Organic Compounds in Water, US Patent 4,792,407 to Ultrox International (1988)

31. H. Taube, Photochemical Reactions of Ozone in Solution *Tran. Faraday Soc.* 53:656 (1956)

32. H. Paillard, B. Legube and M. Dore, Effects of Alkalinity on the Reactivity of Ozone towards Humic Subtances and Manganese, *Aqua* 38:32 (1989)

# NUCLEOPHILIC AND REDUCTIVE REMEDIATION STRATEGIES FOR HAZARDOUS HALOGENATED HYDROCARBONS (PCBs, HCB, PCP, TCE, $CCl_4$, DCBP, EDB, DDT, AND DDE)

Donald T. Sawyer, Paul K. S. Tsang, Seungwon Jeon
and Marjorie Nicholson
Department of Chemistry
Texas A&M University
College Station, Texas 77843

## INTRODUCTION

During the past half-century the world has made use of a large group of halogenated hydrocarbons. The applications include structural polymers [e.g., polyvinylchloride (PVC)], pesticides (e.g., DDT, DCBP, EDB, PCP, DDE, and Lindane), cleaning solvents and degreasing agents (e.g., $CCl_4$, TCE, methylchloroform, and DCE), and non-flammable pump oils, dielectrics, and heat-exchanger fluids (e.g., PCBs, HCB, and related polychlorinated aromatic molecules). In the manufacture and utilization of these materials halogenated wastes result, which have been stored, "dumped and flushed", placed in land-fills, and incinerated. Some of the disposal methods have produced contaminated soils, sludges, harbor muds, and land-fills. Most of the Super-Fund sites contain halogenated-hydrocarbon wastes as a major component and are the most extreme examples of poor "housekeeping" and inappropriate disposal.

Polyhalogenated aromatic hydrocarbons (PCBs, PBBs, and hexachloro-benzene) represent a major environmental problem.[1,2] These materials, which were extensively used as transformer oils and heat-exchanger fluids, are major components of the hazardous-waste disposal problem of the EPA Superfund. The chemistry and toxicology of PCBs, and of their impurities and partially oxygenated products (e.g., dioxins), have been extensively reviewed.[3-5] Clearly the materials contain components that are animal carcinogens and can cause birth defects.[6-8] Hexachlorobenzene (HCB), which is a by-product of the

*Industrial Environmental Chemistry*, Edited by D.T. Sawyer
and A.E. Martell, Plenum Press, New York, 1992

polychloroethylene solvent industry, is as environmentally persistent as PCBs and is a human carcinogen.[9,10]

Although storage and disposal in landfills have been used for PCBs in the past, their long environmental life has lead to the contamination of lakes, rivers, coastal estuaries, and ground water. Currently, incineration is the most utilized technology for the destruction of halogenated aromatic hydrocarbons. The best systems are highly efficient in their conversion of PCBs to HCl, $CO_2$, and $H_2O$, but appear to produce some dioxins.[11]

In addition to the problems of the past, useful materials continue to be produced from the halogenation of hydrocarbons and waste streams from the manufacturing processes are inevitable. The ideal remediation strategy for these wastes should be on-site, on-line processing, which would eliminate the problems and hazards of storage and transport.

Remediation strategies (chemical and biological) for halogenated hydrocarbons seek to destroy all of the carbon-halogen bonds (e.g., $RX \xrightarrow{HO^-} ROH + X^-$) via formation of inorganic halogen. These can be subdivided into (a) oxidative and (b) reductive processes. Within the realm of oxidative chemistry incineration is the most extensively developed technology. The complete mineralization of heavily chlorinated PCBs (Aroclor 1268, $Cl_{7\text{-}10}$), dioxins, and $PhCl_6$ (HCB) requires high temperatures via oxygen-rich combustion. With incineration the production of oxygenated intermediates (dioxins and furans) from PCBs and other halogenated aromatics is a major concern. Also, the volatile products from oxidative combustion and pyrolysis of PCBs include HCl, $Cl_2$, HOCl, $Cl_2O$, and $Cl_2C{=}O$, which must be efficiently removed.

The major chemical oxidants within an oxygen-rich flame are $\cdot O\cdot$, $HO\cdot$, $HOO\cdot$, and $\cdot O_2\cdot$, and their primary chemistry is (a) the abstraction of hydrogen atoms from substrates (oxidation) and (b) formation of bonds with the resultant carbon radicals (oxygenation)

$$CH_4 + HO\cdot \xrightarrow[-H_2O]{} H_3C\cdot \xrightarrow{HO\cdot} H_3COH \qquad (1)$$

Similar oxidants can be produced from $O_2$, $O_3$, HOOH, and $HNO_3$ via (a) catalytic chemical reactors and aerobic biology. Such oxidants are much less reactive with carbon-chlorine bonds than with carbon-hydrogen bonds, which makes oxidative strategies ineffective with fully chlorinated wastes [e.g., $CCl_4$, $PhCl_6$, and $C_{12}Cl_{10}$ (perchloro biphenyl)].

In contrast, reductive technologies for halogenated hydrocarbons invariably transform carbon-bonded halogens to halide ions. This is accomplished with (a) nucleophiles (e.g, $HO^-$, $RO^-$, $O_2^{-\cdot}$, $H^-$, and $e^-$),

$$MeI + HO^- \longrightarrow MeOH + I^- \qquad (2)$$

(b) hydrogen atoms,

$$PhCl + 2\,[H\cdot] + HO^- \longrightarrow PhH + Cl^- + H_2O \qquad (3)$$

and (c) strongly reducing metals via electron-transfer, e.g.,

$$PhCl + 2\,Na + H_2O \longrightarrow PhH + NaCl + NaOH \qquad (4)$$

In the following sections we discuss the current status of various nucleophilic and reductive strategies for the transformation of hazardous halocarbon wastes to harmless materials. Most of the studies have used hexachlorobenzene ($PhCl_6$) and chlorobenzene (PhCl) as models for PCBs, and $CCl_4$ and trichloroethene (TCE) as models for cleaning-solvent and degreasing-agent wastes.

## NUCLEOPHILIC DEHALOGENATION OF HALOCARBONS

Most chemical transformations occur via the reaction of a nucleophile (base, reductant, or electron) with an electrophile (acid or oxidant), which entails a single-electron-transfer (S.E.T.) mechanism[12]

$$Nu\!:^- + RX \xrightarrow{\text{S.E.T.}} R\text{–}Nu + X^- \qquad (5)$$

Because the solvent matrix affects the redox potentials of nucleophiles and electrophiles, it has a major influence on their reactivity. Hence, strong nucleophiles are "leveled" to a reactivity that is equal to that for the conjugate base of the solvent. For example,

$$Nu\!:^- + n\,H_2O \longrightarrow Nu\text{–}H + HO^-(H_2O)_{n-1} \qquad (6)$$
$$(Nu\!:^-;\ H\!:^-,\ RO^-,\ O_2^{-\cdot},\ HOO^-)$$

Although hydroxide ion ($HO^-$) is a highly reactive nucleophile in aprotic solvents ($Me_2SO$, MeCN, DMF, and ethers), it is ineffective with chlorinated aromatics. In the primary step, the chloro-aromatic is transformed to a phenol, which in turn is neutralized with a second $HO^-$ to give a phenolate anion (a weak nucleophile) that is inert to further reaction

$$PhCl_6 + HO^- \xrightarrow[-Cl^-]{} Cl_5PhOH \xrightarrow[-H_2O]{HO^-} Cl_5PhO^- \quad (7)$$

The latter problem is eliminated by the use of alkoxide ions ($RO^-$), which yield an alkyl ether that remains electrophilic

$$PhCl_6 + RO^- \longrightarrow Cl_5PhOR + Cl^- \quad (8)$$

**KPEG Dehalogenation**

The utility of alkoxide nucleophiles for the dehalogenation of PCBs via nucleophilic displacement has been known since the 1980s. In order to avoid the leveling effects of water and to have an aprotic matrix with a low vapor pressure that can withstand high reaction temperatures, Prof. L. L. Pytlewski and co-workers[13] made use of poly(ethyleneglycol) (HPEG) and poly(ethyleneglycol)-monomethylether (HPEGMe) as solvents. They discovered that their conjugate-base salts (KPEG, NaPEG, KPEGMe, and NaPEGMe) are able to dechlorinate PCBs. Subsequent work at General Electric developed this technology for the remediation of PCB-contamined transformer oils.[14]

More recently (1991) the KPEGMe/HPEGMe technology has been successfully applied to PCB-contaminated soil at the Superfund site in Wide Beach, New York. The next paper by Robert Hoch discusses the engineering and technical problems, and the associated economics. Scheme I outlines paths (a, b) for the nucleophilic remediation of PCBs by the KPEG/HPEG reagent, and indicates that in its absence KOH and/or carbonates will produce chlorinated furans and dioxins (paths d-g).

**Superoxide Ion**

Although superoxide ion is a powerful nucleophile in aprotic solvents, it does not exhibit such reactivity in water, presumably because of its strong solvation by that medium ($\Delta H_{hydration}$, 100 kcal) and its rapid hydrolysis and disproportionation. The reactivity of $O_2^{-\cdot}$ with alkyl halides via nucleophilic substitution was first reported in 1970.[15,16] These and subsequent kinetic studies[17–19] confirm that the reaction is first order in substrate, that the rates follow the order primary>secondary>>tertiary for alkyl halides and tosylates,

**Scheme I.** Dehalogenation of PCBs by KPEG.

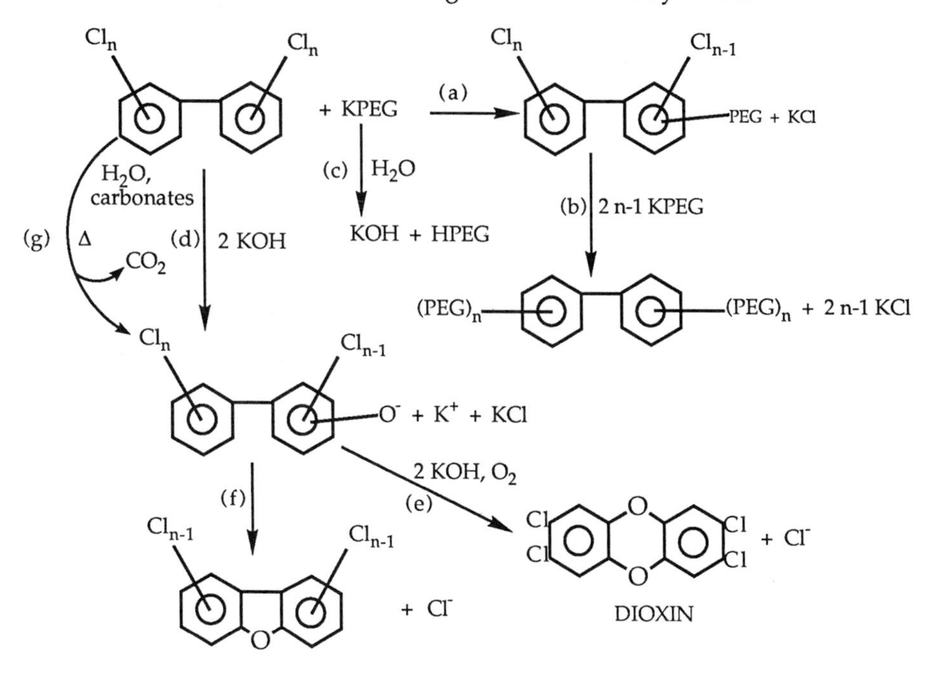


and that the attack by $O_2^{-\cdot}$ results in inversion of configuration ($S_N2$). The stoichiometries and kinetics for the reaction of $O_2^{-\cdot}$ with several halogenated hydrocarbons (alkanes, alkenes, and aromatics) are summarized in Table I.[20–22] [The normalized first-order rate constants, $k_1/[S]$, were determined by the rotated ring-disk electrode method under pseudo-first-order conditions ([substrate] > $[O_2^{-\cdot}]$)].[20]

The nucleophilicity of $O_2^{-\cdot}$ toward primary alkyl halides results in an $S_N2$ displacement of halide ion from the carbon center (Scheme II). The normal reactivity order, benzyl>primary>secondary>tertiary, and leaving-group order, I>Br>OTs>Cl, are observed, as are the expected stereoselectivity and inversion at the carbon center. In dimethylformamide the final product is the dialkyl peroxide. The peroxy radical (ROO·), which is produced in the primary step and has been detected by spin trapping,[23] is an oxidant that is readily reduced by $O_2^{-\cdot}$ to form the peroxy anion ($ROO^-$). Because the latter can oxygenate $Me_2SO$ to its sulphone, the main product in this solvent is the alcohol (ROH) rather than the dialkyl peroxide.

Although formation of the dialkyl peroxide is shown in the prototype reaction (Scheme II), hydroperoxides, alcohols, aldehydes, and acids also have been isolated. The extent of these secondary paths depends on the choice of solvent and reaction conditions. Secondary and tertiary halides also give substantial quantities of alkenes from dehydrohalogenation by $O_2^{-\cdot}/HOO^-/HO^-$.

**Table I.** Stoichiometries and Kinetics for the Reaction of 0.1-5 mM $O_2^{-\cdot}$ with Polyhalogenated Hydrocarbons in Dimethylformamide at 25°C.

| Substrate Concentration, [S] = 1-10 mM | $O_2^{-\cdot}$/S | Products/S | $k_1$/[S] $M^{-1}s^{-1}$ |
|---|---|---|---|
| $CCl_4$ | 5 | $HOC(O)O^-$, 4 $Cl^-$, 3.3 $O_2$ | 3800. |
| $HCCl_3$ | 4 | $HOC(O)O^-$, 3 $Cl^-$, 2 $O_2$ | 0.4 |
| $PhCCl_3$ | 4 | $PhC(O)O^-$(70%), $PhC(O)OO^-$(30%), 3 $Cl^-$, 2.4 $O_2$ | 50.0 |
| $MeCCl_3$ | --- | | <0.1 |
| $HOCH_2CCl_3$ | 4 | $HOCH_2C(O)O^-$, 3 $Cl^-$ | 47. |
| (*p*-ClPh)$_2$CHCCl$_3$(DDT) | 1 | (*p*-ClPh)$_2$C=CCl$_2$, $Cl^-$ | 100. |
| PhCHBrCHBrPh | 2 | 2 PhCH(O), 2 $Br^-$, $O_2$ | 1000. |
| *n*-BuBr | 1 | $Br^-$ | 960. |
| *cis*-CHCl=CHCl | 4 | 2 $HOC(O)O^-$, 2$Cl^-$ | 10.0 |
| (*p*-ClPh)$_2$C=CCl$_2$(DDE) | 3 | $HOC(O)O^-$, (*p*-ClPh)$_2$C=O, 2 $Cl^-$, $O_2$ | 2.0 |

The reaction of $O_2^{-\cdot}$ with $CCl_4$ and $RCCl_3$ compounds cannot occur via an $S_N2$ mechanism because the carbon-atom center is inaccessible. Rather, superoxide ion appears to attack a chlorine atom with a net result that is equivalent to an electron transfer from $O_2^{-\cdot}$ to chlorine. This step is analogous to the single-electron-transfer (SET) mechanism that has been proposed for many nucleophilic reactions; an initial transfer of an electron followed by the collapse of a radical pair (eq. 5).[24]

The rates of reaction for $O_2^{-\cdot}$ with $RCCl_3$ compounds are proportional to their reduction potentials, which is consistent with the SET mechanism.[20] A plot of log $k_1$/[S] (Table II) against the reduction potentials of $RCCl_3$ compounds is approximately linear with a slope of -4.9 decade per volt. Such behavior is consistent with a mechanism that occurs via simultaneous electron transfer and nuclear motion.[25] This correlation indicates that in water, where the $O_2/O_2^{-\cdot}$ redox potential is about 0.44 V more positive, the rate of the reaction for $O_2^{-\cdot}$ with $CCl_4$ would be about 100-200 times slower than in aprotic solvents.

**Scheme II.** Nucleophilic Substitution by $O_2^{-\cdot}$ of Halocarbons.

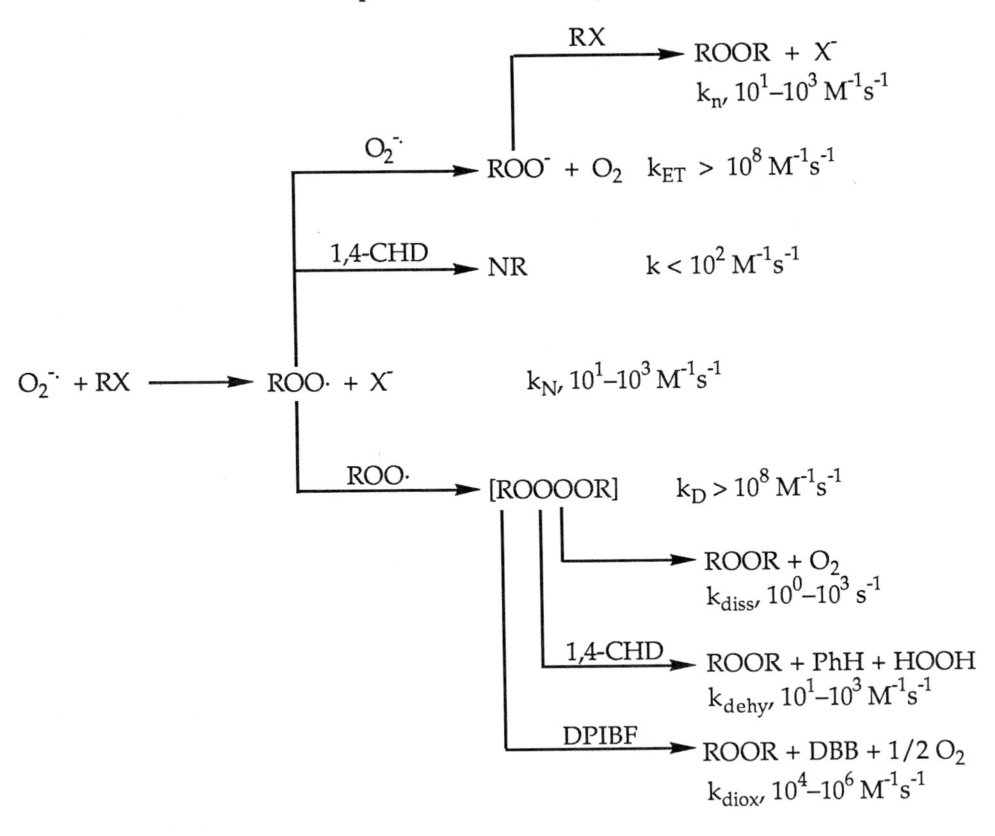


(RX = $CCl_4$, $F_3CCCl_3$, $PhCCl_3$, n-BuBr, n-BuCl)

The stoichiometric data in Table I for $CCl_4$ are consistent with a net chemical reaction that yields bicarbonate ion and four halide ions in the final aqueous workup of the reaction products. When $O_2^{-\cdot}$ reacts with $RCCl_3$ compounds, the *R*–C bond is not cleaved. About 30% of the product from the $PhCCl_3/O_2^{-\cdot}$ reaction is perbenzoate ion.

Superoxide ion reacts with vicinal dibromoalkanes to form aldehydes (Table I). The mechanism proposed for these reactions is a nucleophilic attack on carbon, followed by a one-electron reduction of the peroxy radical and nucleophilic displacement on the adjacent carbon to form a dioxetane that subsequently cleaves to form two moles of aldehyde.[22]

Both *p,p*'-DDT and methoxychlor are rapidly deprotonated by $HO^-$ in aprotic solvents with subsequent elimination of $Cl^-$ to form the dehydro–chlorination products; that is, DDT forms DDE. The same products are formed in their reactions with $O_2^{-\cdot}$. Because the reaction rates that are measured by the rotated ring-disk electrode method are fairly rapid, the primary step must be a direct reaction with $O_2^{-\cdot}$ and not with $HO^-$ [from the reaction of $O_2^{-\cdot}$ with trace water in the solvent ($2\ O_2^{-\cdot} + H_2O \rightarrow O_2 + HOO^- + HO^-$)]. Hence, the initial reaction with $O_2^{-\cdot}$ is deprotonation followed by elimination of $Cl^-$ to form DDE.

The overall reaction and product stoichiometries for the degradation of chloroalkene substrates by $O_2^{-\cdot}$ in DMF are summarized in Table I.[21] Within the limits of a reaction time of 10 min or less, chloroethene, *trans*-1,2-dichloroethene, Aldrin, and Dieldrin are not oxidized by $O_2^{-\cdot}$ in DMF. A reasonable mechanism for these oxidations is an initial nucleophilic addition of superoxide to the chloroalkenes. Subsequent loss of chloride ion would give a vinyl peroxy radical, which can cyclize and decompose to a chloroacyl radical and phosgene.[26] These would undergo subsequent facile reactions with $O_2^{-\cdot}$ to give the observed bicarbonate and chloride ions.

In the case of ($p$-ClPh)$_2$C=CCl$_2$ (DDE) and 1,1-dichloroethene, addition of superoxide can only be followed by β-elimination of chloride if attack occurs at the carbon with the chlorine atoms.

$$R_2C{=}CCl_2 + O_2^{-\cdot} \longrightarrow \left[ {}^{-}\!:\!C(R)(R){-}C(Cl)(Cl)(OO\cdot) \right] \longrightarrow R_2C{=}C(Cl)(OO\cdot) + Cl^- \qquad (9)$$

The sequence would lead to the ketones [RC(O)R] that are observed as products (Table I).

Polyhalogenated aromatic hydrocarbons [e.g., hexachlorobenzene (HCB, $C_6Cl_6$) and polychorobiphenyls (PCBs)] are rapidly degraded by superoxide ion in dimethylformamide to bicarbonate and halide ions (Table II).[27] Because halogen-bearing intermediates are not detected, the initial nucleophilic attack is the rate-determining step. The rates of reaction exhibit a direct correlation with the electrophilicity of the substrate (reduction potential) (e.g., $C_6Cl_6$, $E^{\circ\prime}$ = -1.48 V versus SCE; $k_1/[S] = 1 \times 10^3\ M^{-1}\ s^{-1}$ and 1,2,4-$C_6H_3Cl_3$, $E^{\circ\prime}$= -2.16 V; $k_1/[S] = 2 \times 10^{-2}\ M^{-1}\ s^{-1}$).[27]

Although polyhaloaromatics are degraded by $O_2^{-\cdot}$ in acetonitrile and dimethyl sulfoxide, the rates of reaction are about one-tenth as great in MeCN and 20 times slower in $Me_2SO$. A reasonable initial step for these oxygenations is nucleophilic addition of $O_2^{-\cdot}$ to the polyhalobenzene (e.g., $C_6Cl_6$; Scheme III). Subsequent loss of chloride ion will give a benzoperoxy radical, which will close on an adjacent aromatic carbon center and add a second $O_2^{-\cdot}$ to become a peroxo nucleophile that can attack the adjacent carbochlorine center with displacement of chloride ion to give a highly electrophilic tetrachloro center.

**Table II.** Reactions of Superoxide Ion with Polychloro-aromatics in Dimethylformamide.[a]

| Substrates(S) | $E_{p,c}$,[b] V vs SCE | $O_2^{-\cdot}$/S | $Cl^-$ released/S | base released/S | $k_1/[S]$, $M^{-1}s^{-1}$[c] |
|---|---|---|---|---|---|
| $C_6Cl_6$[d] | -1.48, -1.69 | 12.0±1.0 | 6.0±0.5 | 6.0±0.6 | $1 \times 10^3$ |
| $C_6HCl_5$ | -1.70, -1.98 | 11.0±1.0 | 5.0±0.5 | 6.0±0.6 | $8 \times 10^1$ |
| 1,2,3,4-$C_6H_2Cl_4$ | -1.90, -2.18 | 10.0±1.0 | 4.0±0.5 | 6.0±1.0 | $2 \times 10^0$ |
| 1,2,3,5-$C_6H_2Cl_4$ | -1.95, -2.19 | 10.0±1.0 | 4.0±0.5 | 6.0±1.0 | $1 \times 10^0$ |
| 1,2,4,5-$C_6H_2Cl_4$ | -1.95, -2.19 | 10.0±1.0 | 4.0±0.5 | 6.0±1.0 | $3 \times 10^0$ |
| 1,2,4-$C_6H_3Cl_3$ | -2.16, -2.45 | - | - | - | $2 \times 10^{-2}$ |
| $C_{12}Cl_{10}$ | -1.50, -1.78 | 22.0±4.0 | 10.0±1.0 | 12.0±2.0[e] | $2 \times 10^2$ |
| $CCl_4$[f] | -1.2 | 5.0±0.6 | 4.0±0.4 | 1.0±0.1 | $1 \times 10^3$ |
| $PhCCl_3$[f] | -1.47 | 4.0±0.4 | 3.0±0.4 | - | $4 \times 10^1$ |

[a]Overall reactions: (1) $C_6Cl_6 + 12\ O_2^{-\cdot} \rightarrow 3\ C_2O_6^{2-} + 6\ Cl^- + 3\ O_2$; (2) $C_6H_2Cl_4 + 10\ O_2^{-\cdot} \rightarrow 2\ C_2O_6^{2-} + 2\ HOC(O)O^- + 4\ Cl^- + O_2$; (3) $C_{12}Cl_{10} + 22\ O_2^{-\cdot} \rightarrow 6\ C_2O_6^{2-} + 10\ Cl^- + 4\ O_2$.

[b]First two reduction peaks are listed; a separate peak is observed for each chlorine atom. For $C_6Cl_6$ the other peak potentials are -1.95 V, -2.18 V, -2.44 V, and -2.70 V (also the reduction potential for PhCl).

[c]Apparent pseudo-first-order rate constants, $k$(normalized to unit substrate concentration [S]), were determined from the ratio ($i_{anodic}/i_{cathodic}$) for the cyclic voltammogram of $O_2$ in the presence of excess substrate.

[d]The values of $k_1/[S]$ for $C_6Cl_6$ in MeCN and $Me_2SO$ solvents are 92 $M^{-1}s^{-1}$ and 47 $M^{-1}s^{-1}$, respectively.

[e]Titration curve indicates the presence of a weak base, and a white precipitate forms during the last part of the titration; both consistent with the presence of oxalate ion.

[f]The ring-disc voltammetric technique for reaction rates gave values for $k_1/[S]$ of $3.8 \times 10^3$ $M^{-1}s^{-1}$ for $CCl_4$ and 50 $M^{-1}s^{-1}$ for $PhCCl_3$.

**Scheme III.** Nucleophilic degradation of hexachlorobenzene (HCB) by $O_2^{-\cdot}$

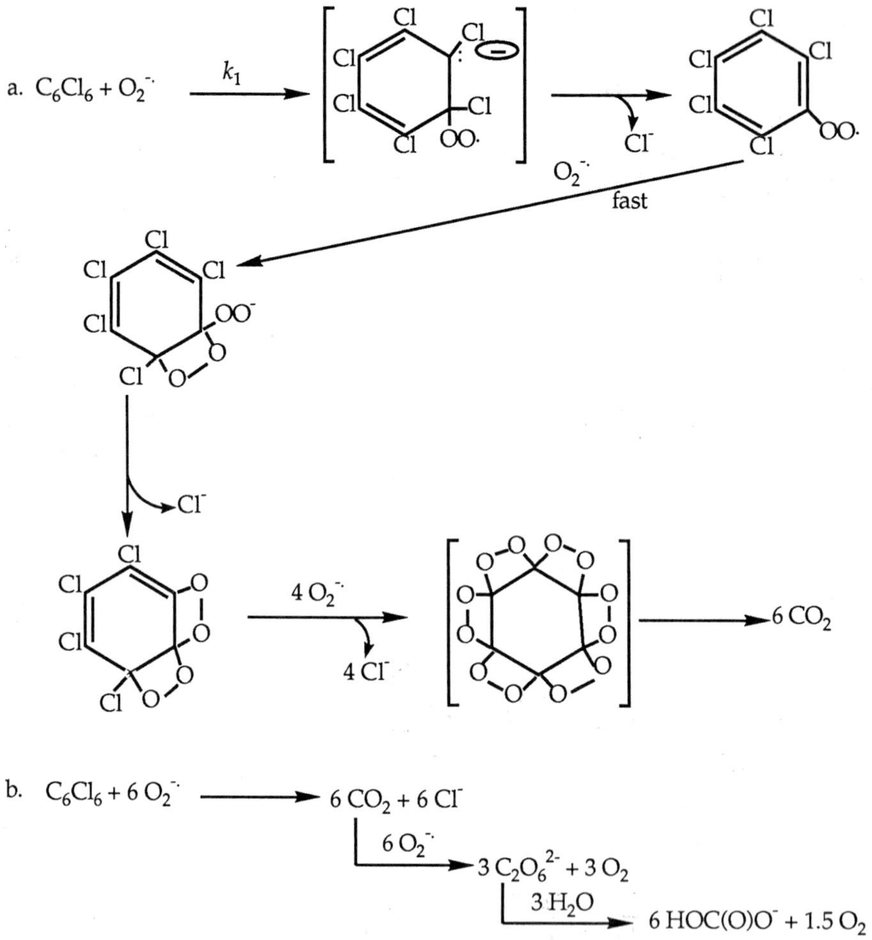

a. $C_6Cl_6 + O_2^{-\cdot} \xrightarrow{k_1}$ [intermediate] $\xrightarrow{-Cl^-}$ ... $\xrightarrow[\text{fast}]{O_2^{-\cdot}}$ ... $\xrightarrow{-Cl^-}$ ... $\xrightarrow[-4\,Cl^-]{4\,O_2^{-\cdot}}$ [intermediate] $\longrightarrow 6\,CO_2$

b. $C_6Cl_6 + 6\,O_2^{-\cdot} \longrightarrow 6\,CO_2 + 6\,Cl^-$

$\xrightarrow{6\,O_2^{-\cdot}} 3\,C_2O_6^{2-} + 3\,O_2$

$\xrightarrow{3\,H_2O} 6\,HOC(O)O^- + 1.5\,O_2$

The latter undergoes facile reactions with $O_2^{-\cdot}$ to displace the remaining chloro atoms. Thus, Scheme III outlines a possible mechanism, but the fragmentation steps are speculative and not supported by the detection of any intermediate species.[27]

When Aroclor 1268 (a commercial PCB fraction that contains a mixture of $Cl_7$, $Cl_8$, $Cl_9$, and $Cl_{10}$ polychlorobiphenyls) is combined with excess $O_2^{-\cdot}$, the entire mixture is degraded (Figure 1). Samples taken during the course of the reaction confirm that (a) the most heavily chlorinated members react first (the initial nucleophilic addition is the rate–determining step) and (b) all components are completely dehalogenated. Tests with other PCB mixtures establish that those components with three or more chlorine atoms per phenyl ring are completed degraded by $O_2^{-\cdot}$ within several hours.

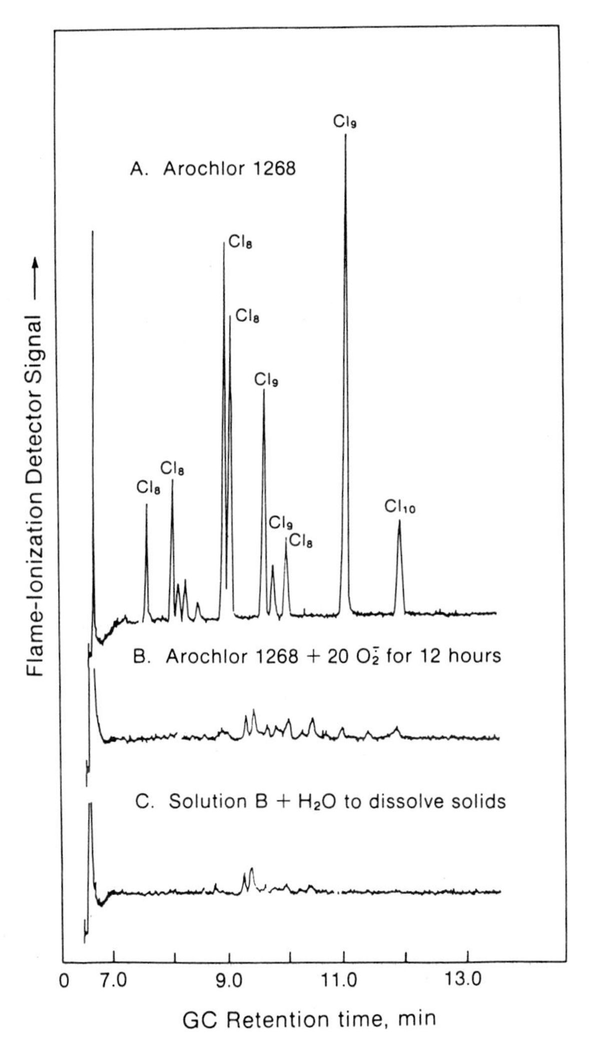


**Figure 1.** Gas chromatograms for Aroclor 1268 PCB mixture ($C_{12}C_{10}$, $C_{12}HCl_9$, $C_{12}H_2Cl_8$): (A) 10.2 mg of Aroclor 1268 in 1 mL of DMF; (B) solution A, 12 hr after the addition of 47.2 mg of $(Me_4N)O_2$ (~20 equiv per PCB); (C) Solution B after the addition of 1 mL $H_2O$ to dissolve any organic solids.

Although superoxide is available as $KO_2$, the material is expensive and almost insoluble in most aprotic solvents. The tetramethylammonium salt $[(Me_4N)O_2]$ is much more soluble, but must be prepared from $KO_2$ and $(Me_4N)OH_{(s)}$ via a solid-phase metathesis reaction.[28,29] A recent report[30] describes the in situ generation of $O_2^{-\cdot}$ from the combination of aniline or hydrazine, $O_2$, and $HO^-$ in a pyridine solvent. However, the most efficient and convenient means of production is in situ reductive electrolysis of dioxygen. The cyclic voltammograms for the reduction of $O_2$ to $O_2^{-\cdot}$ in the absence and presence of $C_6Cl_6$ are illustrated in Figure 2. The enhancement in the cathodic peak current and the decrease in the peak current for the reverse scan that result from the presence of $C_6Cl_6$ indicate a facile multistep reaction between $O_2^{-\cdot}$ and $C_6Cl_6$. The destructive decay of the components of Aroclor 1268 that results from the electrolytic reduction of oxygen to $O_2^{-\cdot}$ in an air-saturated DMF solution is presented in Figure 3. This illustrates a system whereby heavily chlorinated PCBs can be degraded to inorganic ions in a continuous electrolytic reactor. Product assays of numerous related experiments confirm that the initial step is rate limiting, and that organic intermediates do not accumulate in the reactor.

**Relative Reactivity of Oxo Nucleophiles [$HOO^-$, t-BuOO$^-$, $O_2^{-\cdot}$, $MeO^-$, $HO^-$, MePEG$^-$, PEG$^-$]**

The nucleophilicity of oxy anions ($YO^-$) is directly related to their oxidation potentials ($E^{\circ\prime}_{B^-/B\cdot}$, Table III) and the bond energies of their products ($YO$-R) with electrophilic substrates ($RX$)

$$YO^- + RX \rightarrow YO{-}R + X^- \tag{10}$$

Hence, the more negative the oxidation potential and the larger the $YO{-}R$ bond energy, the proportionally greater nucleophilic reactivity that will result. The shift in the oxidation potential of $HO^-$ from +1.89 V versus NHE in $H_2O$ to +0.92 V in MeCN reflects the leveling effect of protic solvents on the nucleophilicity of oxy anions. Likewise, the shift in potential for $HOO^-$ [+0.20 V ($H_2O$) to -0.34 V (MeCN)] is in accord with the exceptional reactivity of $HOO^-$ in aprotic solvents. In aqueous media the reactivity of $HOO^-$ is leveled by extensive anionic solvation,[35] but remains significant with many substrates due to its unique orbital energies[36] and the presence of an unshared pair of electrons on the atom adjacent to the nucleophilic center ($\alpha$ effect).[37]

The reactivity of $O_2^{-\cdot}$ with alkyl halides in aprotic solvents occurs via nucleophilic attack by $O_2^{-\cdot}$ on carbon, or on chlorine with a concerted reductive displacement of chloride ion.[16] As with all oxy anions, water suppresses the

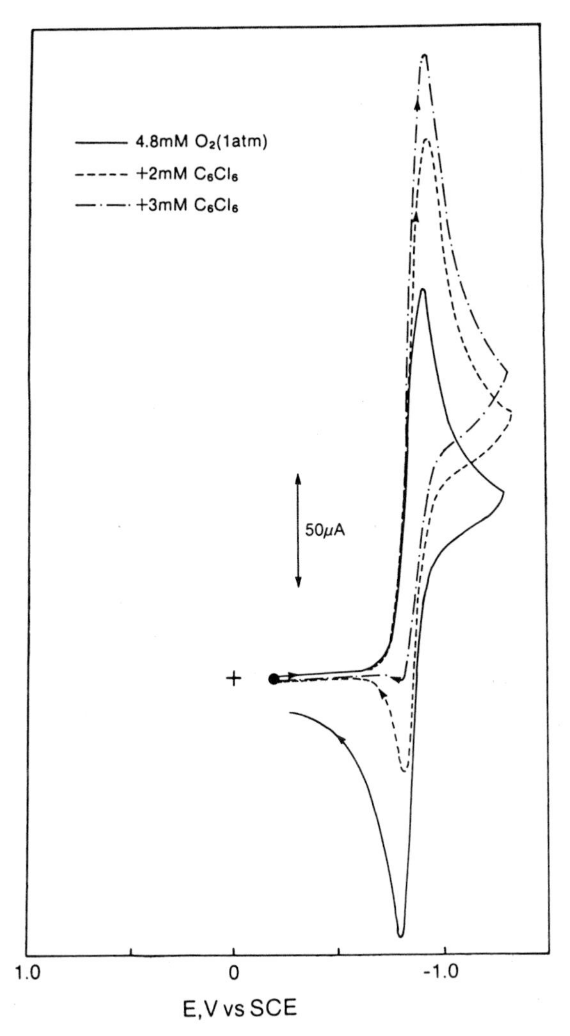


**Figure 2.** Cyclic voltammograms for dissolved $O_2$ (1 atm, 4.8 mM) in dimethylformamide (0.1 M tetraethylammonium perchlorate) at a glassy carbon electrode. The effect of 2 mM $C_6Cl_6$ and 3 mM $C_6Cl_6$ on the peak is illustrated by the dashed (---) and dash-dot (– · –) curves, respectively. Electrode area, 0.062 $cm^2$; scan rate, 0.1 V $s^{-1}$.

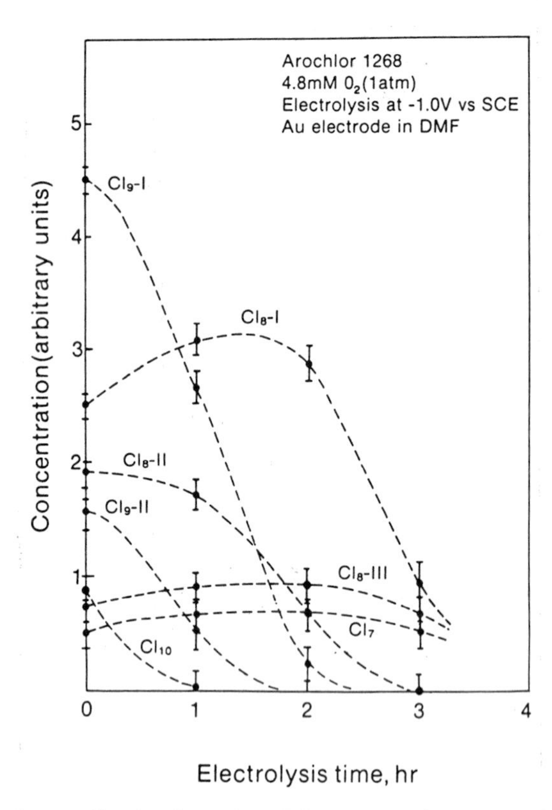


**Figure 3.** Concentration profiles for the various PCB isomers of Aroclor 1268 during the course of an in situ electrolytic reduction (-1.0 V vs SCE) of dissolved $O_2$ (1 atm, 4.8 mM) in DMF (0.1 M TEAP) at a gold-mesh electrode (~20 $cm^2$). Sample concentration, 50 mg Aroclor 1268 in 10 mL DMF.

**Table III.** Redox Potentials for the Single-Electron Oxidation of $HO^-$ and Other Oxy Anion Bases in Water and in Acetonitrile.

| Base ($:B^-$) | $(pK_{HB})_{H_2O}$[a] | $(E^{\circ\prime}{}_B)_{H_2O}$[b] (V versus NHE) | $(pK_{HB})_{MeCN}$[a,c] | $(E^{\circ\prime}{}_B)_{MeCN}$[d] (V versus NHE) |
|---|---|---|---|---|
| $Cl^- \rightarrow Cl\cdot + e^-$ | | +2.41 | | +2.24 |
| $HO^- \rightarrow HO\cdot + e^-$ | 15.7 | +1.89 | 30.4 | +0.92 |
| $PhO^- \rightarrow PhO\cdot + e^-$ | 9.2 | ~+0.7 | 16.0 | +0.30 |
| $O_2^{-\cdot} \rightarrow \cdot O_2\cdot + e^-$ | 4.9 | -0.16[e] | ~13 | -0.63[e] |
| $HOO^- \rightarrow HOO\cdot + e^-$ | 11.8 | +0.20 | ~22 | -0.34 |
| $H^- \rightarrow H\cdot + e^-$ | | -2.20 | | -3.2 |
| $[e^-(H_2O) \equiv H\cdot + HO^-] \rightarrow H_2O + e^-$ | | -2.93 | | -3.90 |

[a] $pK_a$ of the conjugate acid, Ref. 31.

[b] Ref. 32.

[c] Ref. 33.

[d] Ref. 34

[e] Standard state for $O_2$ is 1 M.

nucleophilicity of $O_2^{-\cdot}$ (hydration energy, 100 kcal)[38] and promotes its rapid hydrolysis and disproportionation. The reaction pathways for these compounds produce peroxy radical and peroxide ion intermediates ($ROO\cdot$ and $ROO^-$).

Hydroperoxide ion ($HOO^-$) is unstable in most aprotic solvents, but persists for several minutes in pyridine ($k_{decomp}$, $4.6 \times 10^{-3}$ $s^{-1}$), which allows studies of its nucleophilic reactivity. In pyridine $HOO^-$ is oxidized in an one-electron transfer to give $HOO\cdot$, which is in accord with previous studies in MeCN.[39]

$$HOO^- \rightarrow HOO\cdot + e^-, \; E_{p,a}, \; -0.34 \text{ V versus NHE (MeCN)} \qquad (11)$$

In the case of $HOO^-$ (and $t$-$BuOO^-$), the general leveling of nucleophilic reactivity by protic solvents (water and alcohols) enhances its lifetime such that the net reactions for $HOO^-$ with electrophilic substrates usually are most efficient and complete in $H_2O$ or $t$-BuOH. Almost all other solvents react with $HOO^-$ or facilitate its decomposition. The relative lifetime of $HOO^-$ in various solvents is in the order: $H_2O$>>MeOH>EtOH>>diglyme>pyridine~HPEGMe 350>>MeCN> DMF>>$Me_2SO$.

Table IV summarizes the relative reactivities of $MeO^-$/$HO^-$/$MePEG^-$/$HOO^-$/t-$BuOO^-$/$O_2^{-\cdot}$ with halocarbons and esters in pyridine and the kinetics for the reaction of $O_2^{-\cdot}$ with the substrates in aprotic solvents.[40] The relative reactivity with primary halides ($n$-BuBr and $PhCH_2Br$) in pyridine is $HOO^-$(3.0)>$t$-$BuOO^-$(2.1)>$O_2^{-\cdot}$(1.0)>$MeO^-$(0.3)>$HO^-$(0.2). For $CCl_4$ the relative reactivities are $HOO^-$(1.8)>$t$-$BuOO^-$(1.2)>$O_2^{-\cdot}$(1.0)>$MeO^-$(0.6)>$HO^-(H_2O)$(0.4), and for $PhCl_6$; $HOO^-$(5.0)>$t$-$BuOO^-$(2.9)>$O_2^{-\cdot}$(1.0)>$MeO^-$(0.2)>$HO^-$(0.0).

The relative reactivity of $HOO^-$/$MeO^-$ with primary halocarbons ($n$-BuBr, $PhCH_2Br$, $BrCH_2CH_2Br$) is about 10 in pyridine (Table IV), which compares with a ratio of 13 for their reaction with $BrCH_2C(O)OH$ in $H_2O$[41] and 35 for reaction with $PhCH_2Br$ in 50% acetone/water.[42] Although $O_2^{-\cdot}$ is a powerful nucleophile in aprotic media, it does not exhibit such reactivity in water, presumably because of its strong solvation by that medium and its rapid hydrolysis and disproportionation.[38] Kinetic and electrochemical studies for the reaction of $O_2^{-\cdot}$ with primary halocarbons confirm that the initial step is rate limiting and first order with respect to substrate and $O_2^{-\cdot}$. The other oxy anions of Table IV are believed to react via an analogous nucleophilic displacement.

In contrast, the reactions of $CCl_4$ with oxy nucleophiles ($YO^-$) cannot proceed via an $S_N2$ mechanism because the carbon center is inaccessible. The primary step for the reaction of $O_2^{-\cdot}$ with $CCl_4$ appears to involve a nucleophilic attack of a chlorine to give $Cl_3COO\cdot$ and $Cl^-$.[20] The reactions of

**Table IV.** The Relative Reactivity of $MeO^-/HO^-/MePEG^-/HOO^-/t\text{-}BuOO^-/O_2^{-\cdot}$ with Halocarbons and Esters in Pyridine [0.1 M $(Et_4N)ClO_4$], and the Kinetics for the Reaction of $O_2^{-\cdot}$ with Substrates.

| Substrate | $[k_X/k_{O_2^{-\cdot}}]$ (±20%)[a] | | | | | | $k_{O_2^{-\cdot}}/[S]$,[b] |
|---|---|---|---|---|---|---|---|
| (1-10 mM) | X=MeO⁻[c] | HO⁻ | MePEG⁻[d] | HOO⁻[e] | t-BuOO⁻ | $O_2^{-\cdot}$ | $M^{-1}s^{-1}$ |
| $CCl_4$ | 0.6±0.3 | 0.4 | 0.05 | 1.8±0.6 | 1.2 | 1.0 | $1.4\pm0.5 \times 10^3$ |
| n-BuBr | 0.3±0.2 | 0.2 | 0.00 | 3.0±1.0 | 2.1 | 1.0 | $1.0\pm0.1 \times 10^3$ |
| $CH_2Br_2$ | 0.3±0.2 | | | 3.1±1.0 | | 1.0 | $2.3\pm0.5 \times 10^2$ |
| $BrCH_2CH_2Br$ | 0.3±0.2 | | | 2.7±1.0 | | 1.0 | $1.8\pm0.3 \times 10^3$ |
| $PhCH_2Br$ | 0.3±0.2 | | | 3.0±1.0 | | 1.0 | $>3 \times 10^3$ |
| Br(Me)CHCN | 0.3±0.2 | | | 2.8±1.0 | | 1.0 | $1.4\pm0.3 \times 10^3$[f] |
| $PhCl_6$ | 0.15±0.1[g] | 0.0 | 0.00 | 5.0±1.5 | 2.9 | 1.0 | $1.0\pm0.2 \times 10^3$ |
| MeC(O)OPh | 0.2±0.1 | | | 4.5±1.5 | | 1.0 | $1.6\pm0.5 \times 10^2$[h] |
| MeC(O)OEt | 0.2±0.15 | | | 4.7±2.0 | | 1.0 | $1.1\pm0.2 \times 10^2$[h] |

[a]Determined from the impact of X upon the rate of disappearance of $O_2^{-\cdot}$ in the presence of excess substrate (ref. 12); rate of disappearance of $O_2^{-\cdot}$ monitored by linear sweep voltammetry.
[b]Determined from the ratio of $i_{p,a}/i_{p,c}$ for the cyclic voltammogram of $O_2$ in DMF in the presence of excess substrate (ref. 40).
[c]$MeO^-$ from $(Bu_4N)OH$ in MeOH. [d]$MePEG^-$ (anion of monomethylpolyethyleneglycol 350). [e]$HOO^-$ prepared either from $O_2^{-\cdot}$ plus $PhNHNH_2$ or HOOH plus $HO^-$.
[f]$k/[S] = 3.5 \times 10^2\ M^{-1}s^{-1}$ in MeCN. [g]The primary product is $Cl_5PhOMe$. [h]Kinetics in pyridine.

$CCl_4$ with other oxy anions appears to be similar (e.g., $CCl_4 + HO^- \rightarrow Cl_3COH + Cl^-$).[20] However, chloroform is deprotonated by $HO^-$ to give dichlorocarbene ($HCCl_3 + HO^- \rightarrow :CCl_2 + H_2O + Cl^-$).[20] Thus, the relative reactivity of oxy anions with $HCCl_3$ depends on their relative Brønsted basisicity ($pK_a$; $MeO^- > HO^- > HOO^- > O_2^{-\cdot}$) rather than their nucleophilicity.

The reactions of hexachlorobenzene ($PhCl_6$) with the oxy anions are unique. Thus, $HOO^-$ is five times as reactive as $O_2^{-\cdot}$ and 33 times as reactive as $MeO^-$ (Table IV). Superoxide ion reacts with $PhCl_6$ via an initial nucleophilic addition followed by displacement of $Cl^-$.[27,43] The other oxy anions probably follow the same pathway. In the case of $HOO^-$, the displacement of $Cl^-$ gives $PhCl_5OOH$, which deprotonated by a second $HOO^-$ (or $HO^-$) to give $PhCl_5OO^-$. The latter displaces an adjacent chlorine to give the *o*-quinone.

**Hydride Substitution**

The hydride ion ($H\!:^-$) is the penultimate nucleophile (the electron is a stronger nucleophile and reductant), and has the advantage that it transforms halocarbons to halide ions and the parent hydrocarbon, e.g.,

$$\text{n-BuCl} + H\!:^- \longrightarrow \text{n-BuH} + Cl^- \tag{12}$$

Although there are numerous approaches for hydro-dehalogenation in the organic literature[44] most are ineffective toward chlorinated and fluorinated aromatics. Here we describe a new and unique method for the hydrogenolysis of halogenated aromatics by sodium borohydride ($NaBH_4$) mediated by alkoxide and a nickel complex. Chlorobenzene ($C_6H_5Cl$) has been used as the model substrate for most experiments because it is readily available and has the largest aromatic carbon-chlorine bond energy; ($\Delta H_{DBE}$, 106 kcal mol$^{-1}$); as such it is the least reactive for this class of molecules.

Figure 4 illustrates the product yields for the reduction of $C_6H_5Cl$ to $C_6H_6$ by (a) the combination of $NaBH_4$, $MX_n$, and $MePEG^-$ in MePEGH 350 and (b) the combination of $NaBH_4$ and $MX_n$ in monoglyme. In the case of $Pd(OAc)_2$ in Figure 4a, 1% of biphenyl also is produced. Curves c and d of Figure 4 indicate the product yields for the reduction of $C_6H_5Cl$ to $C_6H_6$ by the combination of NaH, $MX_n$, and $CH_3(OCH_2CH_2)_2O^-$ in monoglyme (c) and the reduction of 1-bromonaphthalene ($C_{10}H_7Br$) to naphthalene by the combination of NaH, $MX_n$, and *tert*-amylate in monoglyme (d).[45] With the latter system $NiCl_2$ also yields 10% naphthalene dimer (R-R), and $PdCl_2$ yields 1% R-R.

The reduction of $C_6H_5Cl$ to $C_6H_6$ by the mixtures of systems a, b, and c of Figure 4 also have been evaluated with $CrCl_2$, $CrCl_3$, $MnCl_2$, $Mn(OAc)_2$, $Mn(OAc)_3$, $FeCl_2$, $FeCl_3$, $CoCl_2$, $Co(OAc)_2$, $Ni(OH)_2$, $PtCl_2$, CuCl, Cu(OAc),

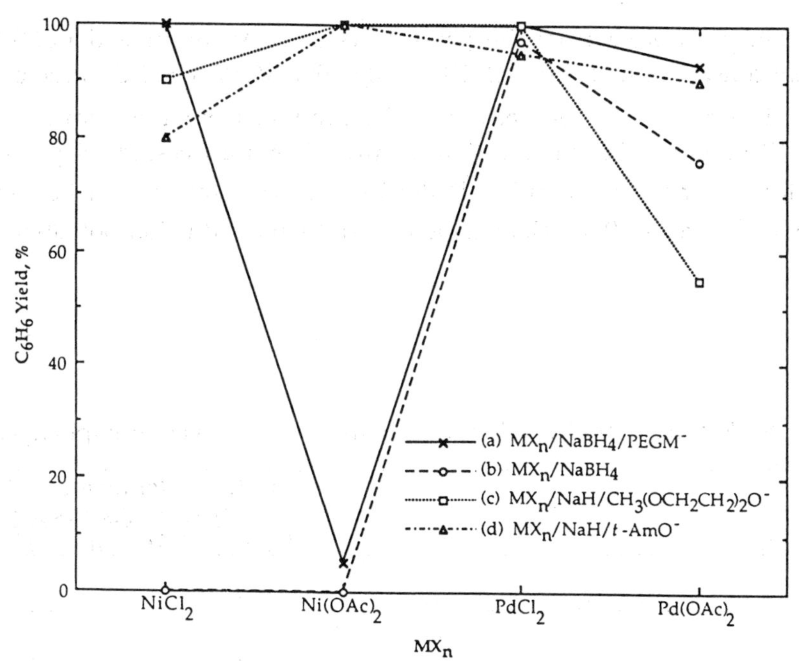


**Figure 4.** Yields of $C_6H_6$ with various transition-metal catalysts from the hydrogenolysis of $C_6H_5Cl$ by the combination of (a) —✕—, $NaBH_4/MX_n/PEGM^-$ in PEGM 350; (b) --o--, $NaBH_4/MX_n$ in monoglyme; and (c) .....□....., $NaH/MX_n/CH_3(OCH_2CH_2)_2O^-$ in monoglyme; and of naphthalene for the reduction of 1-bromonaphthalene by the combination of (d) ---Δ---, $NaH/MX_n/t$-$AmO^-$ in monoglyme[45]. In the case of (d), reactions were carried out under argon or nitrogen for various durations [$NiCl_2$, 17 h; $Ni(OAc)_2$, 5 min; $PdCl_2$ and $Pd(OAc)_2$, 30 min] at 63°C.[45]

$CuCl_2$, $Cu(OAc)_2$, $ZnCl_2$, and $Zn(OAc)_2$ as the catalyst; the yield of benzene for each of these complexes is less than 30%. Control experiments establish that all three components (transition-metal complex, $NaBH_4$, and alkoxide) are necessary for effective hydrogenation of $C_6H_5Cl$.

The product profiles (gas chromatograms) from the combination of Aroclor 1242 (in isooctane) with $NiCl_2$, $MePEG^-$, and $NaBH_4$ are presented in Figure 5. The reactions of this combination with other samples of PCBs are summarized in Table V.

The results of Figure 4 indicate that the mixture of transition-metal complex, $NaBH_4$, and alkoxide is effective for the hydrogenation of $C_6H_5Cl$. Substitution of $NaBH_4$ by NaH gives comparable reactivity (Figure 4c); however, NaH is pyrophoric, more expensive, and has fewer reducing equivalents than $NaBH_4$. A similar system, $MX_n/NaH/t$-amyl alcohol[45], for the hydrogenation of 1-bromonaphthalene (Figure 4d) exhibits similar reactivity to the $MX_n/NaH/CH_3(OCH_2CH_2)_2O^-$ system (Figure 4c), but is completely different from the present system. An optimized ReMED A&M system consists of MePEG⁻ MePEGH⁻ 350 with a MePEG⁻ : $NiCl_2$ ratio of 12 : 1.

The absence of biphenyl as a by-product from the hydrogenation of $C_6H_5Cl$ indicates that the reaction pathway is not a free-radical process. The counterion for nickel is critical [$Ni(OH)_2$ and $Ni(OAc)_2$ cannot be substituted for $NiCl_2$]. The MePEG⁻ alkoxide ion (a) shifts the reduction potential of nickel

**Table V.** Reactions of $NiCl_2$/MePEG⁻/$NaBH_4$ with Polychlorobiphenyls (PCBs).

| | % PCB's Removed on the Basis of Assays for | |
|---|---|---|
| Conditions | PCBs (±5%)[a] | PhPh (±3%)[a] |
| Aroclor 1242 in isooctane (500 μg/g)[b] | 90 | 72 |
| Aroclor 1242 (500 μg/g)[c] | 38 | 23 |
| Aroclor 1242 in Diala AX oil (500 μg/g )[b] | --[d] | 35 |
| Aroclor 1254 in untreated condensate (291 μg/g)[b] | --[d] | --[e] |
| Aroclor 1254 in pretreated condensate (291 μg/g)[b] | 51 | 36 |

[a]PCBs and PhPh were analyzed by GC/MS using SIM; samples were not pretreated except as noted. $d_{10}$-Biphenyl was used as an internal standard and surrogate for the quantitative assays of PCBs and PhPh.

[b]Reaction time, 16 h; $NiCl_2$ : MePEG⁻ : $NaBH_4$ = 0.1 : 0.2 : 0.8 mmol equiv.

[c]Reaction time, 3 h; $NiCl_2$ : MePEG⁻ : $NaBH_4$ = 0.1 : 1.2 : 0.4 mmol equiv.

[d]Matrix interferences precluded detection of PCBs.

[e]Matrix interferences precluded detection of PhPh.

to more negative values, (b) neutralizes the released HCl, and (c) acts as a ligand for nickel.

The results of Figure 5 and Table V indicate that this system reacts with PCB-contaminated wastes as well as $C_6H_5F$ and $C_6Cl_5OH$. The latter substrates usually are not subject to hydrogenation under mild conditions, which further demonstrates the uniqueness of the system and its ability to hydrogenate a wide range of halogenated aromatic substrates.

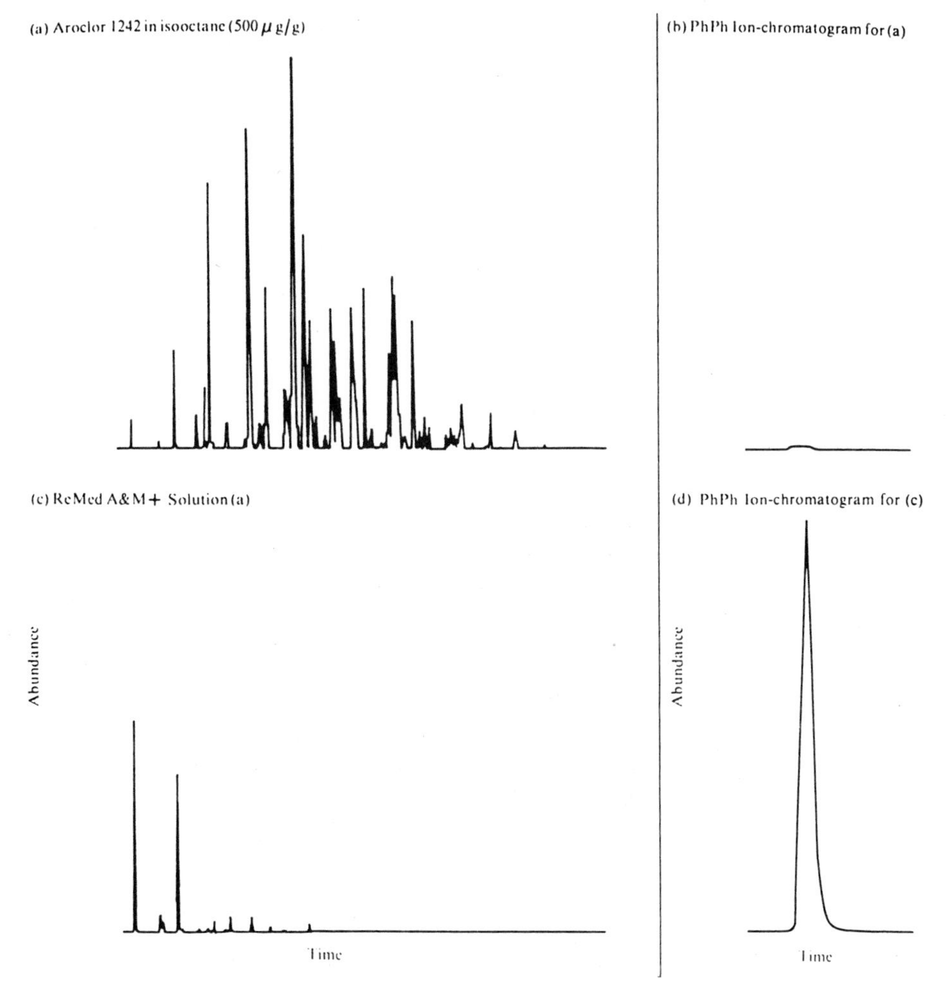


**Figure 5.** Gas chromatograms for (a) Aroclor 1242 in isooctane (500 μg/g); (b) extracted ion-chromatogram of PhPh (154.1 amu) for solution (a); (c) the product solution from the reaction of solution (a) with $NaBH_4/NiCl_2/PEGM^-$ in MePEGH350 after 16 h; and (d) extracted ion-chromatogram of PhPh for solution (c). See Table II for more experimental details.

## The Ultimate Nucleophile ≡ The Electron; Reductive Electrolysis

Although electrical energy is expensive, the stoichiometric introduction of a universal nucleophile (the electron) whose nucleophilicity is determined by the potential of the electrolysis cathode.

Figure 6 illustrates the electrolytic reduction of $C_6Cl_6$, 1,2,3,4,-$C_6H_2Cl_4$, $C_{12}Cl_{10}$, and Arochlor 1268 mixture in DMF at a glassy carbon electrode. Similar experiments for the other chloro derivatives of benzene and bi-phenyl confirm that each of the six peaks of Figure 6a is due to the successive reduction of the six chlorine atoms of $C_6Cl_6$. The decay and product profiles for

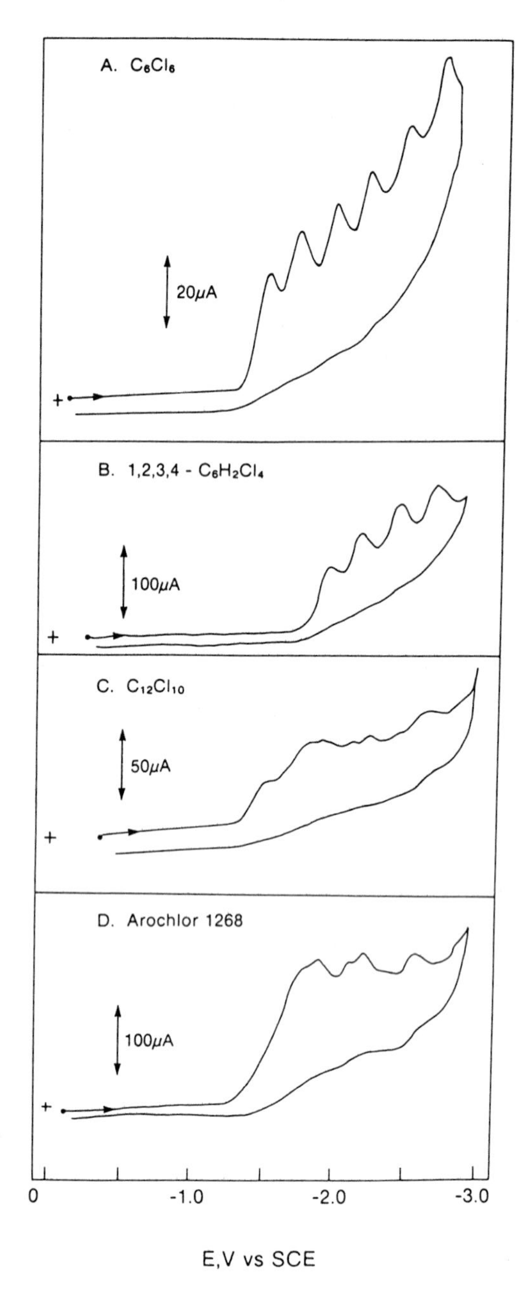


**Figure 6.** Cyclic voltammograms for chlorinated aromatic molecules in dimethylformamide (0.1M TEAP) at a glassy-carbon electrode (area, 0.062 $cm^2$): (A) 1.1 mM $C_6Cl_6$; (B) 2.3 mM 1,2,3,4-$C_6H_2Cl_4$; (C) 1.3 mM $C_{12}Cl_{10}$; and (D) 1 mg/mL Aroclor 1268. Scan rate, 0.1 V $s^{-1}$.

the electrolytic reduction of $C_6Cl_6$ are illustrated by Figure 7 and indicate that reduction produces the hydro-derivative of the substrate molecule (e.g., $C_6HCl_5$ from $C_6Cl_6$). If the electrolysis of Figure 7 is controlled at -2.8 V vs. SCE instead of -2.5 V, the final product is benzene rather than chlorobenzene (PhCl).

Figure 8 illustrates the change in the composition of the Arochlor 1268 PCB mixture (average chlorine content, 68% by weight) during the course of its controlled potential (-2.5 V vs. SCE) reduction in dimethyl formamide. The most heavily chlorinated derivatives of biphenyl are reduced first to the lighter members of the series; all are ultimately reduced to biphenyl. This experiment and that of Figure 7 demonstrate that halogenated-aromatic-hydrocarbon wastes (PCBs, PBB's, and halobenzenes) can be completely dehalogenated to their parent hydrocarbons by electrolytic reduction in dimethylformamide. Such a system provides an efficient means for the safe transformation of these toxic materials into useful products.

In the absence of dioxygen, halogenated aromatic hydrocarbons can be converted to their parent hydrocarbon by electrolytic reductions at a graphite or gold cathode with $Me_4NCl$ as the supporting electrolyte. The data from Figures 6-8 indicate that the overall electrolysis reactions for $C_6Cl_6$ and $C_{12}Cl_{10}$ yield $C_6H_5Cl$ and $C_{12}H_{10}$, respectively.

$$C_6Cl_6 + 5\,H_2O + 10\,e^- \xrightarrow[\text{C, Au}]{\text{-2.5 V vs SCE}} C_6H_5Cl + 5\,Cl^- + 5\,O^- \qquad (13)$$

$$C_{12}Cl_{10} + 10\,H_2O + 20\,e^- \xrightarrow{\text{-2.5 V}} C_{12}H_{10} + 10\,Cl^- + 10\,HO^- \qquad (14)$$

The anodic process involves the oxidation of the chloride ion that is produced in the cathode compartment

$$Cl^- + HO^- \longrightarrow HOCl + 2\,e^- \qquad (15)$$

Again the overall process is equivalent to an electrostimulated oxidation of water (eq 10 plus eq 11)

$$C_{12}Cl_{10} + 10\,H_2O \longrightarrow C_{12}H_{10} + 10\,HOCl \qquad (16)$$

with the anionic products produced at the cathode transfered via diffusion to the anode and oxidized to HOCl.

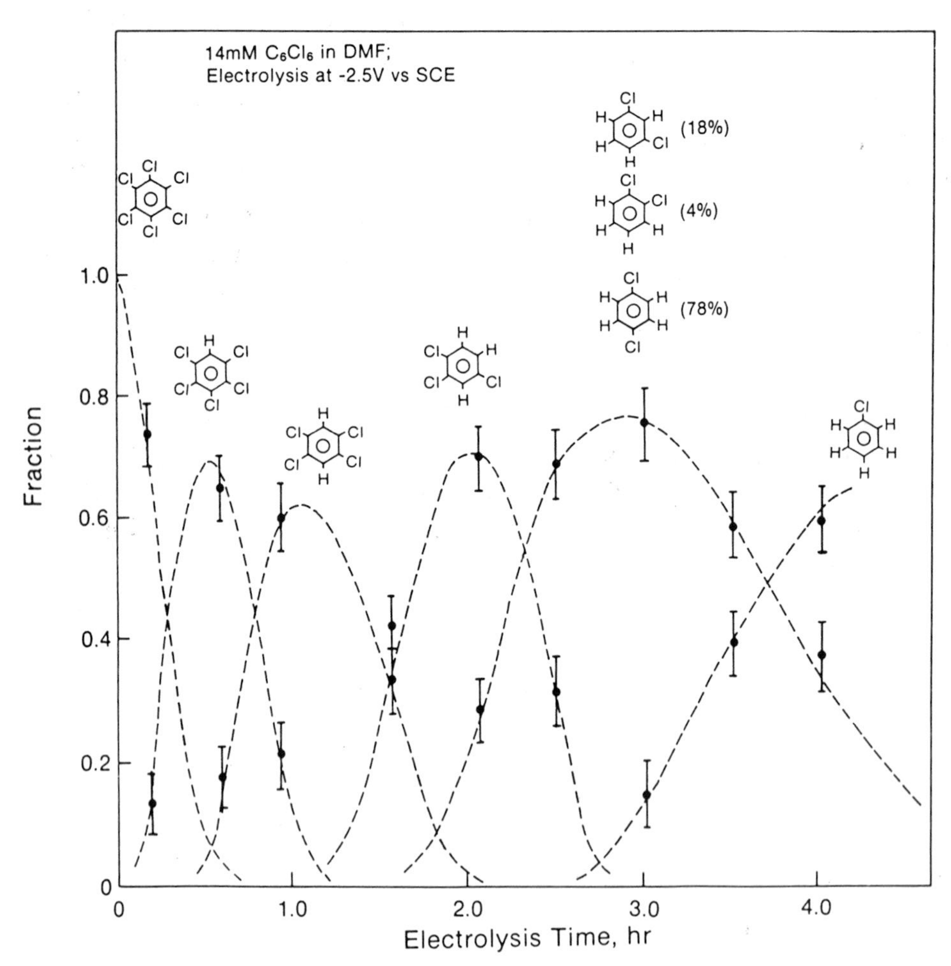


**Figure 7.** Concentration profiles for the controlled-potential (-2.5 V vs. SCE) electrolysis of 14 mM $C_6Cl_6$ in DMF (0.1M TEAP) at a gold-mesh electrode (area, ~20 $cm^2$). The relative amounts of products and reactants were determined by capillary gas chromatographic analysis of samples from the electrolysis solution.

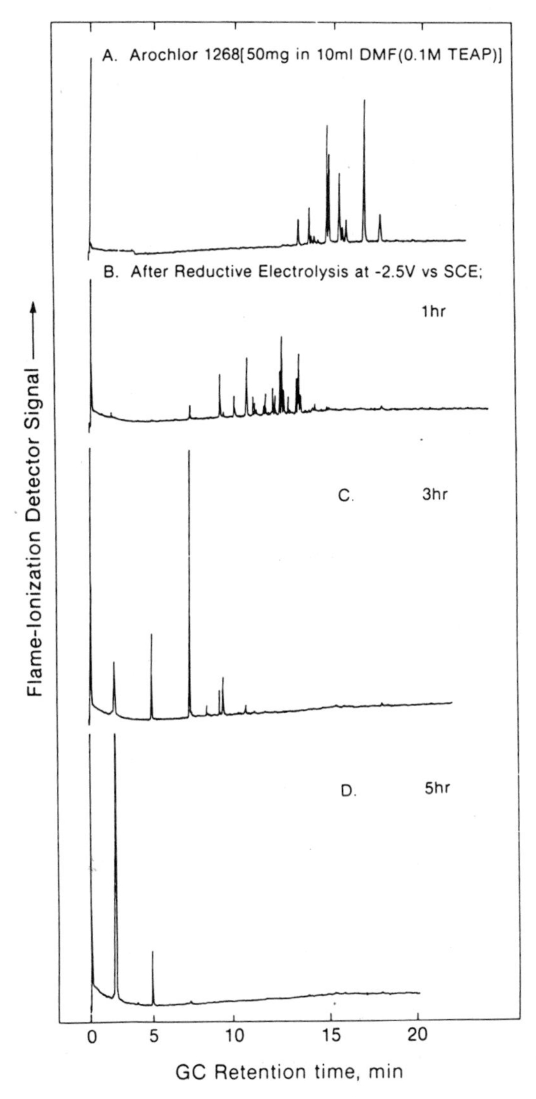


**Figure 8.** Gas chromatograms for Aroclor 1268 in DMF, and for the product solution during the course of its electrolytic reduction at -2.5 V vs. SCE with a 20-$cm^2$ gold-mesh electrode. The peak with a retention time of 2 min is biphenyl, and that at 5 minutes is chlorobiphenyl.

## REDUCTIVE HYDROGENATION BY HYDRAZINE

Conventional hydrogenation with $H_2$ under mild conditions is not effective for chlorinated aromatics.[44] However, anhydrous hydrazine ($H_2NNH_2$) and substituted hydrazines are effective for the dechlorination of polychloro aromatics. Table VI summarizes the reaction efficiencies (expressed as the number of chloride ions released per $C_6Cl_6$ molecule) and the products that result from of hydrazines and hexachlorobenzene (HCB, $C_6Cl_6$, $PhCl_6$).

The results of Table VI indicate that hydrazine is an effective agent for the transformation of $C_6Cl_6$ to $C_6H_6$. The efficiency (expressed as the number of chloride ions released per $C_6Cl_6$) is less than the theoretical 100% (six $Cl^-$) due to the evaporative loss of the lower-chlorinated benzenes. The reaction rate appears to be first-order in $H_2NNH_2$ concentration and first-order in substrate concentration. The relative reaction rates for the various solvents are in the order $Me_2SO$ ~ HPEGMe 350 > pyridine ~ DMF, which is consistent with their relative polarity. Thus, the reaction intermediates are stabilized by the more polar medium.

The reaction products from the combination of $H_2NNH_2$ (or $CH_3NHNH_2$) with $C_6Cl_6$ (Table VI) are consistent with an initial elimination of HCl,

$$C_6Cl_6 + H_2NNH_2 \longrightarrow Cl_5C_6NHNH_2 + HCl \tag{17}$$

with subsequent para- or ortho-elimination of HCl to give the different isomers.

$$Cl_5C_6NHNH_2 \longrightarrow 1,2,3,4\text{- and } 1,2,4,5\text{-}C_6H_2Cl_4 + HCl + N_2 \tag{18}$$

Additional cycles of these steps results in complete reduction via the overall reaction.

$$C_6Cl_6 + 3\,H_2NNH_2 \longrightarrow C_6H_6 + 3\,N_2 + 6\,HCl \tag{19}$$

The proposed mechanism is consistent with that suggested for the reaction of $C_6Br_6$ with hydrazine hydrate.[46] The latter was unreactive with $C_6Cl_6$ in boiling ethanol.[46] The N-H bond dissociation energy [~76 kcal $mol^{-1}$][47] is the lowest of the hydrazine derivatives, and probably is the reason for its superior reactivity. The lack of reactivity for $PhNHNH_2$ and PhNHNHPh is in accord with this proposition.

The present results provide a basis for a reactor to process polyhalogenated aromatic wastes, whereby anhydrous hydrazine transforms

**Table VI.** Reactions of Hydrazine and Substituted Hydrazines with $C_6Cl_6$.

| Conditions[a] | $Cl^-$ Released Per Substrate (±0.5)[b] | Major Products | Minor Products |
|---|---|---|---|
| *A. Hydrazine* ($H_2NNH_2$)[c] | | | |
| 100% $H_2NNH_2$, 25°C | 0.0 | – | – |
| $Me_2SO$, 4% $H_2NNH_2$, 25°C Argon atmosphere | 0.7 | 1,2,3,4-$C_6H_2Cl_4$, $Cl_5C_6NHNH_2$ | 1,2,3,5- and/or 1,2,4,5-$C_6H_2Cl_4$[d] |
| 100% $H_2NNH_2$, 70°C | 4.9 | $C_6H_6$, $C_6H_5Cl$, 1,4-$C_6H_4Cl_2$ | 1,2-$C_6H_4Cl_2$ |
| $Me_2SO$, 20% $H_2NNH_2$, 70°C | 3.7 | 1,4-$C_6H_4Cl_2$ | $C_6H_6$, $C_6H_5Cl$, 1,2-$C_6H_4Cl_2$, 1,2,3-$C_6H_3Cl_3$, 1,2,4-$C_6H_3Cl_3$ |
| $Me_2SO$, 2% $H_2NNH_2$, 70°C | 4.1 | 1,4-$C_6H_4Cl_2$ | 1,2-$C_6H_4Cl_2$, 1,2,4-$C_6H_3Cl_3$, 1,2,3,4-$C_6H_2Cl_4$ |
| HPEGMe 350, 20% $H_2NNH_2$, 70°C | 3.0 | 1,2,4-$C_6H_3Cl_3$, 1,2,3,4,-$C_6H_2Cl_4$ | 1,2,3-$C_6H_3Cl_3$, 1,2,3,5- and/or 1,2,4,5-$C_6H_2Cl_4$[d] |
| HPEGMe 350, 20% $H_2NNH_2$, 150°C | 4.1 | 1,4-$C_6H_4Cl_2$ | 1,2-$C_6H_4Cl_2$, 1,2,4-$C_6H_3Cl_3$ |
| DMF, 20% $H_2NNH_2$, 70°C | 2.1 | 1,2,3,4-$C_6H_2Cl_4$ | 1,2,4-$C_6H_3Cl_3$, 1,2,3,5- and/or 1,2,4,5-$C_6H_2Cl_4$[d] |
| Pyridine, 2% $H_2NNH_2$, 70°C | 2.5 | 1,2,3,4-$C_6H_2Cl_4$ | 1,2,4-$C_6H_3Cl_3$, 1,2,3,5- and/or 1,2,4,5-$C_6H_2Cl_4$[d] |
| *B. Substituted Hydrazines* | | | |
| 100% $CH_3NHNH_2$, 70°C | 2.9 | 1,2,4-$C_6H_3Cl_3$, $Cl_5C_6NCH_3NH_2$ | 1,2,3,4-$C_6H_2Cl_4$, $Cl_4C_6(NCH_3NH_2)_2$ |
| 100% $PhNHNH_2$, 70°C | 0.0 | – | --- |
| $Me_2SO$, 0.6 *M* PhNHNHPh, 70°C | 0.0 | – | --- |

[a]Unless noted otherwise, reactions were carried out by adding $C_6Cl_6$ (3-6 mM in toluene) to 5-10 mL of a hydrazine/solvent mixture. The reaction cell was heated at *ca.* 70°C for 18-24 h, with the cap slightly loosened to prevent pressure buildup.

[b]The reactions were quenched with dilute acid and extracted with toluene. The aqueous layer was analyzed for chloride ion by potentiometric titration with $AgNO_3$ at pH 0.4. Because $H_2NNH_2$ and its derivatives are interferences in the $AgNO_3$ determination of $Cl^-$, they were removed by making the solution alkaline (pH 14) in the presence of $O_2$ (0.2 atm). The equivalence point was determined by the second-derivative method. The organic layer was analyzed by GC/FID and/or GC/MS; evaporative losses during reaction precluded accurate assays of the products.

[c]The reaction of $H_2NNH_2$ with either a condensate that contained Aroclor 1254 or a stock solution of Aroclor 1254 gave erratic analytical results; probably because of inhomogeneities in the condensate or significant solvent loss during the heated reaction cycle.

[d]The GC equipment and parameters used for the analyses could not resolve 1,2,3,5- and 1,2,4,5-$C_6H_2Cl_4$.

toxic organics into $N_2$, $Cl^-$ (under alkaline conditions), and the recyclable parent hydrocarbon.

## ATOMIC METALS VIA ELECTROGENERATION

Research has been initiated in our laboratories to develop an electrocatalytic system that will reduce alkali salts (NaCl, LiCl, or KCl) to their atomic state (Na·, Li·, or K·) within a solution matrix that contains hazardous halogenated hydrocarbons. Although alkali metals react with water, their reaction rates are several orders of magnitude slower than with halocarbons. By an in-situ formation of metal atoms at a carbon-surface electrode, a selective and efficient dehalogenation process should be achieved, e.g.

$$2\,Na^+ + 2\,e^- \longrightarrow 2\,Na\cdot \xrightarrow{RX,\ H_2O} RH + NaX + NaOH \qquad (20)$$

We are evaluating this approach with (a) n-BuCl, $CCl_4$, $HCCl_3$, PhCl, and $C_6Cl_6$ as model substrates, (b) $Me_2SO_2$, DMF, MeCN, THF, and pyridine as solvents, (c) NaCl, LiCl, KCl, $NiCl_2$, and $MnCl_2$, as intermediates for the production of metal atoms, and (d) cyclic voltammetry and controlled-potential electrolysis at glassy carbon electrodes as the means to generate atoms and anions. The effect of various levels of water on the efficiency of dehalogenation is being tested. The advantages of this approach are that (a) the parent hydrocarbon and halide ions are the products (rather than free halogens and $CO_2$ from a oxidative process) and (b) a reductive process does not produce hazardous intermediates.

Although alkali-metal salts are conveniently reduced at platinum and gold electrodes (reduction potentials, ~-2.0 V vs SCE), their metal atoms form metal-metal bonds with Pt and Au, which suppresses their reactivity with halocarbons. However, at carbon the initial reduction deposits metal atoms (~-2.6 V vs SCE), whose reactivity with $C_6Cl_6$ and PCBs is much greater than with residual water because of their radical character (Na·, Li·, and K·). Because PhCl has the strongest C–Cl bond within the unsaturated-halocarbon family, it has been used as a model substrate to optimize the conditions and to test the basic proposition: That in-situ generated alkali-metal atoms will react preferentially and efficiently with halocarbons (PhCl) relative to residual water.

In acetonitrile at glassy carbon (a) chlorobenzene (PhCl) is reduced to a phenyl radical,

$$PhCl + e^- \xrightarrow[MeCN]{C} Ph\cdot + Cl^- \quad E_{red}, -2.8 \text{ V vs SCE} \qquad (21)$$

(b) water to hydrogen atoms,

$$H_2O + e^- \xrightarrow[\text{MeCN}]{\text{C}} H\cdot + HO^- \quad E_{red}, -4.1 \text{ V vs SCE} \tag{22}$$

and (c) sodium chloride to sodium atoms,

$$NaCl + e^- \xrightarrow[\text{MeCN}]{\text{C}} Na\cdot + Cl^- \quad E_{red}, -2.6 \text{ V vs SCE} \tag{23}$$

$$Na\cdot \xrightarrow{H_2O} NaOH + H\cdot$$

which react with residual water. However, with the combination of PhCl, NaCl, and $H_2O$ the initial reduction is at -2.3 V vs SCE,

$$PhCl + NaCl + e^- \xrightarrow[\text{MeCN}]{\text{C}} [Na\text{–}Ph\dot{C}l] \xrightarrow{NaCl, H_2O, e^-} PhH + NaOH \tag{24}$$

and is specific for PhCl.

Scheme IV outlines the reaction paths that are associated with the

**Scheme IV.** Reaction of Electrogenerated Atomic Sodium with PhCl.

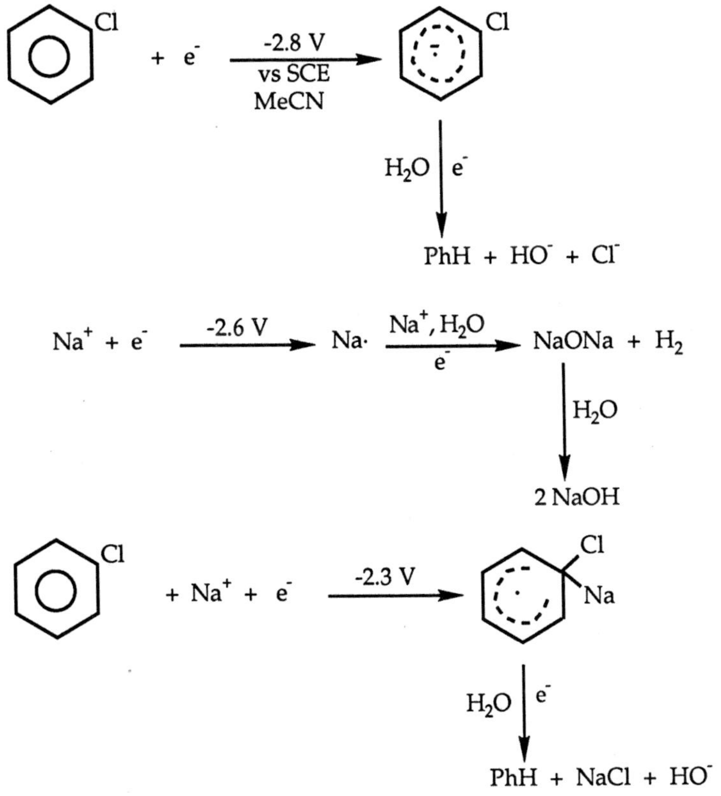

$$PhCl + e^- \xrightarrow[\text{vs SCE, MeCN}]{-2.8 \text{ V}} [PhCl]^{\cdot -} \xrightarrow{H_2O,\ e^-} PhH + HO^- + Cl^-$$

$$Na^+ + e^- \xrightarrow{-2.6 \text{ V}} Na\cdot \xrightarrow[e^-]{Na^+, H_2O} NaONa + H_2 \xrightarrow{H_2O} 2\,NaOH$$

$$PhCl + Na^+ + e^- \xrightarrow{-2.3 \text{ V}} [Ph(Cl)Na]^{\cdot} \xrightarrow{H_2O,\ e^-} PhH + NaCl + HO^-$$

process. A clear understanding of the relative bond energies associated with the substrates, reactants, intermediates, and products is necessary in order to optimize an efficient and selective dehalogenation. Table VII summarizes the dissociative bond energies ($\Delta H_{DBE}$) for a number of relevant diatomic molecules, and is the basis of our current research efforts.

Several reductive approaches for the transformation of carbon-halogen bonds to halide ions have been outlined. No one system is ideal for all problems, and the environment of the hazardous waste often interferes with

**Table VII.** Dissociative Bond Energies for M–X Molecules.

| | $\Delta H_{DBE}$, kcal mol$^{-1}$ | | | | | |
|---|---|---|---|---|---|---|
| Metal(EA)[a] | M | Cl(3.62) | H(0.75) | O(1.46) | S(2.08) | N(0.0) |
| Li(0.62) | 26 | 112 | 57 | 80 | | |
| Na(0.55) | 17 | 99 | 44 | 61 | 75 | |
| K(0.50) | 14 | 104 | 42 | 66 | | |
| Cu(1.24) | 42 | 92 | 66 | 64 | 66 | |
| Ni(1.16) | 49 | 89 | 60 | 91 | 82 | |
| Co(0.61) | 40 | 93 | 54 | 92 | 79 | |
| Fe(0.15) | 24 | 84 | 43 | 93 | 77 | |
| Mn(0.0) | 6 | 86 | 56 | 96 | 72 | |
| Cr(0.67) | 37 | 88 | 67 | 103 | 79 | 90 |
| V(0.52) | 58 | 114 | | 150 | 115 | 114 |
| Mo(0.75) | 97 | | | 134 | | |
| Ru(1.05) | | | 56 | 126 | | |
| C(1.26) | 145 | 95 | 81 | 257 | 171 | 184 |
| H(0.75) | 104 | 103 | 104 | 102 | 82 | 81 |

[a]EA, electron affinity in electron-volts (eV).

effective remediation. Hence, effective treatment will require careful analysis and intelligent design of the treatment strategy for each problem. For most contamined soils, sludges, muds and water, some separation process will be required prior to the chemical destruction of carbon-halogen bonds.

**Acknowledgement:** This work was supported by the National Science Foundation under Grant CHE-8516247 and CHE-9106742, the Welch Foundation under Grant A-1042 with a Robert A. Welch Graduate Fellowship (S.J.), and a Grant-in-Aid from Captiva Capital, Ltd., Colorado.

## References

1. J. Waid. "PCBs and The Environment," CRC Press, Boca Raton, (1986), Col. I-III.
2. F. M. D'Itri and M. A. Kamrin, "PCBs: Human Environmental Hazards," Ann Arbor Science Book, Boston, (1983).
3. O. Hutzinger, S. Safe and V. Zitko, "The Chemistry of PCBs," CRC Press: Cleveland, (1974).
4. S. Safe, S. Bandiera, T. Sawyer, L. Robertson, L. Safe, R. Parkinson, P. E. Thomas, D. E. Ryan, L. M. Reik, W. Levin, M. A. Denomme and T. Fujita, "EHP, Environmental Health Perspect," 60, 47-56(1985).
5. Safe, S. *CRC Crit. Rev. Toxicol, 13,* 319-396(1984).
6. R. D. Kimbrough. *CRC Crit. Rev. Toxicol* 2, 445-498(1974).
7. L. Fishbein. *Annu. Rev. Pharmacol. 14,* 139-156(1974).
8. E. E. McConnell and J. A. Moore, *Ann N.Y. Acad. Sci, 320,* 138-150(1984).
9. R. S. Takazawa and H. W. Strobel *Biochemistry, 25,* 4804-4809(1986).
10. J. R. P. Cabral, P. Shubik, T. Mollner and F. Raitano, *Nature (London), 269,* 510-511(1977).
11. F. W. Karasek and L. C. Dickson, *Science(Washington, D.C.), 237,* 754-756(1987).
12. D. T. Sawyer and J. L. Roberts, Jr., *Acc. Chem. Res., 21,* 469-476(1988).
13. L. L. Pytlewski, F. J. Iaconianni, K. Kevitz, and A. B. Smith, U.S. Patent 4,417,977, Nov. 29, 1983.
14. D. J. Brunelle, A. K. Mendiratta, and D. A. Singleton, *Environ. Sci. Technology. 19,* 740-746(1985).
15. R. Dietz, A. E. J. Forno, B. E. Larcombe and M. D. Peover, *J. Am. Chem. Soc. 102,* (1970).
16. M. V. Merritt and D. T. Sawyer *J. Org. Chem. 35,* 2157(1970).
17. F. Magno, R. Seeber and S. Valcher, *J. Electroanal. Chem. 83,* 131(1977).
18. J. San Fillipo, Jr., C.-I. Chern and J. S. Valentine, *J. Org. Chem. 40,* 1678(1975).
19. R. A. Johnson and E. G. Nidy, *J. Org. Chem. 40,* 1680(1975).
20. J. L. Roberts, Jr., T. S. Calderwood and D. T. Sawyer, *J. Am. Chem. Soc. 105,* 7691(1983); D. T. Sawyer and J. L. Roberts, Jr., U.S. Patent 4,410,402, Oct. 18, 1983.
21. T. S. Calderwood, R. C. Newman, Jr. and D. T. Sawyer, *J. Am. Chem. Soc. 105,* 2337(1983).
22. T. S. Calderwood and D. T. Sawyer, *J. Am. Chem. Soc. 106,* 7185(1984).
23. M. V. Merritt and R. A. Johnson, *J. Am. Chem. Soc., 99,* 3713(1977).

24. L. Eberson, *Adv. Phys. Org. Chem. 18*, 79,(1982).

25. C. L. Perrin, *J. Phys. Chem. 88*, 3611(1984).

26. J. K. Kochi, In *Free Radicals* (J. K. Kochi, ed.). New York: Wiley, (1973).

27. H. Sugimoto, S. Matsumoto and D. T. Sawyer *J. Am. Chem. Soc., 109*, 8081(1987).

28. D. T. Sawyer, T. S. Calderwood, K. Yamaguchi and C. T. Angelis, *Inorg., Chem, 22*, 2577-2583(1983).

29. K. Yamaguchi, T. S. Calderwood and D. T. Sawyer, *Inorg. Chem, 25*, 12- 1290(1986).

30. S. Jeon and D. T. Sawyer, *Inorg. Chem., 29*, 4612-4615(1990); D. T. Sawyer, S. Jeon and P. K. S. Tsang, U. S. Patent Pending (to be issued, June, 1992).

31. R. G. Pearson, *J. Am. Chem. Soc., 108*, 6109(1986).

32. A. J. Bard, R. Parsons and J. Jordon, "Standard Potentials in Aqueous Solution, Marcel Dekker, New York(1985).

33. W. C. Barrette, Jr., H. W. Johnson, Jr., and D. T. Sawyer, *Anal. Chem., 56*, 1890(1984).

34. P. K. S. Tsang, P. Cofré and D. T. Sawyer, *Inorg. Chem., 26*, 3604(1987).

35. C. D. Ritchie. *Acc. Chem. Res. 5*, 348(1972).

36. S. S. Shaik and A. Pross, *J. Am. Chem. Soc., 84*, 16, 1962.

37. J. O. Edwards and R. G. Pearson, *J. Am. Chem. Soc., 84*, 16(1962).

38. D. T. Sawyer and J. S. Valentine, *Acc. Chem. Res., 14*, 393(1981).

39. P. Cofré and D. T. Sawyer, *Inorg. Chem., 25*, 2089(1986).

40. J. L. Roberts, Jr. and D. T. Sawyer, *J. Am. Chem. Soc., 103*, 712(1981).

41. J. E. McIssac, Jr., L. R. Subbaraman, J. Subbaraman, H. A. Hulhausen and E. J. Behrman, *J. Org. Chem., 37*, 1037(1972).

42. R. G. Pearson and D. N. Edgington, *J. Am. Chem. Soc. 84*, 4607(1962).

43. H. Sugimoto, S. Matsumoto and D. T. Sawyer, *Environ. Sci. Technol. 22*, 1182(1988).

44. J. March, "Advanced Organic Chemistry," 3rd. ed., Wiley & Sons: New York, 1985; and references therein.

45. J. J. Brunet, R. Vanderesse and P. Caubere, *J. Organometal. Chem. 157*, 125-133(1978).

46. I. Collins, S. M. Roberts and H. Suschitzky, *J. Chem. Soc. (C)*, 167-174(1971).

47. E. W. Schmidt, "Hydrazine and Its Derivatives," Wiley & Sons: New York, 1984; pp 247-251.

# PROCESS TECHNOLOGY FOR HAZARDOUS WASTE REMEDIATION - (KPEG$^{sm}$ AND REDUCTIVE TECHNOLOGIES)

Robert Hoch
SDTX Technologies, Inc.

## ABSTRACT

A number of technologies have been proposed and promoted for remediation of media contaminated with chlorinated hydrocarbons. While some of these approaches are flawed on first principles, potentially leading, for instance, to production of more hazardous substances, others lack a realistic engineering assessment of what an overall system embodying the chemistry would comprise.

Unlike untried technologies, alkaline glycolate dechlorination, commonly called KPEG$^{sm}$, has a long and successful history, having been extensively used in the destruction of PCB contaminated transformer oil. This chemistry has now been extended to solids and has been applied to the remediation of over 30,000 tons of PCB contaminated soil at the Wide Beach, NY Superfund site.

This paper will briefly review the history of KPEG$^{sm}$, and present experimental data on the complete dechlorination of chlorobenzene and trichlorobenzene as model compounds. The importance of complete dechlorination to project economics will be discussed and the significance of the Wide Beach Project to technology selection for PCB contaminated soils will be assessed.

The total cost of the national remediation effort is sufficiently large that economics, until now quite a minor factor in technology selection, must eventually become a key driving force. Based on this assumption and first principle arguments about efficacy and with non-quantifiable, but quite obvious public prejudices in mind, one can define the design basis of a cost-effective remediation process, where the term process is used in the usual context of the chemical industry. It is shown that the cost of processing in such a system is relatively bounded. Furthermore, the regime of applicability of KPEG$^{sm}$ can be identified as can generic critical features of its integration with other technologies, for remediation of more complex wastes.

## OVERVIEW

Chlorinated hydrocarbons present some of the most vexing problems of hazardous waste contamination. Their environmental persistence coupled with their general and wide

*Industrial Environmental Chemistry*, Edited by D.T. Sawyer
and A.E. Martell, Plenum Press, New York, 1992

spread use persisting well into the 1970's leaves us today with many chlorinated hydrocarbon contaminated media; soil, water, river sediments and various industrial wastes.

The spectrum of technologies that have been proposed for remediation of these situations is staggering. In a rational, economics-driven business, this would suggest that the appropriate combination of remedy and problem is quite specific. In the remediation industry, however, it rather represents a significant opportunity attracting technology regardless of relevancy.

This situation is a microcosm of the hazardous waste marketplace where irrationalities thrive. Among the prominent reasons for this are:

- extensive regulation and attendant red tape
- the relative newness of the industry
- the unwillingness to acknowledge a fundamental base in chemistry
- the tendency to leap from concept to hardware without the intermediacy of a process engineer

Given that the industry lacks a rational basis for technology selection, the following criteria are proposed:

- acceptable economics
- technical acceptability
- social acceptability.

It is the thesis of this paper that many of the technologies suggested for remediation of PCBs fail to meet one or more of the above criteria.

While EPA regulations and procedures, cumbersome as they may be, address the issues of technical and social acceptability, much entropy has been generated by the relegation of economics to the bottom of, for instance, the list of criteria for the selection of superfund remedies and by persistent government support for technology development without *a priori* indication of economic superiority to existing technology for any contaminant in any media. Much more could be written about the unnecessary degree of confusion surrounding the choice of remediation technology.

This paper, however, will focus specifically on issues related to chlorinated hydrocarbons and most especially PCBs and will show that many categories of technologies can be ruled out on first principles. In fact, for sound chemical reasons, reductive systems of destruction are preferred to oxidative systems. The term dechlorination, in the remediation art, is usually applied to these reductive systems. Among such reductive systems, the reaction with sodium or potassium polyethylene glycolate, frequently called NaPEG, KPEG$^{sm}$ or APEG, has a long and successful history and clearly represents the most fully demonstrated such chemistry.

## EARLY HISTORY OF REDUCTIVE DECHLORINATION SYSTEMS

In 1978, Dr. Louis Pytlewski of Drexel University, working at the Franklin Institute Research Center, discovered that aromatic chlorinated compounds like PCBs could be completely dechlorinated by reaction with alkaline glycolates (Pytlewski, et al, 1982). Before this invention, the only practical chemical reaction for destruction of PCBs was the relatively uneconomic reaction with dispersed metallic sodium originally developed by Goodyear (Berry, 1981).

At the time of Dr. Pytlewski's invention, government regulations, responding to the availability of technology, the ease of treatment and the pervasiveness of the problem identified transformer oil as the initial remediation target. The alkaline glycolate treatment (called NaPeg by the Franklin Institute) lent itself readily to this application. Work in this area by Franklin, including pilot studies with electric utilities, led to a number of broad patent filings (Pytlewski, etal 1983 a,b, 1984 a,b,c, 1986). General Electric (Brunelle 1982,1983) and others, eventually fully proved the viability of the glycolate chemistry by successful commercial application in transformer oil. The focus of commercial interest in the United States has now shifted away from relatively well defined systems like transformer oil to the remediation of PCB and dioxin contaminated soils, sludges and sediments.

## UNDERLYING CHEMISTRY

It is generally accepted that the initial reaction (between, for example, potassium polyethylene glycolate and a PCB) is a nucleophilic substitution to produce an aromatic ether, as shown in Reaction **1** of Scheme One.

When this reaction is used for the remediation of transformer oil, further reaction is unnecessary (and may, in fact, be undesirable) because the ether is insoluble in the dielectric fluid and thus easily separated. Conventional practice is to landfill or incinerate the recovered ether.

Although numerous studies commissioned by the EPA and others show various of these polychloromonoethers to be nontoxic, biodegradable and not to bioaccumulate (DeMarini and Simmons, 1989), there remains resistance to leaving chlorinated organics in the soil. The nonhazardous nature of this material is clearly important in soil systems, since recovery of this reaction product from the contaminated medium is impractical. It becomes difficult to argue for on-site re-interment of treated soil containing chlorinated organics.

Early work at the Franklin Institute showed that PCBs can be completely dechlorinated via the alkaline glycolate process. Reaction **2** of Scheme One shows a likely path for complete dechlorination. No analytical protocol now exists for analysis of the reaction products, but complete recovery of the chlorides in inorganic form has been demonstrated (King and Hoch, 1991).

Complete dechlorination is the key to producing a decontaminated solid phase not requiring treatment as a hazardous waste. The implied structure of the products of reaction **2** strongly supports such arguments. Not only is polyethylene glycol (PEG) non-toxic, but, in the biotechnology industry, PEG adducts similar in structure to the products of Reaction **2** are intentionally manufactured to detoxify otherwise dangerous medicinals. (Davis, etal.,1979) Thus, a substantial body of data, including human clinical trials, have shown that reaction with PEG can detoxify otherwise toxic compounds.

The sensitivity of these reactions to moisture level is an issue in extension of this chemistry to soil systems. Scheme One shows chemically why the presence of water is incompatible with alkaline glycolate dechlorination chemistry. In the presence of water, reactions **1** and **2** must compete with the quite facile hydrolysis of the glycolate as shown in reaction **3**. Thus, in order to guarantee the desired reaction path, soils, sludges and sediments should be dried before reaction.

Both theory and experimentation have shown that the hydroxy-chloro-PCBs formed by reaction **4** are less reactive than the ether products of reaction **1** to further dechlorination by either KPEG$^{sm}$ or KOH. (Doubly hydroxy substituted products would theoretically be essentially completely resistant to further dechlorination.) Any significant opportunity for reaction path **4** to occur therefore will prevent complete dechlorination.

Perhaps even more troublesome, these hydroxy species are on the reaction path that converts PCBs to dioxins and polychlorodibenzofurans (PCDFs). Reaction path **6** shows the formation of PCDF by simple ring closure. Production of dioxins, is shown formally

Scheme I.

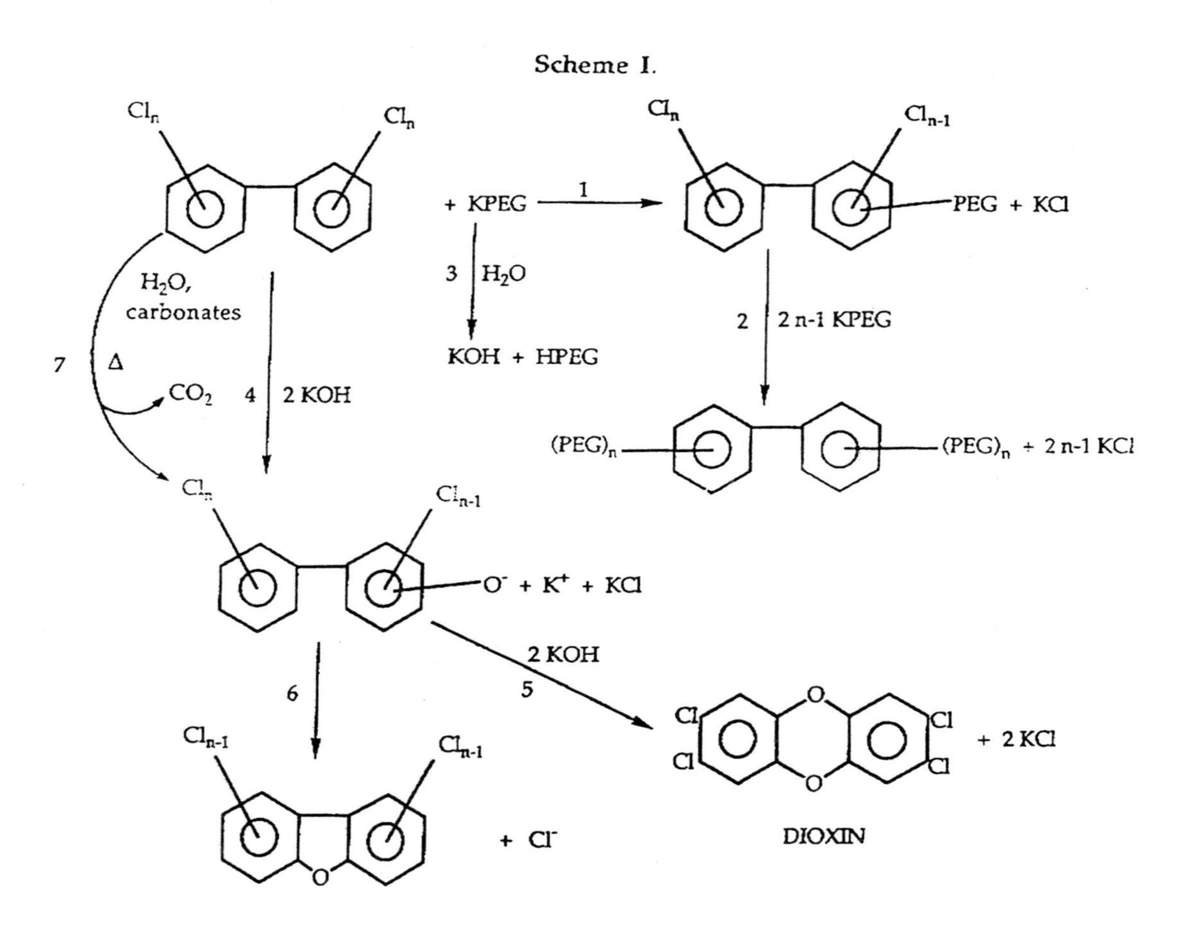

$Cl_n$
$Cl_n$
+ KPEG
1
$Cl_n$
$Cl_{n-1}$
PEG + KCl
3
$H_2O$
KOH + HPEG
2
2 n-1 KPEG
$(PEG)_n$
$(PEG)_n$ + 2 n-1 KCl
$H_2O$,
carbonates
7
Δ
$CO_2$
4
2 KOH
$Cl_n$
$Cl_{n-1}$
$O^-$ + $K^+$ + KCl
2 KOH
5
6
$Cl_{n-1}$
$Cl_{n-1}$
+ $Cl^-$
Cl
Cl
Cl
Cl
O
O
+ 2 KCl
DIOXIN

by reaction **5**. The generation of PCDFs and dioxins will significantly complicate any remediation.

Another chemical issue relates to the potential *in situ* formation of the glycolate reagent. Although frequently practiced, this apparent economy also leads to loss of certainty as to reaction path, because at practical reaction (or desorption) temperatures, KOH (or NaOH) will react with a PCB as shown by reaction **4**. This reaction is much less rapid than reaction **1**. Furthermore, the products of reaction **4** are particularly undesirable.

Note that alkaline glycolates react as readily with dioxins and PCDFs as with PCBs. Reactions **5** and **6** become of concern only in the absence of glycolate. As long as the system has been dried, reaction will proceed along path **2**.

Scheme One also shows why thermal desorption of PCB contaminated soils and sludges without integrated dechlorination is a chemically and economically risky undertaking. In the presence of normal moisture levels and natural soil bases like carbonates, at elevated, ie. desorber, temperatures, hydration according to reaction **7** can lead to the same kind of hydroxy-PCBs as reaction with KOH. Although the simple hydration is much less favored than reaction **4**, theory suggests that at desorber temperatures in the range of ca. 300°C, the reaction becomes possible.

## PROPOSED TECHNOLOGIES FOR CHLORINATED HYDROCARBON REMEDIATION

Based on the arguments in the preceding section, one can understand why essentially any non-reductive remediation strategy is technically flawed. Two such approaches: incineration and bioremediation, however, continue to be proposed.

### Incineration

The emotion with which the public frequently greets the idea of incineration is well known. As applied, however, to incineration of chlorinated wastes, there are sound technical reasons why this hysteria should not be too hastily dismissed.

In 1987, an EPA investigator (Oppelt, 1987) stated:

> "...what appears outwardly to be a straightforward, simple process is actually an extremely complex one involving thousands of physical and chemical reactions, reaction kinetics, catalysis, combustion aerodynamics, and heat transfer. This complexity is further aggravated by the complex and fluctuating nature of the waste feed to the process. While combustion and incineration devices are designed to optimize the chances for completion of these reactions, they never completely attain the ideal. Rather, small quantities of a multitude of other products may be formed, depending on the chemical composition of the waste and the combustion conditions encountered. These products along with potentially unreacted components of the waste become the emissions from the incinerator."

Most risk assessment for incineration ignores these truths. Specifically, risks associated with products of incomplete combustion (PICs) are frequently ignored as is the startlingly high probability of operation at off-normal conditions. Such mal-operation both exaggerates the PIC risk and can lead to emission of hazardous uncombusted feed material.

**Products Of Incomplete Combustion** The potential for incomplete combustion, especially in systems fed with chlorinated materials, has long been recognized. Studies of PCB combustion have focused on dioxins and polychlorodibenzofurans (PCDFs) as PICs. The production of dioxins in incineration has been much studied, (Siebert, etal. 1989; Altwicker, etal. 1989; Shaub and Tsang, 1983), but, like the generation of all products of incomplete combustion, mechanism is poorly understood. An excellent summary of all the uncertainties surrounding generation of PICs was recently given by Dellinger and co-workers. (Dellinger, etal. 1990)

It is well known, however, that the production of dioxins and PCDFs is more serious, all other things being equal, in municipal solid waste (MSW) incinerators than in industrial liquid stream incinerators. This is generally attributed to catalysis by fly ash (Vogg and Stieglitz, 1986; Lane and Hinton, 1990; Altwicker, 1988; Altwicker, etal. 1989). Although the nature of the catalytic sites of the fly ash has apparently never been determined, or even sought, there is evidence of a surface catalytic reaction. (Altwicker, etal. 1989).

In the pyrolysis of soils, sludges and sediments, the volume of solid material per unit of chlorinated contaminants far exceeds that in the MSW incinerators. Furthermore, because of the variability of natural soils and the presence in sewage sludge of a variety of metal residues, the opportunity for adventitious catalysis is ripe. We know of no literature describing PICs from combustion of PCB contaminated soils, sludges or sediments. It is significant, however, that test burns of PCB in liquid injection incinerators (Oppelt, 1987) produced measurable quantities of PCDFs.

Unlike dioxins and PCDFs, some PICs probably remain unidentified. The inability to close chlorine material balance in even the most careful incinerator analyses better than 70% (Dellinger, etal. 1990) strongly suggests that such unidentified PICs exist. Of particular concern is the potential for formation of higher molecular weight analogs of dioxins and PCDFs by sequential occurrence of the same type of condensation reactions that are known to form dioxin from pentachlorophenol. These products may be themselves toxic or may be the antecedents of toxic products produced by the action of natural systems on the discharged soil, for instance, by cleavage back to dioxins and PCDFs.

An alternative mode of exit for PICs is via the gas phase. These materials then enter either the air or the scrubber water. This risk is frequently understated.

Oppelt (Oppelt 1986) found that the performance of incineration deteriorated when less than 500 ppm levels of contaminated materials are present. In fact, a survey of many tests (Oppelt 1987) showed that no compound below 200 ppm met the RCRA 99.99% destruction limit in any incinerator. This phenomenon is frequently attributed to recombination reactions in the effluent. An alternative or contributory mechanism, however, may be by transport of contaminated aerosols. Biswas, (Biswas and Sethi, 1990) found the formation of aerosols to be exacerbated in incinerators containing metals. We know of no fundamental work in soil or sludge filled incinerators, but one could anticipate a similar enhancement of aerosol formation.

**Probability of Off-Normal Operation** Perhaps, an even more serious criticism of the risk commonly assigned to incineration concerns the probability of off-normal operation of the incinerator. Techniques for probabilistic assessment of the significance of such off-normal operation are only beginning to emerge (Zeng and Okrent, 1989).

The ease with which such off-normal operation can be encountered is startling. Okrent and co-workers (Zeng and Okrent, 1989) point out that a 100°C change in operating temperature can increase emission of POHC's by a factor of 5 to 500. Such a temperature anomaly can easily result from thermocouple drift. A recent survey article (Lewis, 1990) highlights the fact that drift is particularly sensitive to the system atmosphere. Because of the inherent variability of feed soils, sludges and sediments, selection of a temperature measurement device resistant to drift cannot be assured. This drift leads to a slow change in actual operating condition which may not be detected for some period of time. The effect of such maloperation on a feed containing PCBs, and metals only heightens the concern over the issues of dioxin, PCDF and PIC generation.

That such incidents can occur, even in carefully supervised facilities, is shown anecdotally by Stumbar and co-workers (Stumbar, etal. 1990) who describe a PCDF formation excursion in an EPA operated test incinerator.

To recapitulate, there are significant gaps in understanding of both the chemistry and engineering of incineration of chlorinated compounds. Because of the variability of the feed, the probability of off-normal operation and the chronic inability to close the chlorine material balance, these uncertainties remain, even in the face of nominally successful test burns. Incineration is simply not an acceptable disposal option for chlorinated organics and will not become acceptable until much more fundamental understanding and analysis has been amassed.

**Bioremediation**

Biotechnology, in contrast to incineration, frequently encounters less public hostility. Furthermore, for reasons unclear, it is, or has been, in the good graces of the EPA. It is well known, however that chlorinated substrates provide a challenge to bioprocessing. In fact, many historic applications of PCBs relied upon the resistance of the material to oxidative and natural attack. The persistence of PCBs in the environment is testimony to this.

It is important to differentiate between *in situ* and *ex situ* bioremediation. *In situ* biological systems are frequently an effective option for readily degraded substrates. In its simplest form, it consists of merely assisting and accelerating naturally occurring biosystems. Not only is such an approach environmentally friendly, it is cheap.

*Ex situ* bioremediation, on the other hand, suffers from the same malady that limits biosynthesis to production of human consumables and extremely complex organics: processing temperature is limited. When alternative synthetic chemistry is available, extension of such chemistry to elevated temperatures will result in a rate and cost benefit that the biologically mediated process simply cannot overcome.

Thus, to be economically attractive, a bioremediation must be an *in situ* process. This effectively rules out genetically engineered organisms. Most proposals for bioremediation of PCBs rely either on engineered organisms or on an undefined combination of observations involving both anaerobic and aerobic organisms.[1] There is no basis to believe that a practical bioremediation process for PCB destruction can be developed.

There is, on the other hand, no doubt that by applying the full power of modern genetic engineering one can, in the fullness of time, create organisms that will consume PCBs. Even then, however, a process utilizing such organisms will still fail to meet all three of the criteria for an effective technology. Rates will be too low to produce acceptable costs. As in all biological systems, very low levels of substrate will adversely effect population dynamics and therefore very strict "clean levels" will not be achievable.

## EARLY EXTENSION TO SOIL SYSTEMS

Extension of the alkaline glycolate technology to solid phase media has been slow to develop. As early as 1982, workers from Franklin under contract to the EPA applied

---

[1] It is generally accepted that aerobic bacteria do not attack heavily chlorinated PCBs and dechlorinate the others impractically slowly if at all. Although it has been shown that anaerobic bacteria can partially dechlorinate fully chlorinated Aroclors, most of these are not, to our knowledge, facultative anaerobes, ruling out *in situ* use and mixed cultures with aerobes. Anaerobic processes are also quite slow and cultures are notoriously unstable. Practical methods for working with mixed cultures of aerobes and anaerobes simply do not exist. Products of reaction are unknown.

NaPEG to soil. At this time, however, it was still hoped that an *in situ* chemical dechlorination could be developed. Results were unacceptable, for reasons having to do with the fundamental irrationality of trying to carry out synthetic stoichiometric chemistry in unmixed solid phase media exposed to uncontrolled conditions of temperature and dilution, ie., rain. Nevertheless, the idea that alkaline glycolate does not work in soils gained momentum.

A series of field trials sponsored by the Environmental Protection Agency through the 70's extended the glycolate chemistry to a wider range of contaminants and contaminant levels, although reporting of the results frequently confounded site specific and other non-chemistry factors with the efficacy of the reagent system. Also, because the experimental approach was evolving from earlier *in situ* efforts reaction temperature was often limited by equipment constraints.

Finally, in 1988, EPA sponsored a field test in Guam (Taylor,etal., 1990) which showed the chemistry to be capable of achieving any contemplated level of remediation in a soil system. In reporting these results, however, no differentiation was made between the costs incurred in the demonstration unit and the costs of full scale remediation which could be reasonably inferred from the successful test.

This history left a residue of skepticism as to the efficacy of alkaline glycolate systems and the attendant cost. Clearly what is needed to fully understand the chemistry is a body of data on model systems. Why was this not done long ago? The answer lies in the system by which technology is selected for Superfund sites which relies on what are called Treatability tests. These tests are site specific and the emphasis is on quick demonstration of efficacy, both to rule out totally inappropriate technologies and because the environmental arena carries an undercurrent of thinking that the chemistry is secondary to the engineering and to the hardware to be used. Doing research under the Treatability system is nearly impossible.

Finally, a model system study was carried out at the National Environmental Technology Applications Corporation (NETAC) (King and Hoch, 1991). Because these data are not readily available, it is useful to review the highlights here.

## NETAC PROGRAM

This privately funded contract program had as its objective identification of process conditions for economical and total dechlorination by determining the rate of removal of chlorine from model aromatic compounds. In contrast to many other environmental studies, this work determined the total amount of chloride ion released as a function of reaction temperature, time and stoichiometry. This approach differs from the usual environmental tests which measure the fraction of the chlorinated aromatic that has reacted - presumably measuring therefore only the first chlorine removal.

Several distinct reactor configurations were employed. At low temperature, i.e. less than 150° C, and below the boiling point of the solvent, a commercial stirred heater block using six dram vials was used. At higher temperatures, but below the boiling point of the reaction solvent, the heater block was placed on a hot plate or a three neck flask with nitrogen purge was used. For the highest temperature runs a 3/16 stainless steel high pressure bomb immersed in a fluidized sand bath and agitated by an internal stainless steel ball served as the reactor.

After reaction, the mixtures were quenched by acidification, diluted with water and chloride released was measured using a specific ion electrode. These data represent the only systematic and self-consistent published study of the dechlorination of well characterized aromatics by alkaline glycolate chemistry.

## TABLE I. HEXACHLOROBENZENE-KPEG RUNS

| Temp deg C | Time hr | moles PEG/C1 | moles KOH/C1 | Solvent | Conversion |
|---|---|---|---|---|---|
| 25 | 16 | 4.01 | 4.40 | Toluene | 51.8 |
| 60 | 2 | 3.91 | 3.87 | Toluene | 52.3 |
| 60 | 4 | 3.89 | 3.90 | Toluene | 61.3 |
| 60 | 8 | 3.90 | 3.86 | Toluene | 57.1 |
| 60 | 16 | 3.89 | 3.90 | Toluene | 62.7 |
| 80 | 16 | 4.00 | 4.38 | Toluene | 58.2 |
| 80 | 16 | 4.01 | 4.39 | Toluene | 54.7 |
| 80 | 16 | 2.01 | 2.22 | Toluene | 54.6 |
| 80 | 2 | 3.99 | 4.61 | Toluene | 61.0 |
| 100 | 4 | 1.61 | 1.80 | Toluene | 68.09 |
| 100 | 2 | 2.53 | 2.76 | Toluene | 64.6 |
| 147 | 16 | 3.92 | 4.35 | Triglyme | 93.7 |
| 175 | 2 | 4.05 | 4.34 | Triglyme | 101.6 |
| 165 | 2 | 4.02 | 4.36 | Triglyme | 86.4 |
| 175 | 2 | 3.90 | 4.27 | Triglyme | 82.7 |
| 200 | 2 | 3.96 | 4.30 | Triglyme | 100.7 |
| 190-195 | 2 | 3.93 | 4.32 | Triglyme | 54.01 |
| 195-200 | 1 | 4.09 | 4.48 | Triglyme | 92.0 |
|  | 2 | 3.98 | 4.38 | Triglyme | 104.8 |
| 181-185 | 1 | 3.84 | 4.20 | Triglyme | 74.5 |
|  | 2 | 3.93 | 4.30 | Triglyme | 96.2 |
| 185 | 1 | 1.35 | 1.48 | Triglyme | 81.1 |
| 150 | 16 | 1.50 | 1.67 | Triglyme | 113.4 |
| 200 | 2 | 3.96 | 4.51 | None | 103.9 |

From King and Hoch 1991

Table I summarizes hexachlorobenzene dechlorination data in various solvents. Note that these data are given in percent of total chlorine removed; as little as 16 percent conversion may, therefore, correspond to essentially complete disappearance of the starting material.

A strong correlation can be seen between the achievable degree of dechlorination and reaction temperature. Loosely speaking, 50 percent dechlorination is found over-night at room temperature. The fourth chlorine atom can be reacted in two hours at 80°C; the fifth requires 150°C and complete dechlorination is observed in the 185° to 200°C temperature range. Data also show that the effect of reagent stoichiometry is not dominant.

Table II presents data on dechlorination of less substituted chlorobenzenes. Generally speaking these less chlorinated materials require more stringent conditions to achieve the same degree of dechlorination. A similar effect was reported for PCBs (Brunelle and Singleton, 1985)

Again relating the conditions required for complete dechlorination to reaction temperature, 1,2,4 trichlorobenzene requires about 230°C for complete dechlorination, 1,3 dichlorobenzene requires about 250°C and chlorobenzene requires even higher temperatures.

This data shows clearly why earlier attempts to dechlorinate materials like pentachlorophenol, which is deactivated to nucleophilic substitution, at moderate temperature conditions is futile even given extremely long reaction times. On the other hand, the seemingly more stringent goal of complete dechlorination can be readily achieved provided that the system is taken to an adequate temperature.

## TABLE II. REACTION WITH CHLOROBENZENES

| Reactant | Temp deg C | Time hr | moles PEG/C1 | moles KOH/C1 | Conversion |
|---|---|---|---|---|---|
| TriClB | 200 | 2 | 4.00 | 4.58 | 67.3 |
| TriC1B | 225 | 2 | 4.00 | 4.56 | 82.5 |
| TriClB | 230-240 | 2 | 3.98 | 4.57 | 102.5 |
| TriC1B | 230 | 1 | 3.97 | 4.56 | 89.9 |
| TriC1B | 240 | 1 | 3.97 | 4.53 | 109.4 |
| TriC1B | 250 | 1 | 3.98 | 4.47 | 72.3 |
| TriC1B | 240 | 1 | 3.73 | 4.27 | 122.1 |
| TriC1B | 250 | 1 | 4.04 | 4.55 | 90.9 |
| DiC1B | 200 | 2 | 4.26 | 4.46 | 42.3 |
| DiC1B | 225 | 2 | 4.49 | 4.67 | 86.5 |
| DiC1B | 250 | 2 | 4.54 | 4.70 | 95.8 |
| ChloroB | 200 | 2 | 7.77 | 8.31 | 52.5 |
| ChloroB | 225 | 2 | 7.47 | 7.79 | 63.3 |
| ChloroB | 250 | 2 | 7.75 | 8.38 | 62.3 |
| ChloroB | 275 | 2 | 7.41 | 7.68 | 40.3 |

From King and Hoch 1991
All runs: triglyme solvent

Furthermore, given the temperatures required, issues surrounding the presence of water become moot. A back of the envelope calculation easily shows that the energy cost to dehydrate even the wettest sediments is negligible compared to the normal costs of remediation. Equipment configured to obtain the temperatures required for dechlorination is easily capable of simultaneously or sequentially dehydrating the feed.

## PROCESS AND ECONOMIC CONSIDERATIONS IN THE DECHLORINATION OF SOILS, SLUDGES AND SEDIMENTS

The key to acceptable economics in soil systems lies in complete dechlorination. Anything less leaves a material of unknown toxicity. Despite the previously cited studies showing these polychloromonoethers to be nontoxic there remains a quite reasonable resistance to their persistence in the environment. If, therefore, complete dechlorination is not achieved the chemistry results in a product still requiring treatment as a potentially hazardous substance. Complete dechlorination, on the other hand, can be coupled with engineering steps that leave the reaction products in the treated soil. One can argue for on site reinterment of this organic chloride free material. On site reinterment is absolutely critical to competitive economics.

Despite its successful history there remained skepticism that alkaline glycolate chemistry could be successfully extended to soil systems. This skepticism was rooted in both the traditional difficulties of the remediation industry in translating chemistry directly into hardware without an intermediate process engineering step as well as in the confusing results of the EPA studies through the 70's.

All of this was swept away in 1990 and 1991 when the Wide Beach, New York, Superfund Site was successfully remediated using alkaline glycolate chemistry in a thermal desorber. It is hard to overstate the significance of the Wide Beach project. Not only did it provide a full scale and fully successful demonstration of alkaline glycolate chemistry in a soil system, but it represented the first use of an innovative, which in the parlance of the remediation industry means non-incineration, technology for PCBs at a superfund site.

Also significant is the fact that the chemistry was readily integrated into an existing thermal desorber. This is significant both because the availability of such desorbers provides additional systems wherein one might employ alkaline glycolate dechlorinations and further because operation of a desorber on a chlorinated aromatic carries a significant risk as explained above with reference to Scheme One.

It should be noted that other reductive technologies for destruction of PCBs have been proposed. These include the reaction with monoethylene glycol monomethylether (MEGME), the base catalyzed decomposition (BCD process) developed by the EPA and reduction by nascent hydrogen developed by Roy F. Weston for the EPA. Beside lacking the track record of the PEG process, each of these approaches has severe handicaps.

Although, based on published data (Friedman, 1990) and reasoning from first principles, MEGME appears to be a facile dechlorination agent, it suffers from the serious drawback of being itself toxic. As a practical matter, this rules out direct application in soil systems followed by reinterment of the treated soil. As explained above such on site reinterment is critical to the economics.

Although the chemistry underlying the BCD process has never been made entirely clear, what has been published suggests that it may suffer from all of the drawbacks associated with the presence of water that have been attributed in the discussion above to desorption in the absence of NaPeg. Furthermore, flow sheets for BCD which have been presented (Kim and Olfenbuttel, 1990) are among the most egregious examples of technology development without regard to process engineering.

Although only limited data is available (Saha, etal.,1990) it is hard to understand how the nascent hydrogen reduction system can be successfully applied to soils. This is another example where elementary process engineering will clearly rule out any practical application.

## ECONOMICS OF REMEDIATION

Having argued that not only are alternative remediation technologies technically or socially unacceptable and further having shown that the fundamentals of the chemistry as well as a body of organized data support the choice of alkaline glycolate dechlorination, it still remains to demonstrate that this technology meets the primary criteria for remediation, viz. acceptable economics. Table Three below shows illustrative economics for a model case. The underlying data for this table are not arbitrary; these figures reflect actual results on a superfund site soil tested at NETAC. Note that these are shovel-to-shovel economics. Site specific factors are thereby excluded.

**TABLE III. ECONOMICS OF TYPICAL SOIL REMEDIATION USING KPEG[sm] 25000 TONS DRY WEIGHT**

| Cost Element | $/ton |
|---|---|
| Reagents and other chemicals | 86 |
| Utilities | 25 |
| Labor | 12 |
| Mobilization/demobilization | 7 |
| Fixed costs, including royalty and return | 103 |
| Total shovel-to-shovel cost | 233 |

Basis: 250 ton/day mobile unit;
3 Hr. reaction time;
3-fold stoichiometric excess of KPEG

It should be noted that this approach of comparing shovel-to-shovel numbers, analogous to use in the chemical industry of the battery limits concept, although not unknown is not stressed in the hazardous waste community. As a result, site specific factors creep into economic data which is imbued with almost thermodynamic universality. It is fair to say that differences in accounting methods, project basis and site specific factors are far more variable than the difference between the economics of *ex situ* processes.

Depending upon the site, the cost elements considered in Table Three may or may not comprise the bulk of the total project cost. These are process economics, not project economics. Cost is seen to be dominated by "reagents and other chemicals" and "fixed costs".

The reagent consumption is, perhaps surprisingly, not a strong function of the type or degree of contamination. Rather, it represents uptake of alkali by the soil media itself. To the extent that representatives samples can be obtained, it can be determined by prior experimentation in fact, determination of factors such as this is the rational basis for the Treatability concept. Experience has shown that alkali uptake can vary widely and can in fact become so large that it forces one to consider other remedies. Many of these, however, may also be adversely affected by a soil or other media with high alkali uptake potential. It is important to make such a technology comparison using consistent data.

Fixed costs, which in the above accounting are all related to unit capital, its amortization and reward, can also vary widely. Here, the reason for inconsistency is not site specific but vendor specific. Neither EPA rules, nor industry practice, provides any guidance on how to charge the capital cost of a mobile unit against specific projects. This reflects the project by project orientation of the remediation industry where, in the extreme case, a contractor could assign a fixed cost of zero. Unfortunately, unrealistically assigned fixed costs are not uncommon. Not only does this always lead to disastrous results for the project at hand, but such numbers find their way into the shoptalk of the industry and by Gresham's law of economics drive the realistic numbers out of circulation.

The figure of $103.00 in Table Three reflects the assumption that the 250 ton/day mobile unit stands idle 50% of the time and is either remediating or being mobilized or demobilized the other 50%. Based on this "activity factor" normal capital weights and 25% simple return are included.

The utilities cost figure needs little discussion; it reflects fuel at $ 3.00/MMBTU and soil moisture of approximately 30 weight percent. It is easy to calculate, then that were the soil water level as high as 90 weight percent utilities cost would rise by less than $5/ton. Although one often hears about the adverse effect of soil moisture on dechlorination, this simple calculation shows that, in systems where the soil is dried before reaction, the utilities cost penalty is relatively minor.

Both the labor cost and the mobilization-demobilization cost, although minor, are related to the size of the unit employed. Matching or mismatching of process unit size to project size can be a further source of strange economic computation. Clearly, the larger the unit employed, the lower the labor cost becomes and the lower certain elements of the fixed cost become. On the other hand, mobilization-demobilization rises and assumptions as to unit utilization become more critical. Assuming the availability of an unreasonably large unit becomes an alternative path by which the unwise or the unscrupulous can compute extremely low project costs.

A point should also be made in this section on the comparison between dechlorination and separation technologies like extraction with various solvents and soil washing. All of these methods in common concentrate, but do not destroy, the contaminant. Furthermore, if the process is not very selective, a number of different contaminated media may be generated. Additionally, driving forces for separation are smaller than for reaction, making strict clean levels more difficult to reach. Finally, integration of many of these separations with alkaline glycolate dechlorination is feasible

## RECAPITULATION

Based on all of the above one can argue that alkaline glycolate dechlorination is the process of choice for remediation of chlorinated soils, sludges and sediments. The simple argument is based on the heirarchy of criteria listed in the "Overview" section above and the technical factors reviewed in this paper. Key points of the logic are the following:

1) Incineration will not be publicly acceptable as well as having serious technical shortcomings.

2) Biotechnology has the same pair of problems.

3) Other oxidative systems and high temperature anaerobic systems run a significant risk of producing dioxins or similar compounds.

4) Landfill should be conserved for wastes that cannot be treated or for mixed wastes that cannot be treated simply and economically.

5) Separation systems merely concentrate the contamination and produce a contaminated product which still requires treatment and/or disposal.

Finally, in this quite conservative industry, the track record of KPEG$^{sm}$, culminating in the Wide Beach Project, is simply unmatched.

## REFERENCES


Altwicker, E.R., 1988, Formation of polychlorodibenzo-p-dioxins and polychlorodibenzofurans during heterogeneous combustion, *3rd Chemical Congress of North America and 195th ACS National Meeting*, Toronto

Altwicker, E.R. etal., 1989, Polychlorinated dioxin/furan formation in incinerators, *Presented at the AIChE Annual Meeting*, San Francisco

Berry, R.I., 1981, *Chem. Eng.*, August, 10, p37

Brunelle, D.J.,1982, U.S.Patent 4,351,718

Brunelle, D.J.,1983, Destruction/removal of polychlorinated biphenyls from non-polar media, *Chemosphere*,12:183

Brunelle, D.J. and Singleton, D. A., 1985, *Chemosphere*, 14:173

Davis,F.F.,etal.,1979, U.S.Patent 4,179,337

Dellinger, Barry, etal., 1990, PIC formation - research status and control implications, *Presented at EPA 16th Annual Hazardous Waste Research Symposium*, Cincinnati

DeMarini,D.M., and Simmons,J.E., 1989, *Chemosphere*, 18:2293

Friedman,A., 1990, *paper presented at 10th International Conference on Organohalogen Compounds*, Bayreuth

Kim,B.C. and Olfenbuttel,R.F., 1990, Demonstration of the BCDP process at USN PWC site in Guam, *presented at EPA Technology Transfer Conference*, Cincinnati

King,A.B. and Hoch,R., 1991, Complete dechlorination of aromatics with alkaline PEG chemistry, *prepared for Air & Waste Management Association Meeting*, Vancouver

Lane, A.M. and Hinton, S.W., 1990, Characterization of municipal waste incinerator fly ash to identify mechanism of dioxin formation, *Presented at AIChE Spring National Meeting*, Orlando

Lewis, C.W., 1990, Measure remote temperatures efficiently, *Chem. Eng.* 97:114

Oppelt, E.T., 1986, Performance Assessment of Incinerators and High Temperature Industrial Processes for Disposing of Hazardous Waste in the United States, in Lorenzen, etal. eds. "Hazardous and Industrial Solid Waste Testing and Disposal", ASTM # 933, Philadelphia

Oppelt, E.T., 1987, Incineration of hazardous waste; a critical review, *JAPCA* 37:558

Pytlewski,L.L., etal., 1982, U.S.Patent 4,337,368

Pytlewski, L.L., etal.,1983a, U.S.Patent 4,400,552

Pytlewski, L.L., etal.,1983b, U.S.Patent 4,430,208

Pytlewski, L.L., etal.,1984a, U.S.Patent 4,417,977

Pytlewski, L.L., etal.,1984b, U.S.Patent 4,460,797

Pytlewski, L.L., etal.,1984c, U.S.Patent 4,471,143

Pytlewski, L.L., etal.,1986, U.S.Patent 4,602,994

Saha,A.K. etal., 1990, Thermocatalytic hydrodechlorination of chlorinated pesticides left in soil medium, effect of temperature, catalyst,acid, and solvent on conversion, *presented at AIChE Summer National Meeting*, San Diego

Sethi, V. and Biswas, P., 1990, Fundamental studies on particulate emissions from hazardous waste incinerators, *Presented at EPA 16th Annual Hazardous Waste Research Symposium*, Cincinnati

Shaub, W. and Tsang, W., 1983, Dioxin formation in incinerators, *Env. Sci. Tech.* 17, 721-730

Siebert, P. C., etal., 1989, Toxic trace pollutants from incineration, *Presented at AIChE Summer National Meeting*, Philadelphia

Stumbar, J.P. etal., 1990, Factors affecting the reliability of operations of the epa mobile incineration system, *Presented at EPA 16th Annual Hazardous Waste Research Symposium*, Cincinnati

Taylor, M.L. etal., 1990, "Comprehensive Report on the KPEG Process for Treating Chlorinated Wastes", EPA Contract No. 68-03-3413

Vogg, H. and Stieglitz, L., 1986, Thermal behavior of PCDD/PCDF in fly ash from municipal incinerators, *Chemosphere*, 15:1373

Zeng, Y. and Okrent, D., 1989, Analysis of off-normal emissions from a hazardous waste incinerator, *Presented at the SRA Annual Meeting*, San Francisco

# A NEW MEMBRANE PROCESS FOR RECOVERING ORGANICS FROM AQUEOUS WASTES

S. V. Ho

Monsanto Company
800 N. Lindbergh Blvd.
St. Louis, MO 63167

## INTRODUCTION

Membrane technology offers great potential for treating industrial wastes. Membrane methods that could potentially be applied to wastewater treatment have been reviewed extensively in a recent Department of Energy report on membrane separation systems[1] and by various workers in the field[2,3]. These methods include reverse osmosis, nanofiltration, ultrafiltration, microfiltration, coupled/facilitated transport (liquid membrane), pervaporation, and electrodialysis. Membranes can also be used as contactors to carry out gas stripping and liquid-liquid extraction[4,5].

A common characteristic of most aqueous wastes discharged from chemical plants is that they are fairly dilute in organic content. These streams typically contain up to several percent low molecular weight (MW) organic compounds such as carboxylic acids, alcohols, nitriles, etc. Most streams also contain inorganic salts with concentrations that can reach 30 wt% or so. Examples of aqueous wastes that belong to the high-salt category include one that contains ~5% highly-charged organics and 20% NaCl; another containing 2% various organics of MW<200 and 6-35% ammonium sulfate; and one stream containing 1% phenolics and 15-20% potassium chloride. From a separation standpoint, a plausible initial treatment of these streams is to remove and/or concentrate the organics either for recycle or to facilitate further treatment. In principle, dilute streams of organics can generally be handled adequately with existing methods such as reverse osmosis or evaporation to reduce the volume, followed by incineration. However, the presence of high concentrations of salt greatly complicates treatment options. For instance, reverse osmosis is not practical for treating these streams due to the very high osmotic pressures involved, and existing nanofiltration membranes generally do not have sufficient rejection for organic compounds with MW below 200 or so. As a result, the costs for treating such streams are usually much higher than for similar ones with no salts present.

Using internal waste streams as models we have developed proprietary composite membranes with unique properties that enable us to concentrate the low MW organics as well as separate them from salts. These membranes operate in a dialytic mode where the driving force is concentration and solubility difference rather than pressure difference. Results obtained with model compounds as well as with two actual waste streams are presented in this paper.

*Industrial Environmental Chemistry*, Edited by D.T. Sawyer and A.E. Martell, Plenum Press, New York, 1992

## EXPERIMENTAL PROCEDURES

All the chemicals used in this study, other than those in actual waste streams, were purchased from either Sigma or Aldrich. The actual wastes streams were obtained from Monsanto plants. Phenol, nitrophenol, nitroaniline, benzoic acid as pure components in aqueous solutions were quantified using spectrophotometric absorption at wavelengths 287, 400, 380, and 224 nm, respectively. Total organic content (TOC) was measured with a TOC machine (Rosemount Analytical, Dohrmann DC-190. Carboxylic acids present as single components were analyzed either by titration with 0.1N NaOH (phenolphthalein indicator) or using the TOC machine. As mixtures, they were analyzed using ion chromatography, which was also the method used for measuring chloride ions in solution. Solution conductivity was measured with a Cole-Parmer conductivity meter.

Separation experiments were carried out using both flat sheet and hollow fiber membranes. These are composite membranes consisting of a proprietary separating layer supported on various microfiltration and ultrafiltration membranes. Due to the patent situation, we are not allowed to disclose the nature of the membranes at this time. Flat sheet microporous membranes from Hoechst-Celanese (polypropylene, trade name Celgard) and from W. L. Gore (polytetrafluoroethylene, trade name Gore-Tex) were used in this study. Some physical characteristics of these membranes are shown below[6,7].

| Type | Thickness | Pore Size | % Porosity |
|---|---|---|---|
| Celgard 2400 | 25 μm | 0.050 μm | 38 |
| Celgard 2500 | 25 | 0.075 | 45 |
| Gore-Tex | 60 | 0.100 | 75 |

Two types of hollow fibers were also used: polypropylene (Celgard) from Hoechst-Celanese, and polysulfone from Permea (now a subsidiary of Air Products). The polypropylene hollow fibers are symmetric, i.e., the pore structure is the same throughout the wall of the fibers. Polysulfone hollow fibers are asymmetric, that is, they have a very thin, dense skin on the surface, supported by a much thicker and more open structure underneath. Physical characteristics of these fibers are shown below.

Celgard X20 400: symmetric; OD = 460, ID = 400 μm; $d_{pore}$ = 0.065 μm; porosity = 40%

Polysulfone : asymmetric; OD = 1100, ID = 866 μm; 160K MWCO; porosity = 76%

These microfiltration hollow fibers were potted into our standard testing modules (polycarbonate tubes, 1 cm inside diameter and 25 cm long with openings on the side of the tube for shell-side fluid circulation), then modified in situ to introduce the separating layer forming the proprietary composite membranes. The membrane area available for transport depends on the number of fibers used; typically the area ranged from 100 to 400 $cm^2$ in our tests.

Figure 1 shows the schematic diagram of various experimental systems used in this study. Transport measurements with flat sheet membranes were conducted using either stirred cells or flow cells. Stirred cells are made up of two glass compartments, each about 30 ml in volume, separated by a sheet of membrane and held together by a clamp. Liquids in the compartments were magnetically stirred. Membrane area in contact with the liquids was 8 $cm^2$. These stirred cells are convenient for rapid membrane screening as well as for determining the effects of organic concentration and temperature on membrane transport. Flow cells are similar to stirred cells except that each compartment has an inlet and an outlet port for circulating liquid through the cells. Flow cells were used for processing large volumes of liquid, which were circulated with pumps from external reservoirs through the cell compartments. With hollow fiber membranes, fluids were circulated through the bore side and the shell side. Organic concentrations of the feed and strip solutions were measured

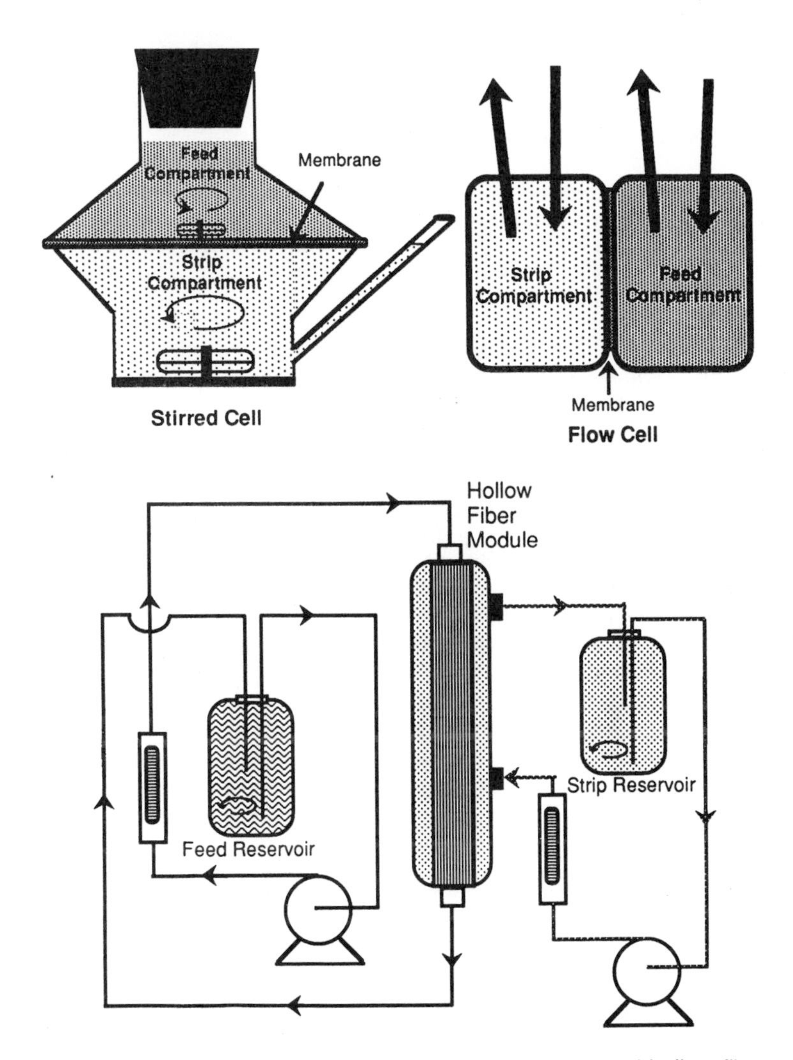


Figure 1. Schematic diagram for stirred cell, flow cell, and hollow fiber membrane systems

as a function of run time. The data were used to calculate the overall mass transfer coefficient using the equations derived in the appendix.

## RESULTS AND DISCUSSION

Rate, selectivity, and stability are three critical elements in determining the value, cost/performance and ultimate utility of a membrane for any separation. Permeability and selectivity are determined by the solubility (interaction) and diffusivity of each component in a system. A general description of the properties of the composite membranes is given below, followed by examples of the application of the technology to waste minimization.

## PROPERTIES OF COMPOSITE MEMBRANES

The transport through our composite membranes is by chemical interaction, not by size difference. A key property of these membranes is that they effectively rejects charged species in solution. This includes both small ions (ions generated by salts, $H^+$ and $OH^-$) and larger charged molecules. Only neutral compounds can pass through. Thus, the membranes can be used to recover organics from an aqueous saline waste as illustrated in Figure 2. For effective separation across such a membrane, a driving force should be provided and the recovery (strip) solution should provide an environment (a "sink") which limits the back diffusion of the compound across the membrane into the waste solution. This could be accomplished in several ways including pH change, reaction, chemical complexation, biodegradation, pervaporation, etc.

To date small ions which have been shown to be effectively rejected by the membranes are $H^+$, $OH^-$, $Na^+$, $K^+$, $Cl^-$, $NH_4^+$, $NO_3^-$, $SO_4^=$. We have not yet found an ion that could readily pass the membrane. This is true for larger charged molecules as well. Data in Table 1 for PNP and formic acid show the significance of the ionization state of a compound on its transport rate through the membrane. At solution pH several units above their pKa's, practically all PNP and formic acid molecules are ionized, resulting in a 20-40x reduction in transport rates across the membrane compared to the rates obtained when these compounds exist predominantly in their neutral forms (pH < pKa).

**Table 1.** Effect of solution pH on transport of organic compounds for composite Celgard 2400 membranes at room temperature.

| Compound | pKa | Solution pH | Initial Rate (mg/hr-cm$^2$ membrane) |
|---|---|---|---|
| p-Nitrophenol | 7.0 | 10 | 0.15 |
| (at 0.5 wt%) | | 5 | 3.5 |
| Formic Acid | 3.75 | 5 | 0.026 |
| (at 4.4wt%) | | 1.6 | 1.0 |

The rate of transport through the composite membrane for non-ionized compounds is very sensitive to their chemical nature. Table 2 shows the transport rates for various compounds such as phenol, amines and carboxylic acids. The rates were obtained with the stirred cell and are reported as the overall mass transfer coefficient (units: cm/sec), which lumps mass transfer resistances of the aqueous liquid films on both sides of the membrane with the resistance due to the membrane itself. A mathematical model of the stirred cell developed in the appendix shows that plot of ln(concentration change) vs. time yields a straight line, the slope of which is proportional to the overall mass transfer coefficient, $K_f$. Such a linearized plot is presented in Figure 3 for phenol. Due to the intentional vigorous liquid mixing in the cell compartments, which minimizes the resistance contributions of the

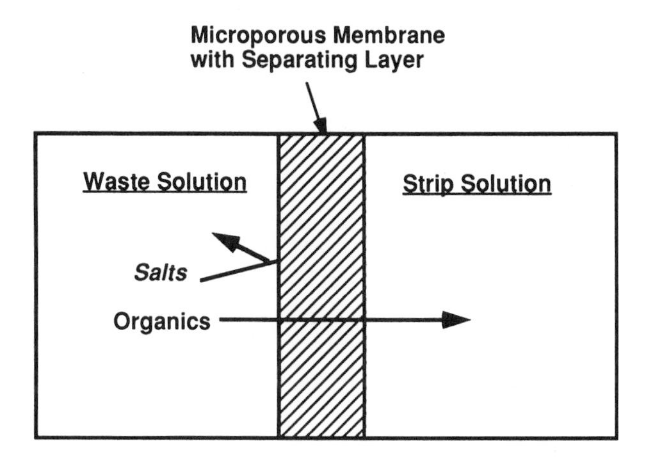


e.g. with NaOH as strip solution forming organic salts

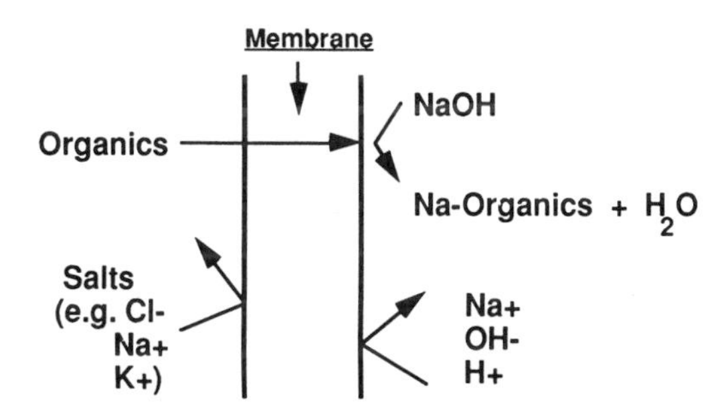


Figure 2. Composite membrane for organic/salt separation

**Table 2.** Transport of organic compounds through a composite Celgard 2400 membrane at room temperature.

| Compound | Initial Feed Con. mg/L | Overall Mass Trans. Coeff. cm/sec |
|---|---|---|
| Phenol | 4000 | $6.0 \times 10^{-4}$ |
| p-Nitrophenol | 4000 | $3.4 \times 10^{-4}$ |
| p-Nitroaniline | 500 | $6.5 \times 10^{-4}$ |
| Benzoic acid | 1700 | $4.0 \times 10^{-4}$ |

*Strip solution: 0.1N NaOH (0.1N HCl for p-nitroaniline)*

**Table 3.** Phenol permeability through various polymeric membranes [(a)]

| Polymer | Film thickness µm | Effective diffusion coefficient, $cm^2$/sec |
|---|---|---|
| Polyethylene | 14.1 | $0.11 \times 10^{-7}$ |
| Polyvinylidene chloride | 10.0 | $0.06 \times 10^{-7}$ |
| Styrene-butadiene copolymer | 41.0 | $0.71 \times 10^{-7}$ |
| XD-7 (Silane) | 68.6 | $2.06 \times 10^{-7}$ |
| Ethyl cellulose | 35.2 | $0.96 \times 10^{-7}$ |
| Composite Celgard 2400[(b)] | 25.0 | $15.0 \times 10^{-7}$ |

*(a) from Klein et al.[8]; (b) This study*

liquid boundary layers, the values obtained for $K_f$ are essentially the membrane permeability values. In these experiments, the pH of the feed solution was such that the organic compounds are un-ionized (pH about 2 units below pKa) and NaOH was used on the strip side for weak acids, and HCl or $HNO_3$ for weak bases.

The practical significance of the overall mass transfer coefficients obtained with the composite membrane shown in Table 2 can be appreciated by comparison with other membrane approaches for phenol transport. Table 3 shows the results obtained by Klein et al.[8] for phenol extraction using various polymeric films as membranes. Transport rates were reported as effective diffusion coefficient (D in $cm^2$/sec = permeability x membrane thickness) for phenol through these membranes. With our composite membrane (25µm thick) the permeability of phenol is $6.0 \times 10^{-4}$ cm/sec (Table 2), which gives an effective diffusion coefficient of $15 \times 10^{-7}$ $cm^2$/sec. This is almost an order of magnitude higher than the best rate achieved by these workers (D = $2.06 \times 10^{-7}$ $cm^2$/sec) with the membrane made from XD-7 (a polymethylsilane-polycarbonate copolymer). Another interesting comparison is with the hollow fiber contained liquid membrane approach by Sengupta et al.[9]. In this approach, the shell side of a membrane module that contains two sets of intermingled hollow fiber bundles is filled with an organic solvent that functions as the liquid membrane. The feed solution flows through the lumen of one set of fibers, and the strip solution through the other. With methyl isobutyl ketone as the contained liquid membrane, they reported an overall mass transfer coefficient for phenol of about $3 \times 10^{-4}$ cm/sec. This is about half of the rate obtained with our composite membrane, which, incidentally, would utilize all of the hollow fiber membrane area available for transport rather than only half of it as in the contained liquid membrane approach.

Transport rates through a composite membrane for a series of carboxylic acids ranging from C1 to C6 are shown in Table 4. Interestingly, the rate increases two orders of magnitude from formic acid (C1) to caproic acid (C6). Thus, hydrophobicity and permeability seem to be closely correlated and the more hydrophobic compounds enjoy faster transport rates. We also found that in addition to hydrophobicity certain polar characteristics (e.g. hydrogen-bonding capability) of an organic compound may also contribute to its enhanced transport rate.

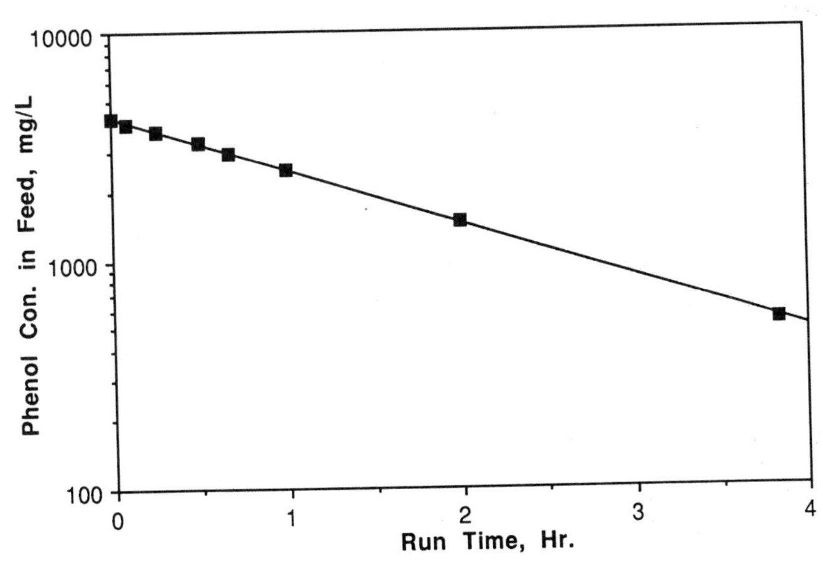


Figure 3. Linearized plot for phenol transport through a composite Celgard membrane. (Feed: 4200 mg/L phenol, pH 5.8; Strip: 0.1N NaOH)

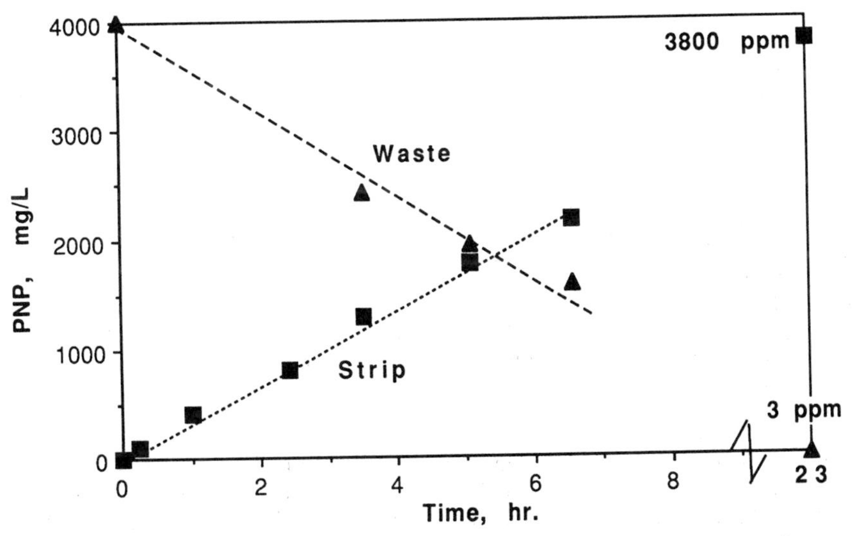


Figure 4. Removal of PNP from an actual phenolics waste stream using a composite Celgard 2400 membrane

**Table 4.** Transport of carboxylic acids through a composite Celgard membranes

| Compound | Overall Mass Transfer Coeff. (cm/sec) |
|---|---|
| Formic Acid ($C_1$) | $0.055 \times 10^{-4}$ |
| Acetic Acid ($C_2$) | $0.10 \times 10^{-4}$ |
| Propionic Acid ($C_3$) | $0.33 \times 10^{-4}$ |
| Butyric Acid ($C_4$) | $0.42 \times 10^{-4}$ |
| Valeric Acid ($C_5$) | $1.6 \times 10^{-4}$ |
| Caproic Acid ($C_6$) | $4.0 \times 10^{-4}$ |

**Feed side: 0.1-5wt% acid in 20wt% KCl, pH = 2; Strip side: 0.1N NaOH; Room Temperature*

Effects of the number of hydroxyl groups on phenolic compounds were explored with resorcinol, hydroquinone, pyrocatechol, and phloroglucinol. More hydroxyl groups attached to the ring make the compound more hydrophilic, which results in lower transport rates. Also very interesting is the behavior of amino acids and its analogs. Being zwitterions, which are always ionized, amino acids even the most hydrophobic ones such as tryptophan, phenylalanine and tyrosine are totally rejected by the composite membrane. Yet hydrocinnamic acid, which is phenylalanine without the amino group, can be transported very fast (comparable to p-nitrophenol) through the composite membrane at low pH where it is non-ionized.

The above properties of our proprietary composite membranes make them very suitable for organic/salt separation as well as for separating similar organic compounds that may be difficult to achieve with other separation methods

## APPLICATIONS TO WASTE TREATMENT

The composite membrane process has been used successfully to remove and concentrate organics from two actual aqueous waste streams. One contains phenolics and high concentration of inorganic salts; the other is a dilute stream of carboxylic acids.

### Phenolics Waste

This waste stream is generated from one of our plants and typically contains p-nitrophenol (PNP) as the major organic contaminant at <1 wt%. The waste stream also contains high salt level (~20wt%, mainly KCl). Existing treatment options for this stream include PNP removal followed by destruction or biodegradation. We investigated the feasibility of using our membrane process to remove PNP from the waste solution in a concentrated form for recycle.

Figure 4 shows the results for a run with the actual waste solution using a flat sheet composite Celgard membrane. In this experiment the membrane was in contact with two circulating solutions: the waste solution at pH 5 on one side, and 0.1M potassium hydroxide (pH 13) on the other side. PNP concentration on the waste side steadily declined and dropped to about 3 ppm overnight, whereas PNP level on the caustic side rose to about 3800 ppm, representing a concentration ratio of over 1000 on the strip side over the waste side. Based on conductivity measurements and chloride ion analysis, only about 0.3% of the salts in the waste passed through the membrane despite the very long contact time allowed. pH values for both solutions were fairly constant throughout the run suggesting no significant transport of $H^+$ and $OH^-$ ions. Water transport across the membrane after several days of operation was not noticeable. These results show that the membrane was quite remarkable in its ability to maintain a large pH differential and significantly block the water transport under

**Table 5.** PNP recovery from a phenolics waste*

| | PNP (mg/L) (at equilibrium) | Chloride (at equilibrium) | Overall Mass Transfer Coeff. (cm/sec) |
|---|---|---|---|
| • T = 25°C ........................ | | | 3.4 x $10^{-4}$ |
| Strip side | 3788 | 0.049 g/L | |
| Waste side | 3.1 | 100 g/L | |
| • T = 60°C ........................ | | | 7.0 x $10^{-4}$ |
| Strip side | 4042 | 0.31 g/L | |
| Waste side | 3.7 | 100 g/L | |

**Membrane: Composite-Celgard 2400; Waste solution: 4000 ppm PNP, 20% KCl, pH 4.5; Strip side: 0.1 N NaOH*

the tremendous osmotic pressure difference of several thousand psi exerted by the 20% salt solution while efficiently transporting PNP across. Table 5 summarizes the results obtained at 25 and 60°C. The high temperature doubled the rate without an adverse effect on salt separation and pH differential across the membrane.

Results for PNP/salt separation using composite hollow fiber membranes prepared with polypropylene (Celgard) and polysulfone (Permea) hollow fibers as the supports are shown in Figure 5 & 6. The hollow fiber configuration is preferred in industrial applications due to its more compact form (higher membrane area per unit volume of packing). Hollow fiber membranes are similar to flat sheet ones in their ability to remove PNP from the saline waste. Note in Figure 6 that removal and simultaneous concentration of PNP was accomplished in this batch operation simply by using a strip volume one-tenth that of the feed. In continuous counter- or co-current operation, the relative flow rates of feed and strip solutions determine the concentration factor.

Effects of the types of membrane support and their structure on PNP rates of transport through the composite membranes are shown in Table 6. Of the two different hollow fiber supports, it was interesting that while the polysulfone membrane was over 4 times thicker than the Celgard membrane (140 μm vs. 30 μm), its PNP transport rate was somewhat comparable to that obtained with the Celgard. The asymmetric nature of the polysulfone hollow fiber support along with its significantly higher porosity probably compensate for the thicker wall.

Based on the rate data obtained with a polypropylene-based (Celgard X20-400 from Hoechst-Celanese) composite hollow fiber membrane, we estimated that a membrane area

**Table 6.** Effects of support characteristics on PNP transport *

| Membrane Support | NaOH Con. | Overall Mass Transf. Coeff. (cm/sec) |
|---|---|---|
| • **Flat Sheet** | | |
| Celgard 2400 (polypropylene) | 0.1 M | 3.4 x $10^{-4}$ |
| | 0.5 M | 3.5 x $10^{-4}$ |
| • **Hollow Fibers** | | |
| Polysulfone | 0.1 M | 1.7 x $10^{-4}$ |
| Celgard X20 400 | 0.1 M | 2.0 x $10^{-4}$ |

**Feed: 4000 mg/L PNP in 20% KCl solution, pH = 4.5; Membrane: composite membranes with various supports; Strip solution: NaOH; room temperature.*

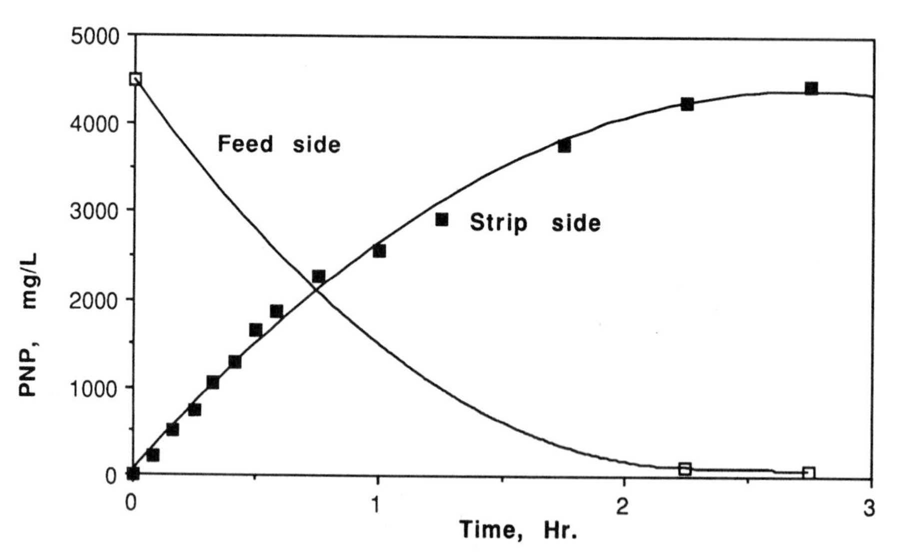


Figure 5. PNP removal with a composite hollow fiber membrane. (Membrane module: Celgard X20-400 fibers, area = 122 cm2; Feed: 4500 ppm PNP, 20% KCl; Strip: 0.1N NaOH

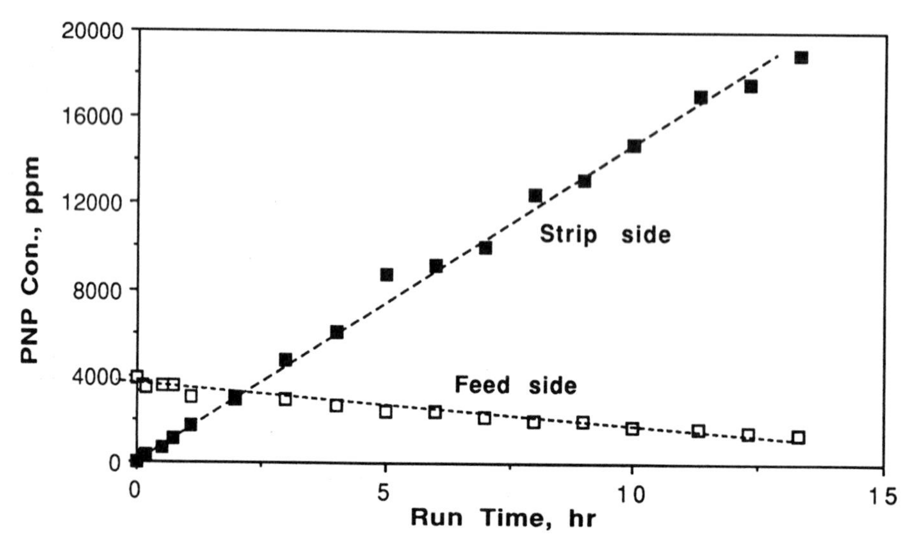


Figure 6. PNP removal with a composite polysulfone hollow fiber membrane. (Membrane module: fibers from Permea, area 120 cm2; Feed: 4000 ppm PNP, 20%KCl, lumen side, 1 liter volume; Strip: 0.1N NaOH, shell side, 0.1 liter volume)

about 15,000 $ft^2$ is required to remove over 99.99% of PNP in a 15 gal/minute waste stream at 60°C (to get below the 200 ppb needed to meet the regulation discharge). On the basis of 6 to 10 dollars per $ft^2$ of membrane installed cost[10], the required membrane area translates into a maximum of $150,000 for the entire membrane module. This is contrasted with an alternate adsorption treatment using activated carbon, which costs from $500,000 to $1 million a year for the carbon. Since the membrane module is likely to last at least a year or so, the membrane process looks extremely attractive, especially if the recovered PNP from the waste could be recycled for some payback value. While PNP is a particularly attractive case, it should be emphasized that the ability to control rate through control of the membrane structure can permit the application of this approach to many cases where the rates based on existing membrane supports are much lower than for PNP.

## Carboxylic Acids Waste

This waste stream is generated from one of our manufacturing plants. The organic content of the waste comprises mainly C2-C6 monobasic carboxylic acids (about 3800 mg/L organics or 2000 mg/L TOC). The stream also contains 1% nitric acid (pH 1.2) and trace metals and is at 60-70°C. Our study involved using the membrane process to simultaneously remove and concentrate the carboxylic acids from the waste solution to facilitate further treatment. At pH 1.2, all the carboxylic acids in solution are in their neutral (uncharged) state and would permeate through the membrane, though at different rates as shown in Table 4. Nitric acid and metal ions also present in the solution exist as charged species, hence can not go through the membrane. The ability of the membrane to remove the organic acids leading to significant reduction in the TOC of the solution is demonstrated in Figure 7. In this run, a 10% sodium carbonate solution was used as the strip solution to pick up the organic acids permeating through the membrane. At the end of the run, TOC in the waste had dropped from the initial value of 2000 ppm to about 50 ppm, equivalent to over 97% TOC reduction.

As shown in Table 4, the carboxylic acids pass through the membrane at different rates, with caproic acid (C6) being the fastest followed by smaller acids in the order of size. In accordance with this, we found that caproic acid, valeric acid and butyric acid in the waste stream passed through the membrane in this order. Because of this characteristic of the process, the rate of TOC removal steadily goes down as more TOC is being removed from the solution. We found, however, that up to 90% TOC reduction the transport process could be approximated by two average rates. Linearized plot of concentration vs. time gives two rate constants: one is termed $K_f$ initial and is valid up to 50-60% removal, the other is $K_f$ final and is good up to about 90% removal. $K_f$ final is roughly half the value of $K_f$ initial. Figure 8 shows such a plot. This approximation allows simple estimation of the membrane area required for a particular degree of TOC removal. Since the rate is directly related to the membrane area required, hence membrane cost, the degree of TOC removal required is a critical parameter in the economics of this process.

The rates of TOC removal from the waste solution using various commercial microfiltration membranes as supports (polypropylene, teflon, polysulfone) are shown in Table 7. Of all the membrane supports tested, the best one in terms of rate appears to be the flat sheet Celgard 2500 (polypropylene). The polysulfone hollow fiber membrane exhibits comparable rate to the polypropylene fiber membrane. Higher temperatures enhance the rate of TOC removal significantly: the rate goes up about 2.5 times as temperature increases from 25 to 65 °C (Table 8). This is an advantageous feature of the membrane process since the waste stream as it exists at the plant is at 65°C.

The organic acids could be recovered in water or a caustic solution. With water as the receiving solution, no concentration of the acids was achieved. With caustic, simultaneous removal and concentration of the recovered organic acids could be done. In an experiment similar to the one shown in Figure 7, concentrating organic acids in the sodium carbonate strip solution to about 6 wt% organics was accomplished along with TOC reduction in the waste solution. This is equivalent to 15-fold concentration of the feed. The composite membrane appears to be stable at the high temperature and in the presence of nitric acid of the waste solution. Preliminary tests of the stability of the membrane were carried out by treating the waste solution continuously for over two weeks. Separate characterization (in terms of

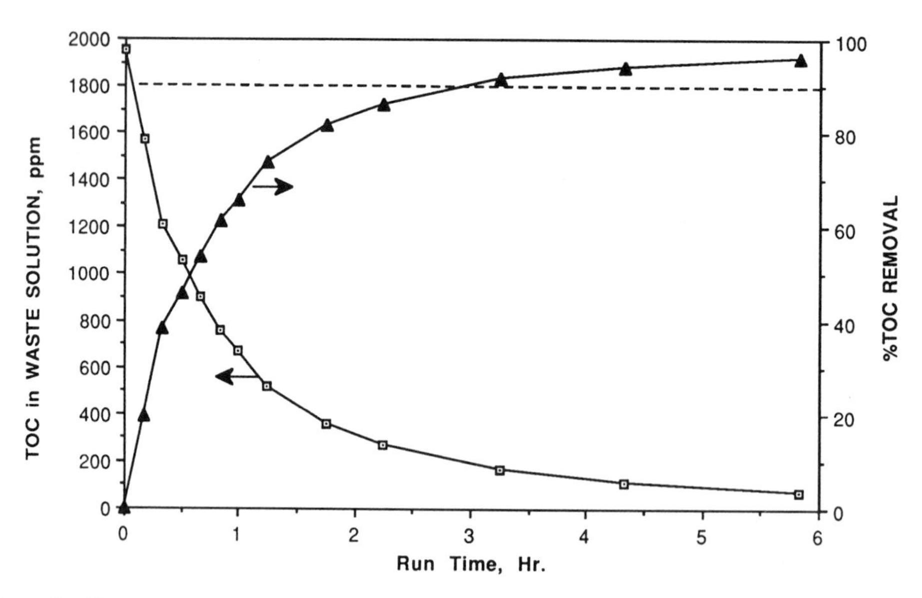


Figure 7. Treatment of a carboxylic acids waste with composite polypropylene hollow fiber membrane. (Membrane module: Celgard X20-400, area = 256 cm2; Feed: 200 ml waste solution, lumen side; Strip: 35 ml 10% sodium carbonate, shell side; T = 60 C)

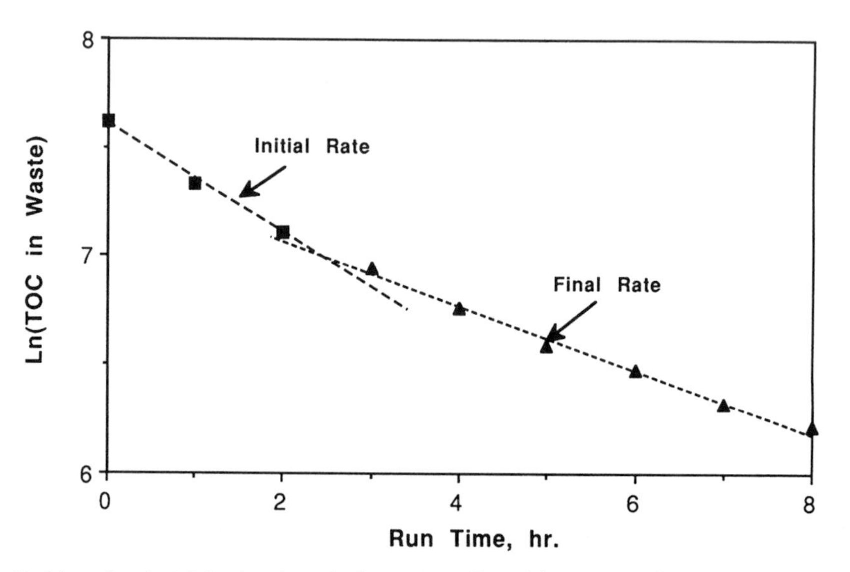


Figure 8. Linearized plot for treatment of a carboxylic acids waste using composite membrane.

**Table 7.** Effects of membrane supports on treatment of a carboxylic acids waste stream using composite membranes

| Membrane Support | Temperature °C | $K_f$ initial [(a)] x $10^4$ cm/sec | $K_f$ final [(b)] x$10^4$ cm/sec |
|---|---|---|---|
| **Flat Sheet** | | | |
| Celgard 2400 | 25 | 0.9 | 0.6 |
| Gore-Tex (0.1μm pore) | 25 | 1.6 | |
| **Hollow Fiber** | | | |
| Polysulfone | 25 | 1.0 | |
| Celgard X20-400 | 25 | 0 .7 | |

*(a) $K_f$ initial is the overall mass transfer coefficient for up to 60% TOC removal; (b) $K_f$ final is the overall mass transfer coefficient from 60 to 80-90% TOC removal.*

**Table 8.** Effect of temperature on rate of TOC removal using a composite Celgard membrane.

| Temperature °C | $K_f$ initial (x $10^4$ cm/sec) | $K_f$ final (x $10^4$ cm/sec) |
|---|---|---|
| 25 | 2.4 | 1.4 |
| 47 | 3.1 | 1.7 |
| 65 | 6.3 | 3.5 |

carboxylic acids transport) of the membrane before and after this extended run shows practically no changes.

Using the rates obtained with the Celgard 2500-based composite membrane at 65°C (Table 8) we estimated that a membrane area approximately 150,000 ft$^2$ is needed to reduce 90% TOC of the 400 gal/min waste stream. Of the 150,000 ft$^2$, less than 50,000 ft$^2$ is required for reduction of the initial 60% TOC, the rest of the membrane area is just for removing the additional 30% TOC. This results from the combined effect of membrane transport (first-order rate) and the chemical makeup of waste solution. On the basis of installed cost of \$10/ft$^2$ with allowance for foundations, pumps, piping, etc., the whole membrane system cost could be about \$2 million, which compares very favorably with other treatment approaches such as distillation, reverse osmosis, and biotreatment. Since a very large membrane system is involved, the actual membrane cost per square foot could be lower. Also, a hybrid system in which membrane is used to remove 60% TOC or so, followed by a less expensive alternative for reducing the remaining TOC (e.g. biodegradation), may result in a lower overall cost.

## ORGANICS SEPARATION

The composite membranes can be used to not only recover but also separate organic compounds based on differences in their physical properties such hydrophobicity, hydrogen-bonding capability, and degree of dissociation as indicated by their pKa's. The example here demonstrates the separation achieved between p-nitrophenol (pKa=7.1) and benzoic acid (pKa=4.25) based on the difference in the pKa's by varying the solution pH. The results are shown in Table 9. As pure component at pH several units below their pKa's, these compounds pass through the membrane at fairly high rates. As a mixture at a solution pH of 5.7, PNP exists in solution mainly in the non-ionized form whereas benzoic acid is practically all ionized. PNP was found in this case to readily pass through the membrane, as

**Table 9.** Separation of p-Nitrophenol and Benzoic Acid

| Solution | pH | Membrane Permeability* | |
|---|---|---|---|
| | | Benz. acid | PNP |
| Pure components | <pKa | 4.0 | 4.7 |
| Mixture (2000 ppm PNP,1700 ppm Benz. acid, 20% KCl) | 2.6 | 3.3 | 5.1 |
| Same as above | 5.7 | 0 | 5.97 |

* *Membrane permeability unit:* $10^{-4}$ *cm/sec*

expected, resulting in about 99% PNP removal from the feed solution overnight. Benzoic acid, however, was virtually blocked by the membrane: overnight only trace amount of benzoic acid (<1% of feed concentration) was detected in the strip solution. It should be noted that this remarkable separation was accomplished in the presence of very high ionic strength (20wt% KCl), which would render ion-exchange methods practically inoperable for this separation. To further demonstrate the effect of pH and ionization on the transport rates, the same experiment was repeated but this time at solution pH of 2.6, which is below both pKa's. Overnight, both PNP and benzoic acid were virtually gone from the feed, having been transported completely to the strip side, and the measured transport rates for the two as a mixture were fairly comparable to those obtained with pure components.

## CONCLUSION

We have developed a proprietary membrane process effective for removing and concentrating low MW organic compounds from dilute aqueous solutions, especially those that also contain high concentrations of inorganic salts which render treatment approaches such as reverse osmosis and ion-exchange ineffective. Our membrane process can handle a broad class of organic compounds including phenolics, aromatic acids and amines, carboxylic acids, alcohols and esters. The membrane appears to exhibit transport rates for common organic pollutants significantly higher than existing membranes. Tested with two actual waste streams, the process appears superior to adsorption using activated carbon, reverse osmosis, and various alternatives (e.g. oxidation, biodegradation). Attractive features of this membrane process include the ability to recover the contaminants in concentrated form for either recycle or more economical disposal, low pressure (ambient) operation, and simple scale-up using commercial hollow fiber modules. We are currently looking into additional applications of this technology such as fermentation products recovery.

## ACKNOWLEDGEMENTS

The author would like to thank J. J. Burke and J. Long for supplying the polysulfone hollow fiber modules; S. Lemp and D. Vinjamoori for analytical support ; and E. Krupetsky for his technical assistance.

## REFERENCES

1. "Membrane Separation Systems - A Research and Development Needs Assessment," Final Report by U.S. Department of Energy, April 1990.

2. S. B. McCray, R. W. Wytcherley, D. D. Newbold, and R. J. Ray "A Review of WasteWater Treatment using Membranes," Proceedings of The 1990 International Congress on Membranes and Membrane Processes, Vol. II, pg 1097-1099.

3. P. S. Cartwright "Membranes for Industrial Wastewater Treatment - A Technical/Application Perspective," Proceedings of The 1990 International Congress on Membranes and Membrane Processes, Vol. II, pg 1131-1133.

4. M. C. Yang and E. L. Cussler, "Artificial Gills," *J. Memb. Sci.*, 42, 273 (1989).

5. R. Prasad and K. K. Sirkar, "Dispersion-Free Solvent Extraction with Microporous Hollow-Fiber Modules," *AIChE J.*, 33, 1057 (1987).

6. Hoechst Celanese Product Literature (1991).

7. W. L. Gore & Associates, Inc,. Publication Literature (1990).

8. E. Kline, J. K. Smith, R. E. C. Weaver, R. P. Wendt, and Shyamkant V. Desai, "Solute Separations from Water by Dialysis. II. Separation of Phenol by Downstream Conjugation," *Separation Science*, 8(5), 585 (1973).

9. A. Sengupta, R. Basu, and K. K. Sirkar, "Separation of Solutes from Aqueous Solutions by Contained Liquid Membranes," *AIChE J.*, 34, 1698 (1988).

10. B. Reid, Hoechst Celanese Corporation, Personal Communication (1992).

11. S. Whitaker, "Fundamental Principles of Heat Transfer," Pergamon Press Inc., New York (1977).

## APPENDIX

## MATHEMATICAL MODELS FOR MEMBRANE STIRRED CELL AND HOLLOW FIBER SYSTEM.

### Membrane Stirred Cell

Rate of organic transport across the membrane in a stirred cell

$$-V_f \, dC_f/dt = A_m K_f \, (C_f - C_s/K_o) \qquad (1)$$

where
- $A_m$ : membrane area, cm2
- $K_f$ : overall mass transfer coefficient in cm/sec
- $C_f$ : organic concentration on feed side, mg/cm3
- $C_s$ : organic concentration on strip side, mg/cm3
- $K_o$ : ratio oi equilibrium concentration (mg/L) of the organic compound in strip solution over that in feed.

Material balance for the organic component, assuming that none was present in the in the strip solution initially and that the amount taken up by the membrane is not significant:

$$V_f C_f^{\circ} = V_f C_f + V_s C_s \qquad (2)$$

where $C_f^{\circ}$ : initial organic concentration in feed

Solving for $C_s$:

$$C_s = (V_f C_f^{\circ} - V_f C_f)/ V_s \qquad (3)$$

then

$$C_f - C_s/K_o = C_f (1 + V_f/V_s.K_o) - V_f C_f^{\circ}/ V_s.K_o \qquad (4)$$

Let $\alpha_1 = 1 + V_f/V_s.K_o$ (5)

$\alpha_2 = - V_f C_f^\circ/V_s.K_o$ (6)

Equation 4 becomes:

$$(C_f - C_s/K_o) = \alpha_1 C_f + \alpha_2 \quad (7)$$

Equation 1 becomes

$$\frac{dC_f}{(C_f - C_s/K_o)} = \frac{- A_m K_f \, dt}{V_f} \quad (8)$$

Substitute in equation 7 and integrate:

$$\frac{dC_f}{(\alpha_1 C_f + \alpha_2)} = \frac{- A_m K_f \, dt}{V_f} \quad (9)$$

$$\ln (\alpha_1 C_f + \alpha_2) = - (\alpha_1 A_m K_f/V_f)\, t + \text{constant} \quad (10)$$

Plotting $\ln (\alpha_1 C_f + \alpha_2)$ vs. time would yield a straight line with slope equal to $- (\alpha_1 A_m K_f/V_f)$, from which $K_f$ can be computed.

With $V_s = V_f$ and $K_o >> 1$:

$\alpha_1 = 1 + V_f/V_s.K_o = 1$

$\alpha_2 = - V_f C_f^\circ / V_s.K_o = - C_f^\circ / K_o$

Equation 10 is simplified to

$$\mathbf{\ln (C_f - C_f^\circ/K_o) = - (A_m K_f/V_f)\, t + constant} \quad (11)$$

The term $C_f^\circ/K_o$ on the left hand side of equation 10 becomes significant when the system gets near equilibrium, i.e. $C_f$ approaching $C_f^\circ/K_o$, at which time the transport process slows down significantly. For large values of $K_o$ ($\geq$100), which is the case for most systems in this study, the actual numerical value of the term $C_f^\circ/K_o$ has little effect on the value of $K_f$ obtained when the system is far from equilibrium.

**Hollow Fiber System**

Assuming plug flow of the feed solution through the lumen side of the fibers (strip solution on the shell side) and no liquid movement through the membrane, mass balance on the hollow fiber unit for counter-current flow gives the following expression for the concentration change in the feed during a single pass[11] :

$$\mathrm{Ln}\left[\left(\frac{C_{fo} - C_{si}/K_o}{C_{fi} - C_{so}/K_o}\right)\right] = - A_m K_f \left(1/Q_f - 1/Q_s K_o\right) \quad (12)$$

Where
$C_{fi}$ & $C_{fo}$ : feed inlet and outlet concentrations
$C_{si}$ & $C_{so}$ : strip inlet and outlet concentrations
$Q_f$ & $Q_s$ : feed and strip flow rates

Note that for large $K_o$, equation 12 can be reduced to

$$\mathrm{Ln} ( C_{fo} / C_{fi} ) = - A_m K_f / Q_f \quad (13)$$

The parameter $A_m K_f/Q_f$ is the ratio of the mass of material transported through the membrane to that passing through the module per single pass. For batch operation in the labs where the amount of membrane area is low relative to the circulation rate, the change in feed concentration per pass through the module is very low, and $K_f$ can not be obtained accurately from equation 12 or 13. Recirculating the feed solution through the module is needed to build up the concentration change.

Combining equation 12 with the unsteady-state mass balance for the feed reservoir (assuming that feed and strip solutions in the reservoirs are well mixed and their concentrations in the reservoirs are the same as those entering the hollow fiber module), the following equation can be obtained relating the concentration in the feed reservoir to time.

$$\text{Ln}\left[\frac{C^{o}_{si}/K_o - C^{o}_{fi}}{(C^{o}_{si}/K_o - C_{fi}) + (V_f/K_oV_s)(C^{o}_{fi} - C_{fi})}\right] = \phi t \qquad (14)$$

Where $V_f$ & $V_s$ : volumes of feed and strip reservoirs
$C^{o}_{fi}$ & $C^{o}_{si}$ : initial feed and strip concentrations

and
$$\phi = (1/V_f + 1/V_sK_o)\,[\,1 - \exp(\beta)]\,/\,[1/Q_f - \exp(\beta)/Q_sK_o] \qquad (15)$$

with
$$\beta = -A_m K_f\,[1/Q_f - 1/Q_sK_o] \qquad (16)$$

Equation 14 can be used to obtain the overall mass transfer coefficient from the data of feed concentration change vs. time.

For the special case of large $K_o$ and with $V_f$ and $V_s$, and $Q_f$ and $Q_s$ being somewhat comparable in values, equation 14 can be significantly simplified to

$$\mathbf{Ln\ (C_f/C^{o}_f) = (Q_f/V_f)\,[exp(-K_f\,A_m/Q_f) - 1]\ t} \qquad (17)$$

And if $A_mK_f/Q_f$ is very small (minute concentration change per pass through the module), equation 17 can be further reduced to the following

$$\text{Ln}\ (C_f/C^{o}_f) = -(A_m K_f/V_f)\,t \qquad (18)$$

which is the same as equation 11 that was derived for the membrane stirred cell.

# APPLICATION OF IMMOBILIZED CELL TECHNOLOGY FOR BIOTREATMENT OF INDUSTRIAL WASTE STREAMS

Daniel E. Edwards[1] and Michael A. Heitkamp[1]

[1]Environmental Sciences Center
Monsanto Company
800 North Lindbergh Boulevard
St. Louis, MO 63167

## INTRODUCTION

Social and regulatory pressures are forcing industry to further reduce environmental discharges of hazardous chemicals. Monsanto, like many other companies, is focusing its waste reduction efforts in three main areas: 1) Improved process efficiency leading to less waste produced. 2) Recycling of waste chemicals and unreacted intermediates. 3) Innovative waste treatment technologies. This chapter will discuss applied research designed to evaluate the fluidized bed reactor (FBR) as one type of innovative technology for biotreatment of industrial wastes.

Once process modifications and chemical recycling have been maximized, some compounds are likely to remain in plant effluents. Whenever possible, biological treatment is the method of choice to remove these waste chemicals, since it is usually less expensive, often results in more complete destruction of compounds, and is sometimes more socially acceptable than technologies such as carbon adsorption and incineration. The most common configurations of biological treatment are activated sludge and trickling filters. In the case of activated sludge, microorganisms are freely suspended in wastewater and aggregate into biomass units called flocs. In trickling filters, microorganisms coat a bed of material (such as gravel or plastic rings) to form a thin biofilm and wastewater is sprinkled over the top of the bed. Over the past sixty or more years as these processes have been widely used, various mechanical improvements and innovations have increased the performance of these systems. However, recent needs for higher levels of chemical removal in wider applications than ever before has created interest in a new generation of biological waste treatment technologies.

There are two main forms that improvements to conventional biotreatment could take: 1) improved performance; 2) reduced cost. Factors improving performance include removal

*Industrial Environmental Chemistry*, Edited by D.T. Sawyer
and A.E. Martell, Plenum Press, New York, 1992

of previously non-degradable compounds, more complete removal of compounds, and tolerance to harsh conditions such as high salt, high temperature, toxic mixtures of chemicals, surge loadings, etc. Factors reducing cost of the systems include resistance to upsets (reduced down time), reduced sludge production (lower sludge handling costs), and reduced reactor size (lower capital and operating expense). One new biotreatment approach which is promising in many of these areas is immobilized cell technology (ICT).

## Configurations of Immobilized Cell Technology

The basic concept of ICT involves bacteria attached to a submerged solid support; waste water is then passed through the bacteria/support complex. There are a number of configurations of ICT in use or under development at this time. The main difference between them is the type of solid support and the degree of motion of the support (i.e. biocarrier). Biocarrier beds may be fixed (i.e non-moving or "packed"), expanded (i.e. lifted slightly by upflow of wastewater), or fluidized (i.e. lifted significantly by upflow of wastewater with each support particle being in motion independent of other particles). The subject of this paper is the configuration of ICT as an FBR.

One common feature of all FBRs is a reaction vessel containing a volume of sand-sized granules which act as the bacterial support. A flow of wastewater passed upward through the biocarrier bed is sufficient to fluidize the granules to a bed height approximately 40% greater than the unfluidized (i.e. collapsed) bed height (Figure 1).

Following inoculation and operation of the bioreactor, each granule becomes coated with a relatively low density biofilm which decreases the overall density of the granules and results in a higher level of fluidization (Figure 1). The biofilm thickness must be mechanically controlled by some type of shearing system in order to prevent excessive biofilm development which may decrease reactor performance or cause the bed to be flushed out of the reactor vessel.

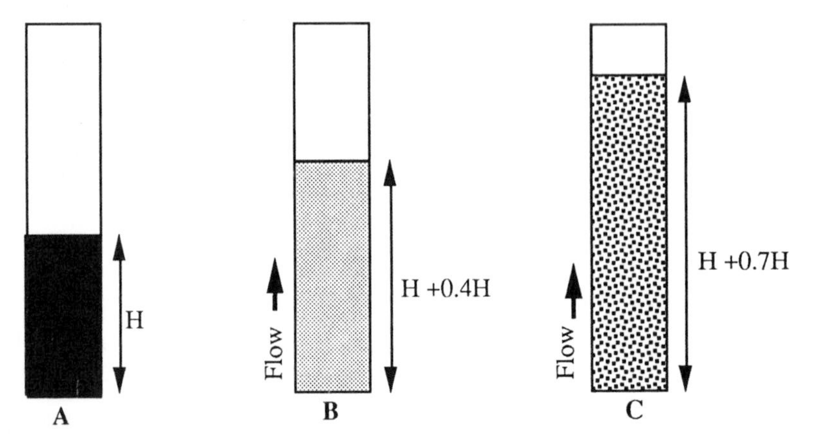


**Figure 1.** Diagram of biocarrier beds (H=bed height): A) before fluidization; B) following initial fluidization; and C) fluidization after full biofilm development.

The primary advantage of the FBR configuration is that small granules provide an expansive surface area for microbial growth within a small reactor volume. An additional advantage is that the constantly moving bed is less prone to the biomass clogging experienced in packed bed reactors. Another significant feature of this system is that oxygen must be dissolved into the waste water before treatment, since excessive bubbling in the bed creates turbulance which can disrupt optimum fluidization conditions. This feature may be a disadvantage for high strength waste streams since the low solubility of oxygen may result in a system which is oxygen limited at high loadings. However, external oxygenation would be an advantage for treating waste streams containing volatile compounds since these materials will not be stripped out of the water as may occur in air-sparged systems. FBRs have been examined for wastewater treatment since the early 1970's.

### Fluidized Bed Reactors

Ecolotrol Inc. developed the Hy-Flo® FBR as a proprietary process to upgrade municipal wastewater treatment (Jeris et al. 1977). This technology used a fluidized sand bed as a support for microbial growth which provided very high available surface area ( about 1,000 $ft^2/ft^3$ of reactor) leading to a maximum of 40,000 mg $L^{-1}$ of mixed liquor volatile suspended solids (MLVSS). This level of bacteria is about 10 times higher than normally found in activated sludge. This high mass of attached cells allowed "...an extremely high rate system which combines the best features of activated sludge and trickling filtration into one process". This system also claimed to provide greater stability in handling shock loads with minimal sloughing of attached biomass. Jeris and co-investigators presented data from three pilot scale units operated on a municipal waste stream for removal of carbonaceous BOD (biological oxygen demand), nitrification, and denitrification.

Results for BOD removal showed that greater than 90 percent of the BOD could be removed with retention times as short as 6 minutes. However, internal recycling of flow was sometimes needed to dilute the waste so that oxygen would not be limited. Nitrifying bacteria were [characteristically] slow to establish in this FBR study. Once the proper bacteria became established, $NH_4$-N was removed from a level of 19.1 mg $L^{-1}$ in the influent waste to less than 0.2 mg $L^{-1}$ in the effluent with a hydraulic retention time of 10.6 minutes. It is noteworthy that the FBR configuration appeared to be especially favorable for denitrification; $NO_3$-N was reduced from 21.5 mg $L^{-1}$ in the influent to 0.2 mg $L^{-1}$ in the effluent with a detention time of 6.5 minutes.

Since sludge production was low from these reactors, the authors concluded that intermediate clarification would not be necessary and final clarifiers may also be eliminated. Since retention times were short, the FBRs occupied about 5% of the space required by conventional technology to treat the same waste. The ability to reduce reactor sizes and eliminate clarifiers led the authors to project major cost savings through use of this technology.

Sutton and Mishra (1991) presented a historical review of FBR usage during the 1970's and 1980's. They noted that at an international symposium in 1980 the FBR was declared "the most significant development in the wastewater treatment field in the last fifty years". However, by 1990 only 65 full-scale FBRs had been installed in North America and Europe. According to the authors, wider use of the technology was hampered by mechanical scale-up issues, slow development of economically attractive commercial systems, and proprietary constraints. The principle mechanical problems affecting scale-up of this technology were: 1) clogging or uneven flow from the influent wastewater distribution system; 2) oxygen transfer and control: 3) removal of excess biofilm growth from the granules.

The authors listed 24 specific cases of full-scale municipal and industrial applications of FBR technology including ammonia removal from fish hatchery water, denitrification of nuclear fuel processing wastewater, carbonaceous oxidation of chemical wastewater, and various other examples of wastewater treatment. The current largest industrial user of this technology was listed as General Motors, which has installed nine aerobic FBR reactors for treatment of synthetic metal cutting fluids. The conclusion of this review was that if mechanical scale-up problems could be resolved economically, FBR technology would be widely used in many different applications (especially in industry).

## OBJECTIVES

Previous research, as documented by Jeris et al., Sutton and Mishra, and others (Gardener et al. 1988, Heijen et al. 1990, Hickey et al. 1991, Holladay et al. 1978, Hosaka et al. 1991, and others), has shown FBR technology to be highly effective and cost-attractive for a variety of low-strength waste streams. The purpose of research at Monsanto is to examine this technology as it applies to waste streams commonly found in the chemical industry. These waste streams are often different from municipal and simpler industrial waste streams in a variety of ways including (but not limited to) the following features:

- Moderate to high organic concentrations (COD up to 15,000-20,000 mg $L^{-1}$)
- Recalcitrant chemicals.
- Combinations of diverse chemical structure.
- Compounds which are toxic at elevated concentrations.
- Compounds which are regulated to very low concentrations.
- Volatile compounds.
- Biologically unfavorable conditions (e.g. high salt, metals).
- Broad fluctuations in strength and composition of wastes.

If FBR technology could be shown to be more effective than conventional waste treatment under many of these conditions, then it would be useful in chemical industry applications. In addition to evaluating this technology in general, three specific objectives were identified for study:

- Comparison of sand and granular activated carbon (GAC) as biocarriers.
- Development of biomass estimation techniques.
- Estimation of sludge production.

Sand is the biocarrier historically used in FBR's. It has the advantages of low cost (usually less than $20/ton), high density (therefore smaller granules may be used providing a high surface area/unit volume), and excellent fluidization properties (e.g. flows well, no clumping, limited flotation, etc.). GAC is increasingly being used in FBRs because it offers adsorption of certain chemicals as a second mechanism for chemical removal and has a rough, porous surface to facilitate microbial attachment. The disadvantages of GAC are higher cost ($0.50-$1.00/*pound* or more), attrition (10-15% of the GAC may be lost annually due to grinding, deterioration, etc), and lower density (larger particles must be used and excessive flotation occurs more readily). The objective of this study was to directly compare performance of these two biocarriers for treating chemical industry wastes in order to determine whether improved performance of GAC is sufficient to offset the cost advantages of using sand.

MLVSS is used as a measure of biomass in design and operation of activated sludge systems. A similar standard method has not been defined for biomass measurement in attached growth systems such as the FBR. Since high biomass is claimed as one of the major advantages of this technology, a reliable biomass estimation technique was desired to measure this quantity and provide better operational control of the reactor. The second objective of this study was to develop a technique that could provide data directly comparable to MLVSS. In order to test this new technique, it was used to track biomass growth trends in experimental FBR units.

The final main objective of this study was to estimate sludge production from FBRs. Sludge processing, handling, and disposal are some of the most costly factors in most biological waste treatment systems. Sludge production was monitored from the experimental FBR units in this study to determine whether it was lower than expected sludge production from an activated sludge system treating a similar waste stream. If reduced sludge production by FBRs could be documented, it would represent significant cost savings for this technology

## EXPERIMENTAL DESIGN

Two laboratory-scale FBRs were set-up and operated side-by-side (Figure 2). One reactor used river sand (0.6-0.8 mm diameter) as its biocarrier and the other used anthracite based GAC (1.2-1.4 mm diameter) as its biocarrier. The bed of sand or GAC was contained in a plexiglass "reactor column" which was 6 ft. in height and 4 in. inside diameter. The initial, unfluidized, bed heights of both reactors were 22 in. Prior to inoculation, the beds were fluidized to a height of 31 in. (i.e. 40% fluidization) by a recycled flow of 1.2 GPM of water. The recycle flow passed out of a port 13 in. from the top of the reactor column and subsequently flowed through a "settling basin" where heavier solids would drop out of the water. The flow then passed through a recycle pump to a venturi which created a pressure drop of about 15 PSI across the device. Following the venturi, pure oxygen was metered into the recycle flow and allowed to bubble through the water in a sealed "oxygenation" column. Any bubbles not entering solution were pulled back into the flow via a "bubble return line" by the vacuum created in the venturi. Bubbles would continue to recycle in this way until becoming dissolved. Oxygenated water exited the bottom of the column and concentrated feed stock solution was metered into the flow by a small feed pump (normal flow was 5-6 mL $min^{-1}$) pump. The flow then entered the bottom of the reactor column, completing the cycle. Excess water would exit the system through an effluent port located 5

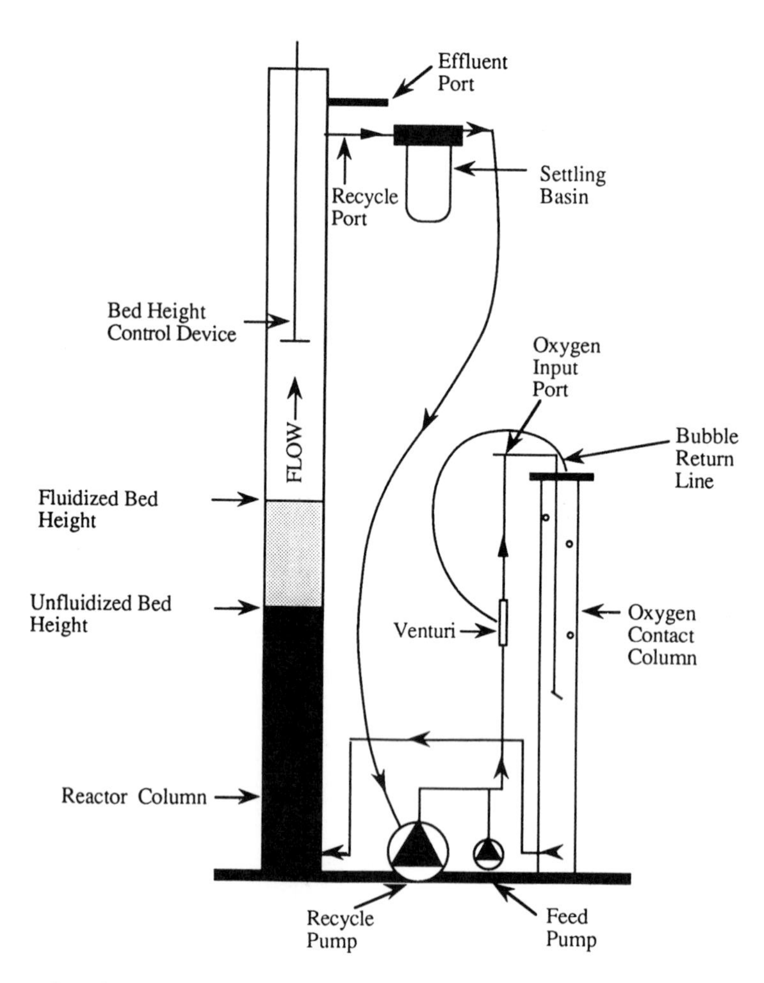


**Figure 2.** Diagram of laboratory-scale fluidized bed waste treatment bioreactor.

in. from the top of the reactor column. The total volume of each system was 20 L.

A synthetic waste stream containing four organic components was chosen for this study. These common industrial waste chemicals were sodium formate (FRM), aniline (ANL), methyl-ethyl ketone (MEK), and paranitrophenol (PNP). FRM was the largest component of this waste stream by weight, being chosen since it would provide much of the COD in a highly degradable form which would not be adsorbed to GAC. ANL is a component of moderate degradability and is highly adsorbed to GAC. MEK is a volatile compound which was added to the mixture to test the ability of the FBRs to degrade it without air stripping. PNP was the component which was least degradable of the four and had the added feature of being highly colored (at $pH \geq 7.0$). The bright yellow coloration of PNP (even at concentrations as low as 1 mg $L^{-1}$) became a useful visual indicator of reactor performance. The appearance of yellow color in the effluent implied that the reactor was suffering a correctable malfunction or was overloaded. During operation, it was possible to eliminate the visible evidence of PNP breakthrough by correcting the malfunction or by reducing feeding until the excess chemical was degraded. The original feed stock was prepared using 5 g FRM, 0.5 g PNP, 0.5 mL MEK, and 0.5 mL ANL to one liter of well water containing ammonium chloride (0.1% w/v) and dipotassium phosphate (0.03% w/v) as inorganic amendments. Loading was increased in seven equal steps (setpoints) by increasing the concentration of each of the organic components until at the highest setpoint feedstock was prepared using 20 g FRM, 5 g PNP, 5 mL MEK and 5 mL ANL. Approximate chemical loadings at setpoints 1-7 were 0.2, 0.3, 0.5, 0.6, 0.75, 0.85 and 1.0 lbs chemical oxygen demand (COD)/ cu ft of fluidized bed/day, respectively. Exact loadings fluctuated slightly according to variations in bed height, feed pump rate speed, and feed concentration. Both reactors were fed from the same container throughout the study.

Each FBR was inoculated with 1 L of fresh mixed liquor from a full size activated sludge plant treating a waste stream containing primarily formate along with various other chemicals. Feeding began at a low rate and was batched (i.e. feeding temporarily suspended) until clear effluent and D.O. (dissolved oxygen) accumulation indicated that the chemicals were removed from solution. As colonization of the granules proceeded, bed height increased gradually. When the beds reached a height of 42 in., a "bed clipping device" was installed (Figure 2). This device consisted of an electric stirring motor with a 36 in. shaft and a horizontal blade placed at the base of the shaft. As the bed reached the level of the blade, the shearing force was sufficient to knock off loose biomass which would float past the blade and exit the reactor column through the recycle or effluent port. This device was effective in maintaining constant bed height at 42 in.

## RESULTS

### Chemical Loading and Removal

Chemical removal (as evidenced by lack of effluent coloring) began almost immediately in the GAC reactor, so loading was quickly raised to the first chemical loading setpoint where it was maintained until about 104 days after start-up (DAS) (Figure 3). The sand reactor required a longer "start-up period" of slow feeding and batching due to visible breakthrough of PNP. At 30 DAS, when COD sampling began, the average chemical loading remained less than 0.05 lbs COD/cu ft/ day. When consistent chemical removal began about 70 DAS, the loading into the sand reactor was gradually increased to setpoint 1 by 100 DAS. At 104 DAS, the carbon reactor's loading was increased to setpoint 2. The sand reactor remained at setpoint 1 for about 20 days to confirm consistent chemical removal and was then raised to setpoint 2. The two reactors were loaded identically (setpoints 2-4) through 125-170 DAS.

At 171 DAS (setpoint 5) a major upset event occurred in the form of a weekend power failure. The loss of electrical power caused the recycle pumps to shut down, resulting in bed collapse and anaerobic conditions. When power was restored, recycle pumps did not re-start, but feed pumps continued to run, resulting in a surge loading of toxic chemicals. Recycle

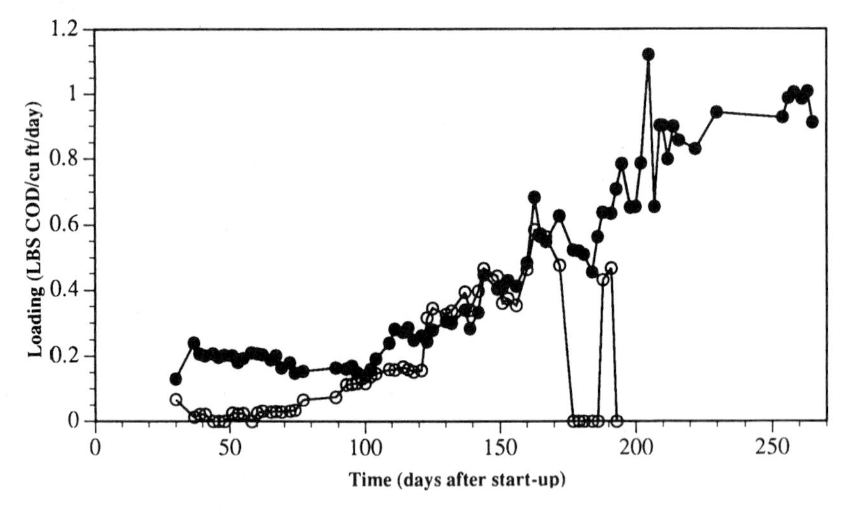


Figure 3. Chemical loading of the GAC FBR (O) and the sand FBR(●).

pumps were re-started manually about 18 hours later. Sand reactor performance was severely damaged by this upset. Loading of the sand reactor was suspended for about 14 days until significant chemical removal resumed. At that point, loading was quickly raised to the pre-upset levels, but continued poor performance led to the shut-down of the sand reactor at 193 DAS. It was understood that the reactor could probably have been brought back to good performance by flushing excess chemical out of the system and restarting under a slower regime of feed increases. However due to time limitations, another lengthy start-up period was not feasible. The GAC reactor was not seriously affected by the power outage, presumably due to adsorption of the toxic compounds in the feed solution. Chemical loading of the GAC reactor was rapidly raised through the last three set-points until being shut down at 265 DAS after having successfully treated a maximum loading of about 1.0 lbs/cu ft/day.

During the middle of the start-up phase (30-60 DAS) of the sand reactor, effluent COD varied between 100-300 mg $L^{-1}$, in response to feed/batch cycles. In the late start-up phase (60-100 DAS) effluent COD was consistently less than 120 mg $L^{-1}$. The average effluent COD from 30-100 DAS was 117 mg $L^{-1}$ which represented 97% reduction in COD of the feed stock concentrate (Table 1). Average COD removal by the sand FBR during setpoints 1-4 was 97%-99% (Table 1). Chromatographic analysis of effluent samples revealed no detectable breakthrough of the four components during setpoint 1 (note that samples were not taken for specific chemical analysis during setpoint 2). At setpoint 3 there were trace

**Table 1.** Average feed stock concentrations, average effluent concentrations, and specific chemical breakthrough during operation of FBRs using sand and GAC as biocarriers.

| Setpoint | Time period<br>Days after Start-up | Feed Stock | Effluent | Chemical Breakthrough[1]<br>Component (mg $L^{-1}$) |
|---|---|---|---|---|
| | | --------COD mg $L^{-1}$-------- | | |
| Sand Reactor | | | | |
| Start-up | 0-100 | 3728 | 117 | Not Sampled |
| 1 | 100-121 | 4741 | 118 | None Detected |
| 2 | 122-142 | 5529 | 92 | Not Sampled |
| 3 | 143-156 | 7322 | 115 | FRM (11.3), MEK (0.46) |
| 4 | 157-172 | 9548 | 191 | MEK (0.49) |
| Post-upset | 173-188 | ----- | 468 | Not Sampled |
| GAC Reactor | | | | |
| 1 | 0-104 | 3745 | 74 | Not Sampled |
| 2 | 105-142 | 5364 | 62 | None Detected |
| 3 | 143-156 | 7322 | 123 | FRM (12.3), PNP (0.05) |
| 4 | 157-191 | 9815 | 128 | None Detected |
| 5 | 192-207 | 12,088 | 155 | FRM (29.7); PNP (0.05) |
| 6 | 208-222 | 14,989 | 315 | FRM (71.3); PNP (0.05) |
| 7 | 223-265 | 17,086 | 243 | FRM (100.7); PNP (0.05) |

[1]FRM= formate, MEK=methyl ethyl ketone, PNP=paranitrophenol, ANL=aniline. Omission of a specific component indicates that chemical was not detected by chromatographic analysis.

concentrations of FRM (11.3 mg $L^{-1}$) and MEK (0.46 mg $L^{-1}$) (Table 1). At setpoint 4, FRM was not detectable, but MEK (0.49 mg $L^{-1}$) was still present in the effluent. COD in excess of quantities accountable by chromatography is presumed to be due to soluble organic products of microbial activity. During the post-upset period (173-188 DAS), average effluent COD remained over 400 mg $L^{-1}$ (Table 1) despite the lack of additional feed input. No specific chemical analysis was performed during the upset period.

Chemical removal by the GAC FBR was superior to the sand FBR during the first 100 DAS due to elimination of the start-up period (Table 1). From 100-172 DAS, when both FBRs were being loaded similarly, COD removal by each was comparable (98-99%). During the last three setpoints (after the upset of the sand FBR), COD removal by the GAC FBR continued at 98-99% through feed stock concentration step-increases from 12,088 mg $L^{-1}$ to 14,989 mg $L^{-1}$ to 17086 mg $L^{-1}$ (Table 1). At the fifth setpoint, FRM was detected in the effluent (29.7 mg $L^{-1}$) along with a trace quantity of PNP (0.05 mg $L^{-1}$). FRM was detected in increasing concentrations at the final two setpoints (71.3 mg $L^{-1}$, 100.7 mg $L^{-1}$ respectively) (Table 1). Trace PNP continued to be detected at the last two setpoints but did not increase in response to the increasing chemical loading. MEK and ANL (both known to be highly adsorbed to GAC) were not detected in the effluent of the GAC FBR.

### Biomass Measurement

Biomass was measured gravimetrically from samples (22 mL) of the sand and GAC beds. Samples from pre-determined bed heights (data shown from 10 in. below top of fluidized bed) were transferred into a 125 mL flask and rinsed three times with distilled water; rinsate was filtered through a 1 μM glass fiber filter for determination of TSS (total suspended solids, Cleseri et al. 1989). This TSS value was used as a measurement of the non-attached solids (i. e. non-attached biomass or NAB) and expressed in mg $L^{-1}$ of fluidized bed. The rinsed bed material was dried at 105°C, weighed, and digested in concentrated nitric acid for 24 hours. The digested sample was rinsed and dried again at 105°C. The change in dry weight following the acid digestion was used to calculate the "attached biomass" of the fluidized bed (mg $L^{-1}$). This quantity was statistically corrected for chemical reaction of the acid with the bed material by treating 10 replicate samples of each virgin bed material by the same technique and determining a correction factor.

Attached biomass in the GAC FBR was found to be approximately 15,000 mg $L^{-1}$ when the first biomass samples were taken after 90 DAS (Figure 4). Biomass remained in equilibrium near that level until the beginning of the fifth setpoint at 190 DAS. When the first samples were taken from the sand FBR after 120 DAS, attached biomass (4,000-5,000 mg $L^{-1}$) was much lower than in the GAC FBR. After 130 DAS when both FBRs were being loaded at the same level, biomass in the sand FBR increased to the same level as the GAC FBR (Figure 4). Attached biomass in the sand FBR decreased to about 5,000 mg $L^{-1}$ following the power failure after 170 DAS (Figure 4), but did recover to near its previous level before being shut down at 193 DAS. Attached biomass in the GAC FBR was not reduced by the upset and increased to a new equilibrium level near 25,000 mg $L^{-1}$ during the fifth and sixth setpoints. Following the final chemical loading increase, attached biomass in the GAC FBR had reached a level of 35,000 mg $L^{-1}$ to 40,000 mg $L^{-1}$.

Non-attached biomass (NAB) in both FBRs was consistently much lower than attached biomass, but was significant (1,000 mg $L^{-1}$ -2,000 mg $L^{-1}$). NAB in both FBRs remained near these same levels throughout the study and did not appear to be responsive to chemical loading increases or system upsets (Figure 4).

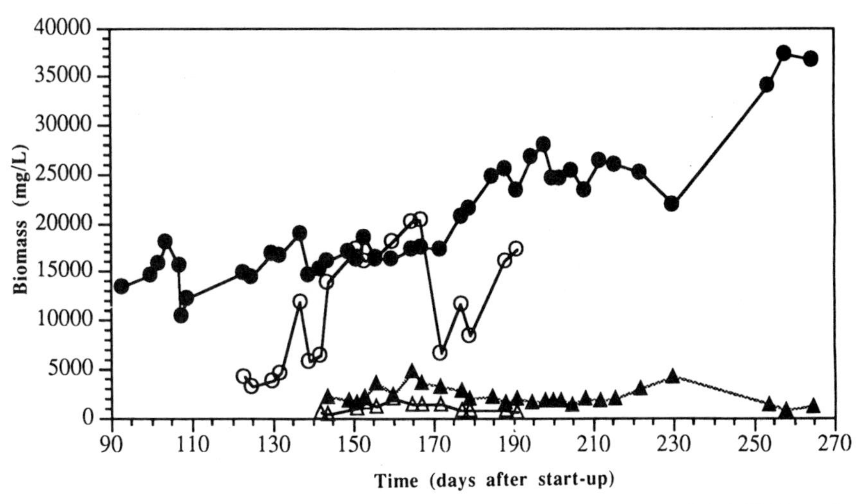


**Figure 4.** Biomass measured from bed samples taken from GAC FBR and sand FBR. Symbols ●, ▲ represent attached and non-attached biomass from the GAC reactor, respectively; symbols O, Δ represent attached and non-attached biomass from the sand reactor, respectively.

## Sludge Production

The two major routes for sludge to exit these laboratory-scale units were: 1) solids suspended in effluent; 2) solids deposited in the settling basins. Throughout most of this study, effluent was pooled into a large carboy for 2-3 days and was sampled at those intervals for TSS. Effluent solids were generally less than 600 mg $L^{-1}$ at steady state (i.e. not immediately after an upset or feed increase) in both FBRs (data not shown). However, since only the lightest solids exited the systems by this route, effluent solids represent only part of the sludge production picture. For several periods of time, both effluent solids and trapped sludge were measured from the systems. Trapped sludge measurement involved allowing the contents of the settling basins to accumulate for discrete periods of time and then sampling the TSS contained within them. Total trapped sludge was determined by calculating the total mass of solids accumulated in the settling basin during that period of time.

An example of data obtained by intensive sampling of effluent solids and trapped sludge is shown in Table 2. This data was collected during the final 11 days in operation of the GAC FBR. Since this was the period of highest loading, it represents the period of highest steady-state sludge production. Total solids exiting the system through the effluent port were calculated to be 27.5 g (dry weight); solids deposited in the settling basin were calculated to be 57.1 g (Table 2). The 2:1 ratio of trapped:effluent solids was representative of data taken during other periods of sampling. Average total attached and non-attached biomass in the bed during this time were calculated to be 315.0 g and 11.2 g respectively (Table 2). By combining the sludge production and total biomass into a single expression over those 11 days, sludge age was calculated to be 42 days (Table 2). COD loaded into the system during the same period totaled 1490 g, leading to a sludge:COD ratio of 0.056 g/g (Table 2).

Experience of Monsanto plants treating formate waste in full-scale activated sludge units has led us to expect sludge age around 20 days and sludge:COD ratios of 0.15-0.2 lb/lb. If sludge production data found in this study is confirmed in pilot-scale testing and full-scale application, this technology will provide high treatment efficiency using biomass which has a long retention time and relatively low sludge production. The economic benefits of reduced sludge production are obvious: reduced sludge handing and disposal costs, downsized or eliminated clarifiers, etc. Increased sludge age (i.e. longer SRT) may allow the use of microorganisms which have intrinsically slow reproductive rates. Processes which use slowly reproducing microorganisms include anaerobic/anoxic systems, nitrification, and recalcitrant chemical degradation.

**Table 2.** Total solids[1] production by the GAC FBR during the period 254-265 DAS.

| | | |
|---|---|---|
| **Sludge exiting system** | | |
| Effluent Solids: | 27.5 g | |
| Trapped Sludge: | 57.1 g | |
| Total: | **84.6** g | |
| **Biomass within fluidized bed** | | |
| Attached biomass: | 315.0 g | |
| Non-attached biomass: | 11.2 g | |
| Total: | **326.2** g | |
| **Total COD treated during period** | **1490** g | |
| **Sludge age:** | | 1). 84.6 g sludge lost ÷ 11 days = 7.7 g $day^{-1}$ |
| | | 2). 326.2 g biomass in bed ÷ 7.7 g $day^{-1}$ = **42 days** |
| **Sludge production:** | | 84.6 g sludge lost ÷ 1490 g COD treated = **0.056 g sludge/g COD** |

[1]All sludge and biomass quantities are expressed as grams dry weight.

## CONCLUSIONS

Aerobic FBRs containing sand or GAC as biocarriers provided excellent COD removal from a moderate-high strength synthetic industrial waste stream (COD up a maximum to 17,000 mg $L^{-1}$ loaded at 1.0 lb COD/cu ft of fluidized bed/day) used. GAC offered major advantages in reduced start-up time and resistance to chemical surge loading. Especially for adsorbable compounds, the advantages of GAC in treating potentially hazardous and/or regulated industrial wastes appear to outweigh high capital cost of the material. Using a mixture of unrelated chemicals did not favor or hinder biodegradation of any specific components.

Biomass in the reactor beds was found at high levels (15,000-40,000 mg $L^{-1}$); the level of biomass was proportional to the level of chemical loading. Biomass data corroborated reports of high MLVSS found in FBRs by Jeris et al. (1977). Sludge production by FBRs was lower and sludge age was longer than would be expected from activated sludge systems treating similar waste.

These laboratory-scale tests provided evidence that Immobilized Cell Technology (ICT) in the form of a Fluidized Bed Reactor (FBR) is suitable for treatment of typical chemical industry waste streams. Due to high treatment efficiency, longer solids retention, and reduced sludge production, FBR technology appears to offer significant cost and technical advantages compared to conventional treatment techniques. Based on data and experience gained from this study, there are no identified biological obstacles to the adoption of this technology for a wide spectrum of industrial uses.

## ACKNOWLEDGEMENTS

This work was funded by the Monsanto Chemical Company. The laboratory-scale FBRs were professionally custom-manufactured by Monsanto Specialty Shop Services. The authors thank William Adams for developing the initial concept of this research and support for this work. We are grateful to the following individuals for their excellent analytical and technical support: Chi Trang (GC-MS), Jon Wehler (HPLC analysis for PNP and aniline), Eileen Hahn (HPLC analysis for formate), and William Garvey (COD, TSS). Inoculating sludge was obtained by Zip Beedle through the coordination of Larry Hallas. Edward Valines added valuable engineering insight and guidance in the early phases of the study.

## REFERENCES

Clesceri, L. S. (Chairman), A. E. Greenberg, R. R. Trussel and M. A. H. Franson (Eds.). 1989. Standard methods for the examination of water and wastewater. 17th Edition. American Public Health Association, Washington, D. C.

Gardener, D. A., M. T. Suidan, and H. A. Kobayashi. 1988. Role of GAC activity and particle size during the fluidized-bed anaerobic treatment of refinery sour water stripper bottoms. Journal WPCF **60**(4):505-513.

Heijen, S. J., A. Mulder, R. Weltevrede, P. H. Hols, and H. L. J. M. van Leeuwen. 1990. Large-scale anaerobic/aerobic treatment of industrial wastewater using immobilized biomass in fluidized bed and air-lift suspension reactors. Chem. Eng. Technology **13**:202-208.

Hickey, R. F., D. Wagner, and G. Mazewski. 1991. Treating contaminated groundwater using a fluidized-bed reactor. Remediation/ Autumn 1991: 447-460.

Holladay, D. W., C. W. Hancher, C. D. Scott, and D. D. Chilote. 1978. Biodegradation of phenolic waste liquors in stirred-tank, packed-bed and fluidized-bed bioreactors. Journal WPCF **50**(11): 2573-2589.

Hosaka, Y., T. Minami, and S. Nasuno. 1991. Fluidized-bed biological nitrogen removal. Water, Environment, and Technology. August: 48-51.

Jeris, J. S., R. O. Owens, R. Hickey, and F. Flood. 1977. Biological fluidized-bed treatment for BOD and nitrogen removal. Journal WPCF **49**:816-831.

Sutton, P. M. and P. N. Mishra. 1991. Biological fluidized beds for water and wastewater treatment. Water Environment and Technology. August: 52-56.

# MEMBRANE-BOUND MICROORGANISMS FOR REMOVING ORGANISMS AND METALS

S. S. Sofer

NJIT Biotechnology Laboratory, Department of
Chemical Engineering, Chemistry, and Environmental Science
New Jersey Institute of Technology
Newark, New Jersey 07102

## ABSTRACT

Many contaminated aqueous industrial streams contain mixtures of organics as well as heavy metals. Bacteria offer the advantage of having the ability to oxidize organics as well as adsorb heavy metals. Membrane-immobilized bacteria have added advantages including improved mass transfer for faster reaction and adsorption rates, as well as the capability to be re-used. Performance characteristics of immobilized bacteria for oxidation and adsorption are herein presented. A hybrid process configuration is proposed for the treatment of heavy metals combined with high concentrations of organics.

## IMPORTANCE OF FLOW RATE

Microbial reaction systems are limited by their physical surroundings. A major limitation is the availability of substrate at the surface of the microorganism. Consider the figure on the right. The curves show that the reaction rate increases as a

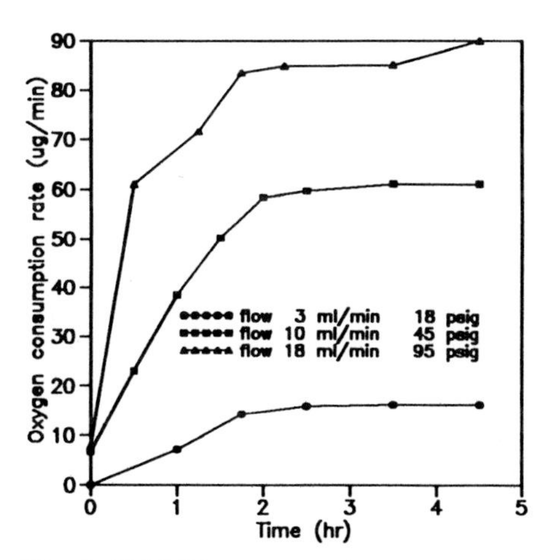


Fig. 1 FMC Bioreactor Performance Effect of flow rate on oxygen consumption rate in single pass configuration.

*Industrial Environmental Chemistry*, Edited by D.T. Sawyer and A.E. Martell, Plenum Press, New York, 1992

function of flow rate – the physical resistance to substrate migration to the cell is drastically reduced by immobilizing the microorganisms and forcing liquid past the reaction surface.

This and other advantages of immobilization are reported by Lakhwala and others (1988), and Mattiasson (1983). The data for this paper are for bacterial consortia designed to oxidize organics. The experimental setup (next page) has been described in detail elsewhere (Lakhwala, Goldberg, and Sofer, 1990 and 1992).

Three runs are illustrated for a single-pass configuration, where contaminated fluid is pumped past the membrane and exits the system. For each run, an initial rise of oxygen consumption is observed upon pumping of 50 ppm phenol in water. This is followed by a plateau at which the maximum steady-state rate is observed. Oxygen consumption rate increases as a function of flow rate because the fluid mixing profile is such that there exists a thinner static layer of water over the bacteria at higher flow, allowing phenol to diffuse more quickly into the active bacterial layer.

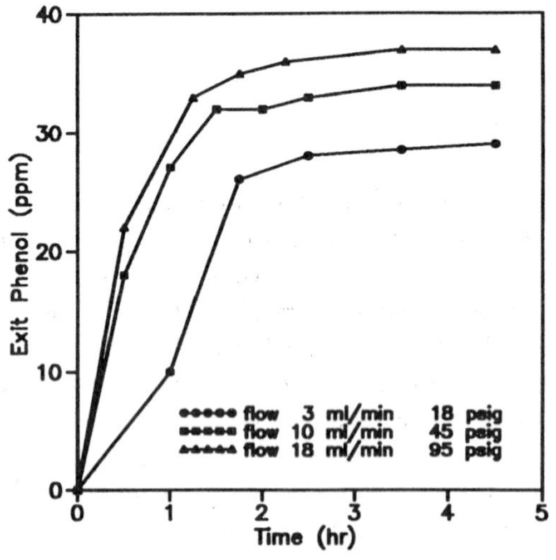


Fig. 2 Effect of flow rate on exit concentration of phenol in membrane reactor (contnuous mode).

Even though the reaction rate increases as a function of flow rate, the single-pass configuration has its limitations. The figure on the right shows the exit phenol concentrations for the same runs. They appear higher because the over-all flow rate is high and carries much substrate through.

This could be a significant disadvantage if the bioreactor is in service for water purification.

One way to solve this problem is to operate under a recirculation flow configuration. Here, the contaminated water is pumped from a reservoir into the bioreactor and flows back into the well-mixed reservoir. This way, the high flow rates are maintained

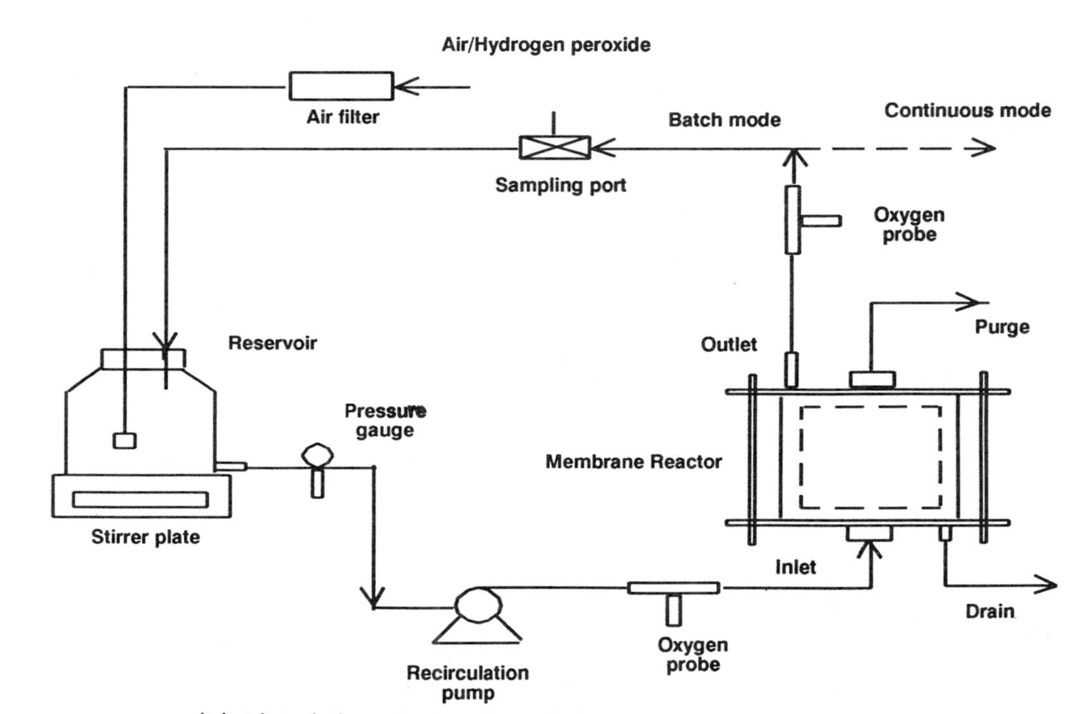


In batch mode the waste stream is recycled between the reservoir and the bioreactor. Oxygen is added to the reservoir in the form of air or hydrogen peroxide. Dissolved oxygen is monitored before and after the bioreactor using probes. In continuous mode the waste stream is pumped at a lower flow through the bioreactor without recycling.

Fig. 3 Experimental Set Up of the Membrane Bioreactor.

to take advantage of the resulting lower mass transfer resistance, and higher associated reaction rates. The liquid is then recirculated until an acceptable contaminant level is reached and the product water is clean.

This is demonstrated in the figures below.

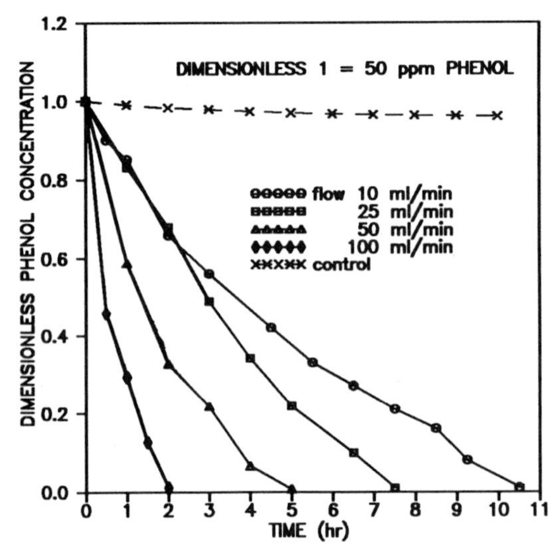


Fig. 4 Variation in Phenol Concentration with Recirculation Flow Rate in Membrane Bioreactor. (batch mode)

In this flow regime, increased flow rate increases biodegradation of phenol. For this membrane bioreactor, a flow rate of 100 ml/min decontaminates the water in less than three hours.

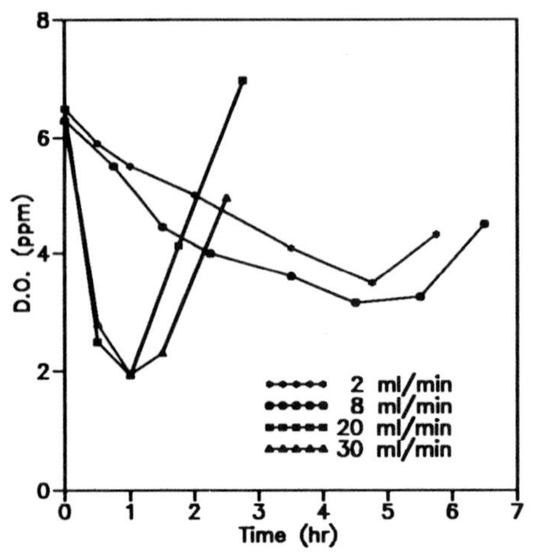


Fig. 5 Variation in Exit DO with Flow in Membrane Reactor (batch mode)

The initial rates for phenol removal at the high flows are in the order of 40 ppm in less than one hour. At these reaction velocities, the exit dissolved oxygen (DO) concentrations can reach very low levels as shown below. When most of the oxidation is accomplished, the exit DO approaches that of the entry level as expected.

Dissolved oxygen concentration is an important parameter to keep in mind throughout these considerations. The bio-oxidations are naturally dependent on oxygen availability.

Therefore when increased flow rate increases reaction rate, a sufficient decrease in DO would have a negative effect on the overall oxidation process due to the lower availability of oxygen.

A recirculation configuration bioreactor is the ideal way to investigate reaction rate as a function of oxygen concentration. Under very high recirculation flow conditions, the change in DO across the reactor lessens, and an accurate assessment of rate versus DO can then be made because the concentration remains essentially constant across the length of the membrane.

## Biomass Availability

A membrane biosupport with too small an amount of bacteria on the surface is not efficient. While the bacteria may remain active, the relative cost of membrane is too high. As the amount of biomass immobilized on the membrane increases, the reaction rate will increase. Beyond a certain point, the biomass layer becomes too thick, and the productivity of the bacteria farthest away from the water/bacterial surface decreases. In this case, the cost of biomass becomes too high. Therefore, an optimum biomass thickness exists for each reaction system. A typical membrane biosupport is wound around itself in a spiral fashion. As bacteria attach to the membrane, they reduce the spacing between each layer of the spiral, resulting in an increased pressure drop across the bioreactor.

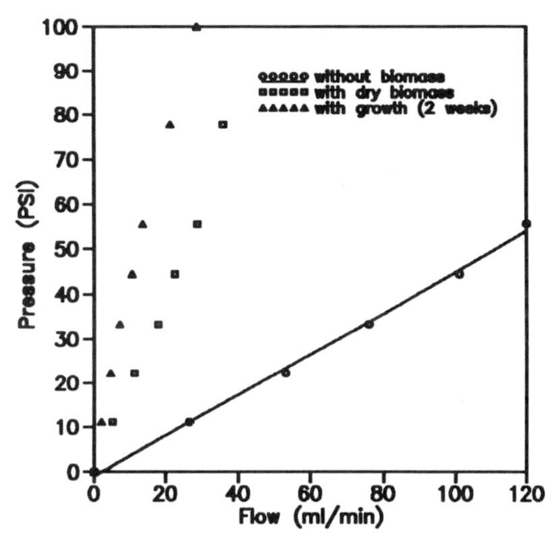


Fig. 6 Effect of Biomass Loading on Pressure-Flow Relationship.

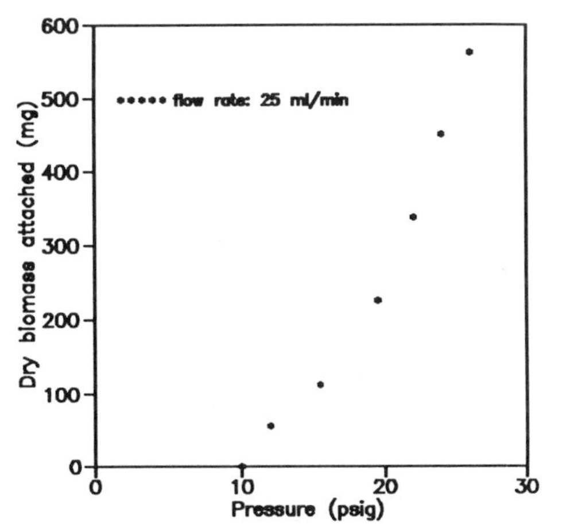


Fig. 7 Effect of Biomass Loading on Operating Pressure.

The figure above demonstrates the effect of biomass loading on the pressure/flow relationship. Pressure at the entrance of the reactor increases upon loading of biomass. If the biomass is allowed to grow, then the pressure increases accordingly. At left is shown the effect of increased biomass loading on operating pressure at constant flow rate.

From these two figures it is apparent that growth increases pressure drop, and high biomass loadings lead quickly to unacceptable pressure drops. Bioreactors for decontaminating wastewaters should be run under no-growth conditions where possible to limit the obstacles posed by this type of rise in inlet pressure.

## OXYGEN AND HYDROGEN PEROXIDE

Oxygen requirement can be met by bubbling with air, using pure oxygen, or using hydrogen peroxide.

Air contains a great deal of nitrogen. This causes the stripping of volatile components into the atmosphere, converting a water pollution problem into an air pollution problem. This justifies, in some cases, the use of pure oxygen or hydrogen peroxide.

Hydrogen peroxide is converted by the enzyme, catalase, into water and oxygen. Since catalase is present in large quantities in bacteria, it becomes a convenient source of enzyme. A hydrogen peroxide driven bio-oxidation has the strategic benefit that it can operate under a closed system—a particular advantage for treatment of certain toxic volatiles.

On the right, the inlet and outlet DO concentrations are shown upon injection of a single bolus of $H_2O_2$ solution into the reservoir of a recirculation mode membrane bioreactor. Oxygen concentration increases dramatically, being higher at the exit. At point B, substrate is injected, and both concentration profiles show an immediate decrease. Within

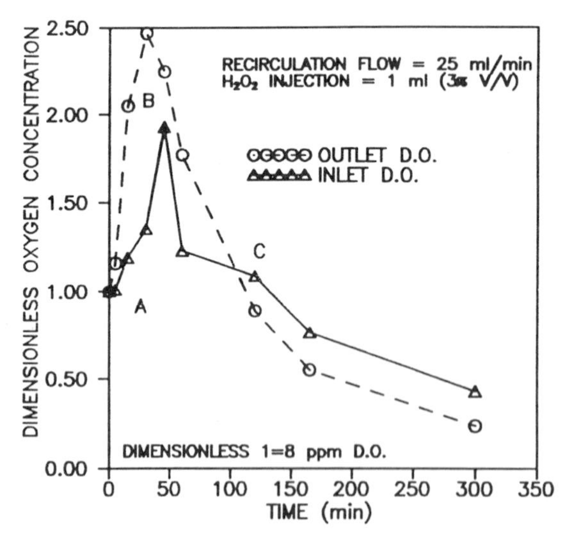


Fig. 8 Release of Oxygen in the Membrane Bioreactor on Addition of Hydrogen Peroxide. (batch mode)

about two hours at point C, the $H_2O_2$ is completely consumed and the outlet concentration becomes lower than the inlet concentration as expected.

Hydrogen peroxide is very expensive, and its use should be carefully controlled. If the oxygen generation rate is too high, oxygen will be lost as the solution goes above saturation.

## SUBSTRATE INHIBITION

Certain substrates such as phenol have an inhibitory and even toxic effect on the bacteria. At high concentrations, rates are drastically reduced. The design of bioreactors dealing with these and other problems is more formally addressed elsewhere (Lakhwala and Sofer, 1991).

A membrane biosupport system has a thin layer of bacterial mass attached to it. As flow rate increases, the effects of substrate inhibition and toxicity increase. Therfore, membrane bioreactors are not recommended for operation at high concentrations of inhibitory substrates.

Entrapping the bacteria within a protective gel for this regime offers significant advantages.

## ALGINATE BEAD BIOREACTORS

Immobilization of bacterial biomass within calcium alginate gel beads yields a powerful moiety for bio-oxidation.

The figure at the top of the next page summarizes the performance of a gel-immobilized bioreactor. The two critical parameters, dissolved oxygen and phenol concentration, are presented. Percent degradation is shown for a single pass mode. As expected, the reaction rates increase as a function of available oxygen. The rates are maximum at 200 ppm phenol – quite high when compared to membrane and especially free-cell systems. Even at 1,000 ppm the reaction is very robust.

The alginate bead forms a protective sphere around the bacteria. As long as oxygen is available, the substrate within the sphere is oxidized. Therefore, in the robust oxidative environment of the gel bead, substrate concentrations remain low and the bacteria are well protected.

In this environment, a difference between oxidation rate and phenol disappearance may be observed as shown below. Oxygen is consumed very quickly, and intermediate products of phenol oxidation build up.

## ADSORPTION OF HEAVY METALS

It has long been known that heavy metals are adsorbed by bacteria. Gourdon, Rus,

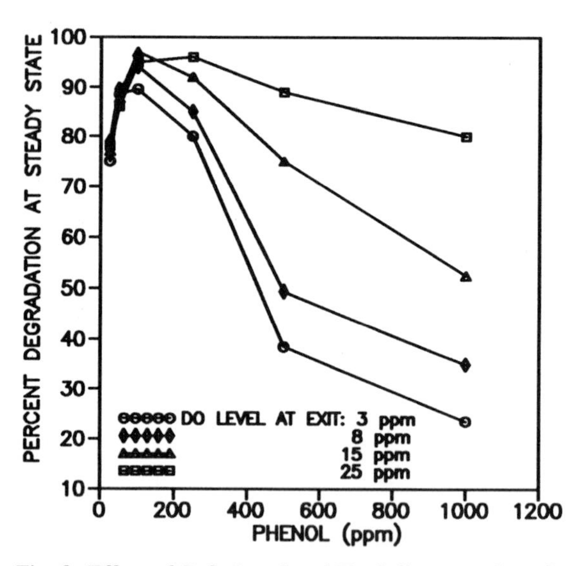


Fig. 9 Effect of D.O. Level and Feed Concentration of Phenol on Continuous Degradation of Phenol at Steady State in the Alginate Bead Bioreactor.

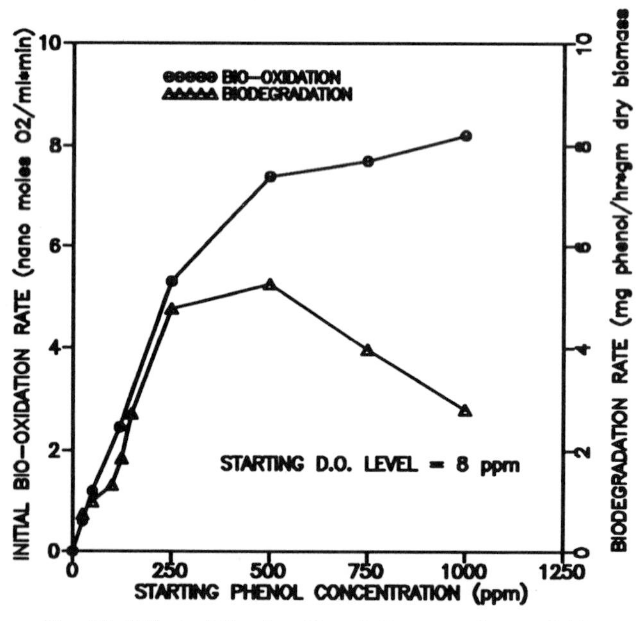


Fig. 10 Effect of Starting Phenol Concentration on Initial Bio-Oxidation Rates and Biodegredation Rates in Calcium Alginate Bead Bioreactor.

Bhende, and Sofer (1990 & 1990) have shown that metals such as cadmium, chromium, lead, and others adsorb very well on immobilized bacteria.

The figure below indicates cadmium uptake as a function of temperature. For lead, the adsorption is in the order of 30 times higher (next page). The adsorption appears to follow well defined behavior, and the cells can be desorbed and re-used.

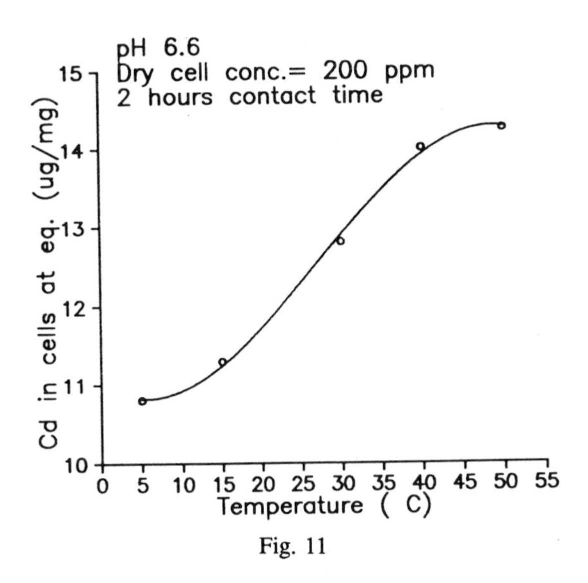

Fig. 11

## HYBRID BIOREACTION SYSTEMS

In our laboratory, we have found that immobilized bacteria remove a great deal of heavy metals before exhibiting any inhibitory characteristics. Heavy metal uptake and bio-oxidation can take place at the same time.

For a system containing organics as well as heavy metals, a primary bioreactor at the start is recommended for the removal of heavy metals. This reactor serves as and adsorber, and may be regenerated by a decrease in pH, operating in cycles.

For high concentrations of organic contaminants, an alginate entrapped system is recommended in recirculation mode. As shown above, gel entrapment offers excellent economy under these conditions. When the concentration of organics is decreased, then a single pass membrane bioreactor is ideal for polishing and cleaning the water to the required specifications.

This difference is even greater amplified for a continous-flow, single pass configuration shown on the next page.

A hybrid gel/membrane system is therefore a potent combination for attacking concentrated organics and heavy metal contaminated waters.

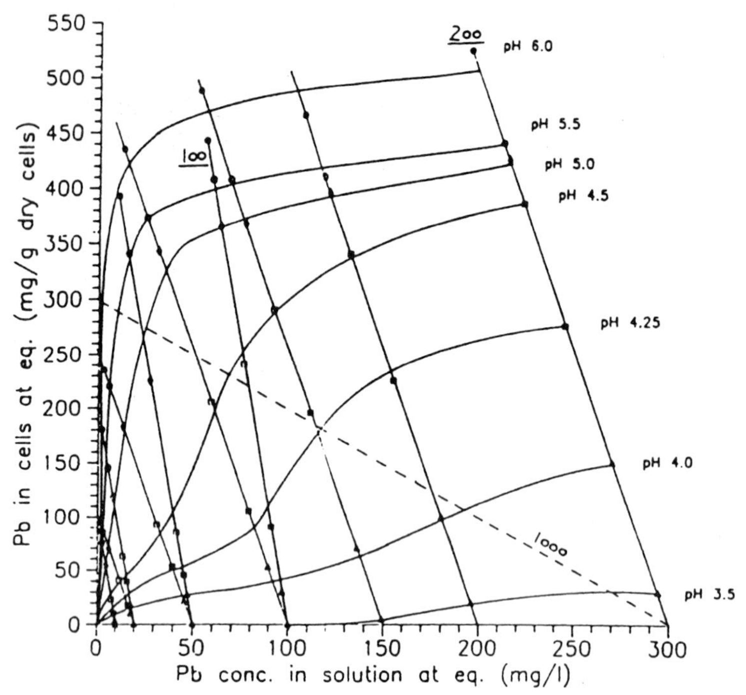


Fig. 12 Isotherm of Lead Biosorption by Free Cells at 30 C
Dry Cell Concentration: 200 mg/l or 100 mg/l.

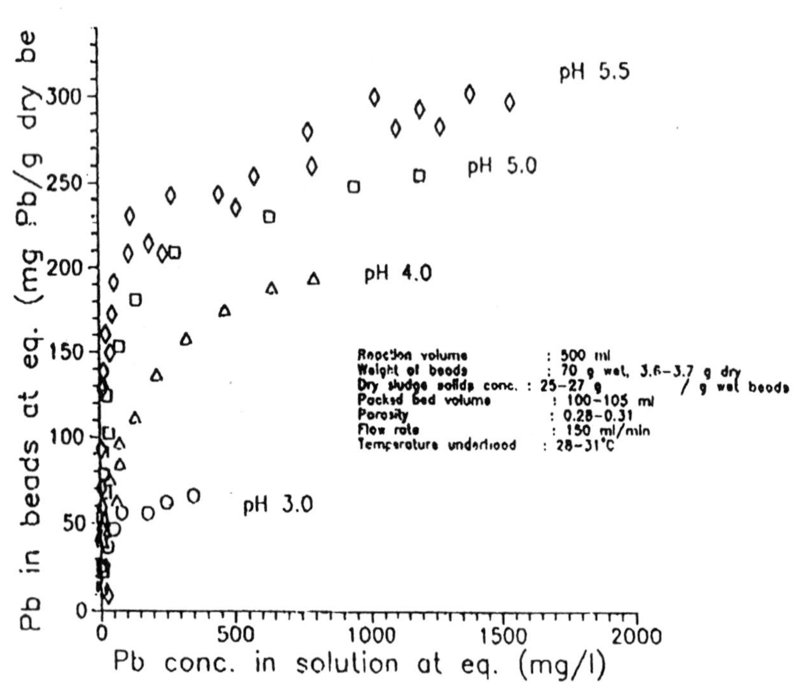


Fig. 13 Isotherm of Lead Biosorption by Bacterial Beads in
Column Reactors Operated in a Recirculation Mode.

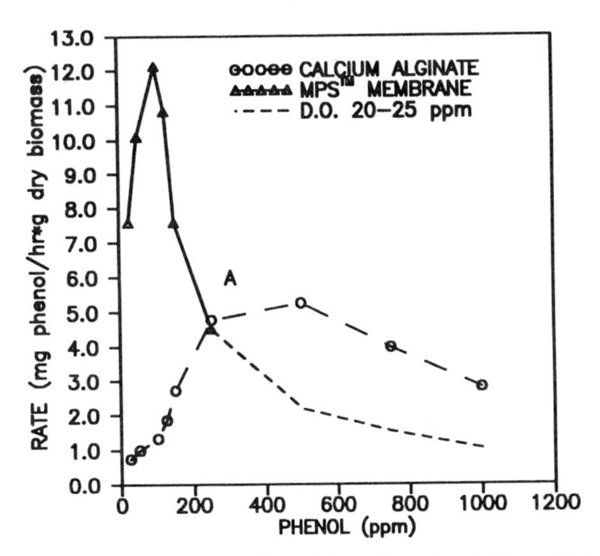


Fig. 14 Comparison of Biodegradation Rates in Calcium Alginate and Membrane Reactors.

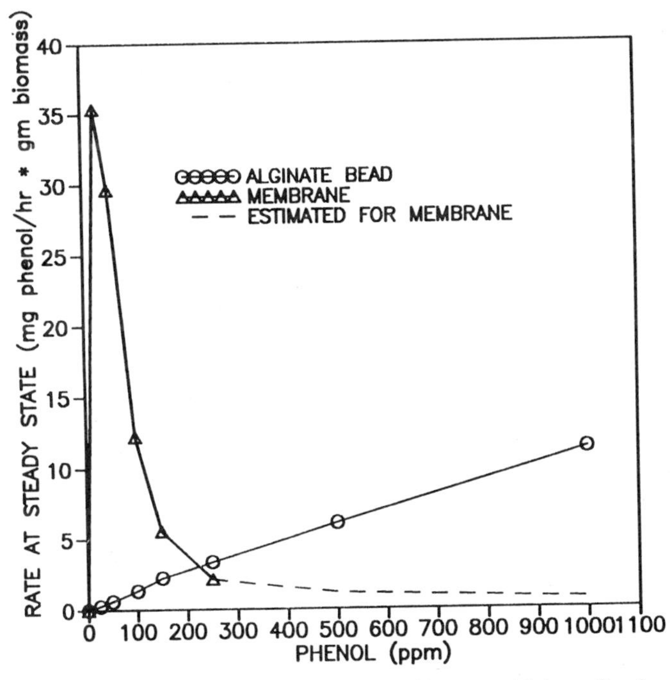


Fig. 15 Comparison of Degredation Rates Between Alginate Bead and Membrane Bioreactor.

## ACKNOWLEDGEMENTS

This work was done at the NJIT Biotechnology Laboratory in Newark NJ, and was funded by the State of New Jersey Sponsored Chair in Biotechnology, the New Jersey Institute of Technology, and the Biosupport Group of FMC Corporation. The excellent technical assistance of Emilia Rus and Fayaz Lakhwala is gratefully acknowledged.


## REFERENCES


1. Lakhwala F.S., Lodaya M.P., Yang Kai Chung, Lewandowski G.A. and Sofer S.S. "Design of Toxic Waste Treatment Bioreactor: Viability Studies of Microorganisms Entrapped in Alginate Gel". Presented at International Conference on Physiochemical and Biological Detoxification of Hazardous Wastes, May 3-5, 1988, Atlantic City, N.J.
2. Mattiasson B., "Immobilization Methods" in Immobilized Cells and Organelles, vol I,. B. Mattiasson, Ed., CRC Press, Boca Raton, FL. Chap 2, 3-26 (1983)
3. Lakhwala, F.S., Sinkar, V., Sofer, S.S., and B. Goldberg "A Polymeric Membrane Reactor for Biodegradation of Phenol in Wastewater" *J Bioactive & Biocompatible Polymers*, vol 5:439-452 (Oct 1990)
4. Lakhwala, F.S., Goldberg, B.S., and S. S. Sofer "A comparative study of gel-entrapped and membrane attached microbial reactors for biodegrading phenol" *Bioprocess Engineering* Vol 8 (1992)
5. Lakhwala, F.S. and S.S. Sofer "Design Considerations for an Immobilized Cell Bioreactor Operating in a Batch Recirculation Mode"*Journal of Chemical Technology and Biotechnology* Vol 52:499-509 (1991)
6. Gourdon, R., Rus, E., Bhende, S., and S. Sofer "Mechanism of Cadmium Uptake by Activated Sludge" *Applied Microbiology & Biotechnology* vol 34, 1990, p 274.
7. Gourdon, R., Rus, E., Bhende, S., and S. Sofer "A Comparative Study of Cadmium Uptake by Free and Immobilized Cells from Activated Sludge" *Journal of Environmental Science and Health*, vol A25:1019 (1990)

# BIOFILTRATION: AN AIR POLLUTION CONTROL TECHNOLOGY FOR HYDROGEN SULFIDE EMISSIONS

Eric R. Allen and Yonghua Yang

Environmental Engineering Sciences Department
University of Florida
Gainesville, Florida 32611

## INTRODUCTION

Traditional air polllution control technologies for pollutant gases, such as adsorption, absorption and combustion, were developed to treat high concentration waste gas streams associated with process emissions from stationary point sources. Although these technologies rely on established physico-chemical principles to achieve effective control of gaseous pollutants, in many cases the control technique yields products which require further treatment before disposal or recycling of treatment materials. In the case of treatment of dilute waste gas streams, however, these traditional methods are relatively less effective, more expensive and wasteful in terms of energy consumption; and identification of alternative control measures is warranted. A suitable alternate air pollution control technology is biofiltration, which utilizes naturally occurring microorganisms supported on a stationary bed (filter) to continuously treat contaminants in a flowing waste gas stream. This application of aerobic biodegradation has received considerable attention and use in Europe, particularly in Germany and Holland during the last 15 years.[1-21] In the United States, however, relatively little attention[22-29] has been paid to this emerging, innovative and versatile 'low' technology approach to solving current environmental problems in air quality.

Biofiltration was originally developed for the control of odorous emissions,[3,7,14,24] an application involving low concentration waste gas streams. More recently, however, biofilters have been shown to have a high potential for successfully controlling volatile organic compounds (VOCs) and air toxics emissions from both traditional and non-

*Industrial Environmental Chemistry*, Edited by D.T. Sawyer
and A.E. Martell, Plenum Press, New York, 1992

traditional sources[12,25,30,31]. Vapor phase organic emissions in simple and complex mixtures have been effectively mitigated using biofilters. Examples of emission control applications at industrial and commercial sources include chemical waste treatment plants, smelters, rendering plants, publicly owned treatment works, bakeries, breweries, etc. Control of VOCs and air toxic emissions is of immediate interest in view of the importance of reactive VOCs in ozone production in non-attainment areas and the requirements to limit toxic air pollutants emissions under Titles I and III, respectively, of the 1990 Amendments to the Clean Air Act. In the latter case, limits of 10 tons/year for individual air toxics or a total of 25 tons/year in mixtures implies that even dilute (<5000 ppm) waste gas stream emissions from process vents and leaks, as well as area and fugitive emission sources will require controls. In this regard biofiltration appears to be an ideal air pollution control technology, provided that the appropriate microorganisms are available in the filter medium. Biofilters have been demonstrated to have the following attributes for air pollution control of dilute waste gas streams. They are simple to construct from readily available materials, inexpensive to maintain and operate, versatile in treatment and flexible in operating conditions, and reliable and effective in their ability to destroy most air pollutants. Biofiltration may be described as biologically catalysed oxidation at ambient temperatures yielding products that require no further treatment, such as $CO_2$ and $H_2O$ from hydrocarbons, or products requiring minimal treatment, such as $H_2SO_4$ from $H_2S$. The current status of the development, applications and use of biofiltration in Europe and the U.S. has been the subject of several recent reviews.[16,26,27,31]

One reason for the limited interest in biofiltration applications in the U.S. is the lack of detailed information concerning the design, operational variables, maintenance procedures and microbial processes involved for specific control applications. Also, systematic compilations of long term operational data and associated problems are lacking. Such information is necessary considering that the operating lifetime of biofilters could be as long as 5 years.

Two distinct types of biofilter systems have been employed. In-ground (soil beds) and on-ground compost beds require undeveloped land to be available for installation, whereas packed biofilter towers have minimal area requirements and are generally more versatile and controllable under varying gas flow and pollutant concentration conditions. High pollutant removal efficiencies are obtained through preliminary conditioning of the biofilter, where the appropriate microbiological populations are optimized through exposure to the pollutants of interest and adjustments to the filter bed physical and chemical properties.

During the last six years the University of Florida has been involved in several biofilter applications projects[27-29,32-34], specifically oriented to the control of hydrogen sulfide ($H_2S$) emissions. Although these studies were designed to address the problem of objectionable odor complaints from communities residing near wastewater treatment plants, considerable quantities of $H_2S$ are released to the atmosphere from other sources. $H_2S$ is emitted as a by-product of industrial processes, such as petroleum refining, rendering, paper and pulp manufacturing, food processing, 'sour' natural gas processing, etc. Human exposure to low concentrations of $H_2S$ in air can cause headaches, nausea and eye irritation and at high concentrations exposure can cause paralysis of the respiratory system leading to fainting and possibly death. In addition, $H_2S$ is corrosive to materials and harmful to crops and vegetation. Control of $H_2S$ emissions is essential, therefore, to protect public health and welfare as well as to mitigate environmental impacts, such as vegetation and material damage. It is interesting to note that $H_2S$ is included in the listing of 190 air toxic compounds where emissions are required to be controlled under Title III of the 1990 Amendments to the Clear Air Act. Thus, the development and application of efficient and economical control technologies for this obnoxious gas are both desirable and necessary.

Laboratory studies have been conducted to determine optimal operating conditions for compost-based biofilters in packed tower systems. Also, field studies have been conducted in association with engineers at a local wastewater treatment plant on the design, construction and operation of a full-scale on-ground compost biofilter bed odor contol system. These projects are described briefly to support the contention that biofiltration is a viable, competitive but underutilized air pollution control technology.

## METHODS

### Laboratory Studies

The laboratory-scale experimental biofilter system used in these studies is shown in Figure 1. The experimental system consists of parallel dual column filters which can be run simultaneously and controlled separately. The biofilter bed material is enclosed in transparent rigid plastic (Acrylic) columns having an inner diameter of 0.15 meter (m) and height of 1.2 m. Each column is packed with the desired compost to a height of 1.0 m. The packed biofilter material is supported by a sieve plate to ensure a homogeneous distribution of the inlet gas across the face of the bed. Six sampling and measurement ports are distributed evenly along the column for inlet, intermediate and outlet gas and compost sampling, as well as temperature and pressure measurements. Room air is forced into a humidification chamber by an air blower (Gast Regenair Model R3105-1).

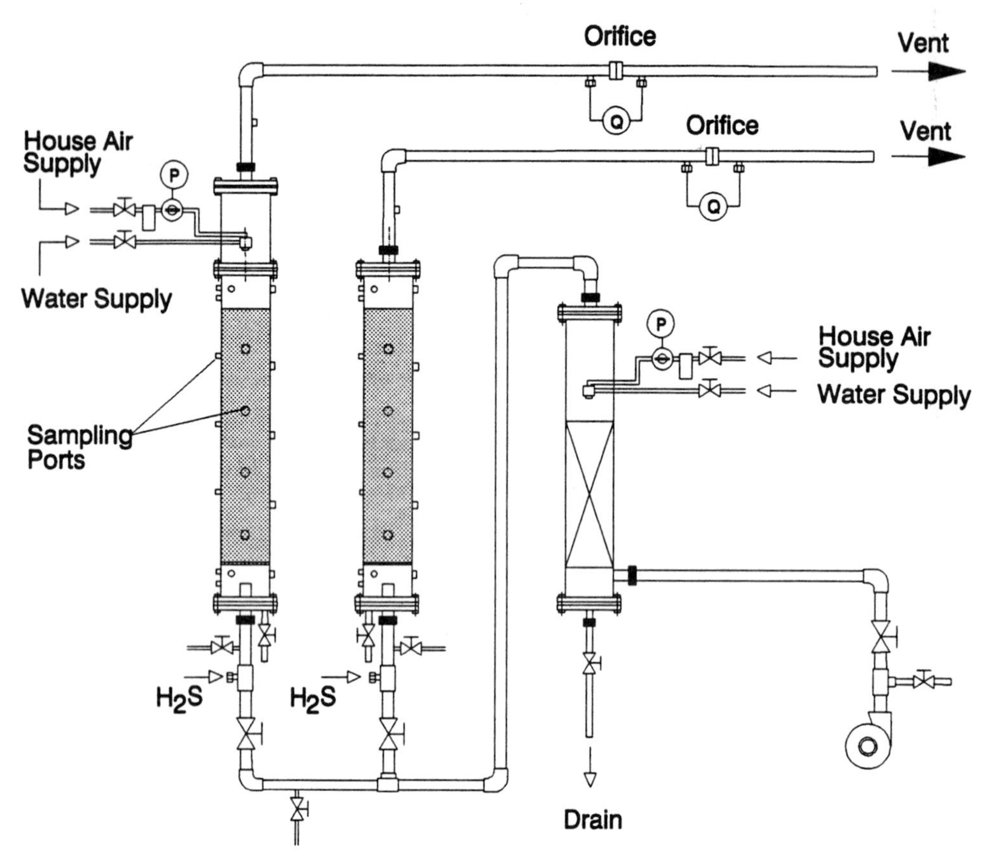


Figure 1. Laboratory Scale Biofilter System.

Water is distributed into the chamber using a spray atomizer and wets the Pall rings packed in the chamber. The latter extend the wetted surface area in contact with the flowing air. Relative humidities in the range from 95 to 100% were routinely and continuously achieved with this system.

Purified hydrogen sulfide (Liquid Air, purity 99+%) was regulated and mixed into the humid air flow prior to the biofilter bed. Flow rates of air and $H_2S$ were controlled by needle valves and monitored by previously calibrated rotameters in order to obtain the desired $H_2S$ concentrations and gas flows through the columns.

Gas samples were taken from various locations on the column by attaching a gas-tight syringe to a sampling port and extracting an aliquot of gas. These samples were accurately diluted with pure nitrogen (Liquid Air, Purity 99.99%) after transfer to a sealed Tedlar bag. Diluted gas samples were analyzed using a commercial gas chromatograph (Tracor Model 250H) equipped with a sulfur-specific flame photometric detector. The $H_2S$ detection limit of the method was determined to be 0.02 parts per million by volume (ppm) with 95% confidence.

The filter materials used in the laboratory studies were composts made from yard trash, pine bark and chips from a variety of sources. Compost samples were stored in sealed plastic bags at room temperature until needed. Some compost samples were stored in this way for several months prior to use. When results for stored compost samples were compared to those for fresh samples, no significant differences in operating characteristics were observed as a result of extended storage time. The sources and principal properties of the composts studied are summarized in Table I.

Table I. Sources and Properties of Laboratory Biofilter Materials.

| | Compost ID# | | | | | | |
|---|---|---|---|---|---|---|---|
| | 3 | 6 | 13 | 14 | 16 | 17 | 17A |
| Bulk Density (g/cc) | 0.18 | 0.20 | 0.27 | 0.18 | 0.22 | 0.20 | 0.20 |
| Pore Volume (%) | | | 84.3 | 89.7 | 88.4 | 88.7 | 88.7 |
| Water Content (wt%) | 62.8 | 64.8 | 54.9 | 62.4 | 56.5 | 62.7 | 62.7 |
| Organic Matter (wt%) | 72.3 | 67.2 | 66.5 | 64.2 | 64.3 | 62.5 | 62.5 |
| pH | 9.22 | 7.21 | 1.60 | 6.44 | 6.66 | 8.10 | 8.10 |

Description of the Composts:

Compost #3: Yard trash compost from Pompano Beach, FL. one year old.

Compost #6: 25% by volume of sewage sludge compost and 75% of yard trash mixed and composted; about 18 months old.

Compost #13: Compost #6 mixed with tree bark and sewage sludge, lime was used to adjust pH before use. Compost obtained from a filter bed which has been used for $H_2S$ removal for 2.5 years.

Compost #14: Compost from Kanapaha WWTP, Gainesville, FL. Yard trash, grass and sewage sludge werre mixed and composted; lime was used to adjust pH; 2.5 years old.

Compost #16: Yard trash compost from Wood resource recovery, Inc., Gainesville, FL. 3.5 months old.

Compost #17: From same source as compost #16, 1:1 by volume of yard trash and grass composted; 3.5 months old.

Compost #17A: Compost #17 mixed with 2% lime ($CaCO_3$), weight percent of dry weight of compost.

## Full Scale Biofiltration Studies

A biofilter system, for control of hydrogen sulfide emissions at a local wastewater treatment plant, was installed in the summer and made operational in the fall of 1988. The system consisted of an air collection system, anti-corrosive blower, ductwork, humidifier, central air flow shaft for flow adjustment and a dual-bed Siebo-stone air distribution system, as shown in Figure 2.

The plant's malodor source was identified to be the grit chamber located in the headworks building. The building was sealed as tight as possible and a negative static

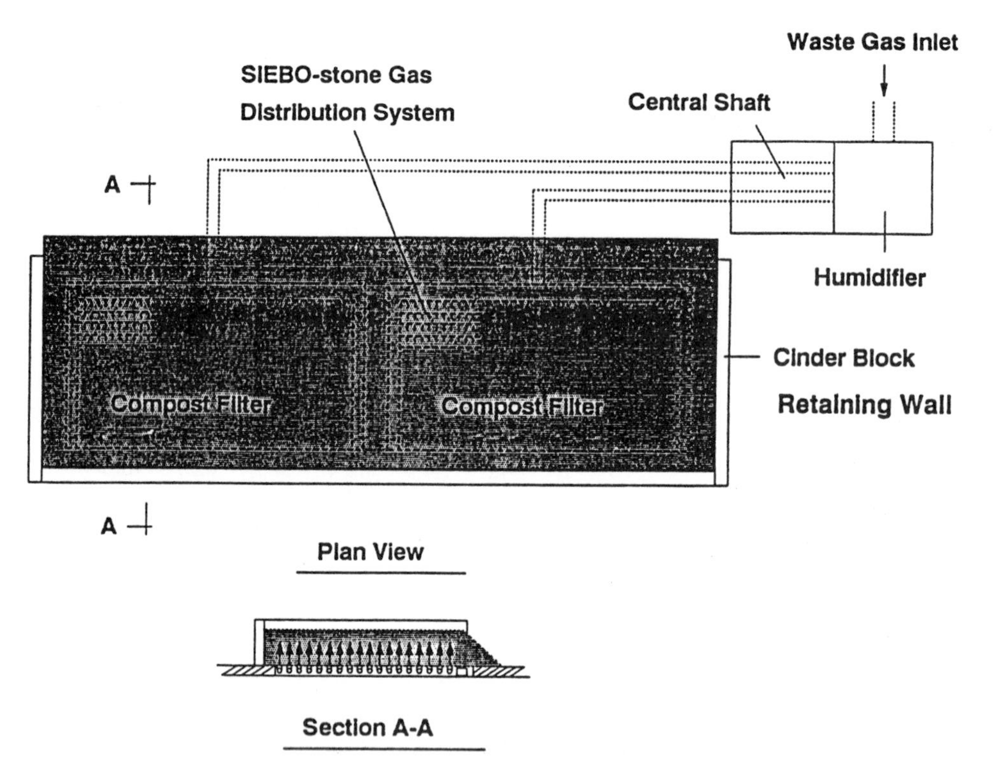


Figure 2. Full Scale Biofilter System.

pressure was maintained by the blower drawing contaminated air at 85 $m^3$/min. through 0.3 m diameter exhaust ducts. The exhaust air was forced to a humidifier where water was dispensed by several spray nozzles to saturate the waste gas stream. All pipes in the waste gas collection and transport system were made from polyvinyl chloride to minimize corrosion problems.

The biofilter bed gas distribution system was constructed from the German patented SIEBO stones. These interlocking sinter block base units of the biofilter bed when installed, were rigid and allowed for heavy vehicles to be driven on the base without damage. The SIEBO-stone base system provided for an even distribution of the inlet air to the filter bed as well as functioned as a drainage system. The overall filter bed area was 100 $m^2$, divided into two equal sections, which could be operated and controlled individuallly. During normal operation each section treated half of the total contaminated gas flow by balancing the separate input flows. The total flow, however, could be diverted to one section if necessary, when repairs or refurbishing were needed, without reducing overall pollutant removal efficiency.

Composts obtained from Pompano Beach and Jacksonville, Florida were mixed and used as the filter bed material. This mixture consisted of yard waste compost, pine bark, wood chips and sewage sludge. Lime was applied to the compost material to buffer the bed acidity (pH) prior to installation. The biofilter material was spread over the SIEBO-stone air distribution system to a depth of at least 1.3 m. A 1.3 m retaining wall was used on three sides of the bed to hold the compost. The fourth side of the bed was left open for access and the compost on this side was overlapped with the edge of the air distribution system by 1.3 m, through incorporation of an incline of 45° in the bed from bottom to top.

The waste air flow rate through the biofilter was in the range from 79 to 96 actual cubic meters per minute, which corresponded to an average gas loading rate of 52 $m^3/m^2$-h and average gas retention time of 88 seconds. The full-scale biofilter design and operating parameters are summarized in Table II.

To monitor biofilter bed performance, influent and effluent gas samples were collected and analyzed for $H_2S$ by procedures described in detail elsewhere.[32,34]

## RESULTS AND DISCUSSION

### Laboratory Studies

The composts listed in Table I have been studied in the laboratory-scale biofilter system for periods varying from 1 to 4 months. Typical physical and chemical properties of these composts are presented in Table III. Most of the composts showed a high removal efficiency (>99%) during the first 24 to 48 hours of operation. This high

Table II. Full Scale Biofilter Design and Operating Parameters

| | |
|---|---|
| Total Flow: | 79 - 96 acmm |
| Filter Area: | 100 $m^2$ |
| Filter Height: | 1.3 meter |
| Gas Loading Rate: | 47 - 57 $m^3/m^2$-hr |
| Retention Time: | 82 - 100 sec. |
| Temperature: | 15 - 30°C |
| Pressure Drop: | 150 - 200 $mmH_2O$ |

efficiency, however, was probably due to physical and chemical, rather than biological, interactions between $H_2S$ and the biofilter materials. During the subsequent 2 to 14 days, the removal efficiency declined to about 90%, presumably as a result of acclimation and development of the optimum population of sulfur-oxidizing microorganisms. After this conditioning period the biofilters approached their final stable condition and the most efficient control (>99%) of $H_2S$ was achieved. If the compost had been exposed previously to $H_2S$ then the initial physico-chemical and conditioning stages did not occur. Seeding of sulfur oxidizing bacteria to the compost was not necessary because the

Table III. Physical and Chemical Properties of Typical Laboratory Filters.

| Property | Range |
|---|---|
| Bulk Density | 0.18 - 0.30 $g/cm^3$ |
| Particle Density | 1.7 - 1.9 $g/cm^3$ |
| Porosity | 0.84 - 0.90 $cm^3/cm^3$ |
| Water Content | 0.45 - 0.63 g/g |
| Organic Matter | 0.59 - 0.67 g/g |
| pH | 1.6* - 8.1 |
| Total Sulfur | 0.74 - 70.4* mg-S/g |
| Soluble Phosphorus | 0.11 - 0.22 mg-P/g |
| Total Carbon | 0.31 - 0.41 g-C/g |
| Total Nitrogen | 0.013 - 0.043 g-N/g |
| C/N Ratio | 7.1 - 32 g-C/g-N |

*Extreme values from 'used' biofilter materials previously exposed to $H_2S$

required bacteria exist naturally in soils and sludges.[35,36] Mixing of small amounts of sewage sludge with the composts, however, was found to increase the initial bacterial population and provide for a shorter conditioning period. Although addition of sewage sludge can provide an additional source of nutrients, unseeded composts were found to be as efficient as sludge-seeded composts once they were conditioned and stabilized.

An important operating variable in biofilter systems is the energy required for air movers to transport the waste gas through the filter bed at the required flow rates. The pressure drop across the bed increased markedly as the flow rate was increased. Since the pressure drop was determined by the bed depth it was necessary to keep the gas velocity as low as possible. One of the most important properties of the bed material, which significantly affected the pressure drop, was the compost small particle content ($d_p$ < 1.2 mm). It was found that for a 1 m bed depth if particles with sizes less than 1.2 mm were removed then gas velocities as high as 10 m/min could be obtained for pressure drops up to 50 $cmH_2O$. It should be noted, however, that during operation of the biofilter to remove $H_2S$, mineralization of bed material and sulfur accumulation occur which increased the biofilter small particle content. Thus, measures should be taken to separate the accumulated small partlicles periodically, otherwise a substantial increase in pressure drop and reduction in gas velocity will be observed during use, which leads to enhanced energy consumption.

The effect of gas retention time on $H_2S$ reduction has been studied by varying the gas flow rate through the biofilters. When the $H_2S$ loading rate was kept constant and below the maximum loading capacity of the bed the removal efficiency remained high (>99%) for residence times as low as 23 seconds. In subsequent tests, at very high $H_2S$ concentrations in the inlet gas the removal efficiency remained high as long as the residence time of the flowing gas was increased. For example, with an inlet concentration of 2651 ppm $H_2S$, a 99.8% removal efficiency was obtained for a residence time of 197 seconds. In this case the filter bed maximum loading rate was determined to be 1.2 g-$H_2S/m^2$-min for a typical compost bed. However, the latter variable is a function of $H_2S$ concentration and gas velocity. A more appropriate measure of biofilter capacity is the concept of $H_2S$ mass loading rate per unit mass of compost (mg-$H_2S$/kg-min). The maximum $H_2S$ loading capacity for typical compost biofilters has been determined to be in the range from 2.3 to 4.8 mg- $H_2S$/kg-min. The long term influence of $H_2S$ loading rates on $H_2S$ removal efficiency is shown in Figure 3.

The effect of varying $H_2S$ concentration in the inlet gas stream on biofilter removal eficiency was investigated at constant waste gas flow rate. No significant differences in $H_2S$ removal efficiencies (>99%) were observed for concentrations in the range from 5.5 to 518 ppm because the $H_2S$ loading rate was always less than the maximum $H_2S$ loading capacity for the bed. However, if the maximum loading capacity was exceeded then the $H_2S$ removal efficiency was observed to decrease significantly. For example, when the $H_2S$ loading rate was 7 mg-$H_2S$/kg-min for a bed with a maximum loading capacity of 2.3 mg-$H_2S$/kg-min the removal efficiency was reduced to 65% and the biofilter was overloaded. The maximum loading capacity of the bed is the maximum amount of $H_2S$ that the optimal microbial population can consume, without inhibition of activity.

The effect of varying compost water content has been studied with respect to $H_2S$ removal efficiency. The data indicated that for water contents greater than 30% by weight, high removal efficiencies (>99%) were obtained consistently. However, when the water content was reduced to less than 30% the efficiency decreased markedly with decreasing water content. It appears that a liquid water film on the compost is necessary

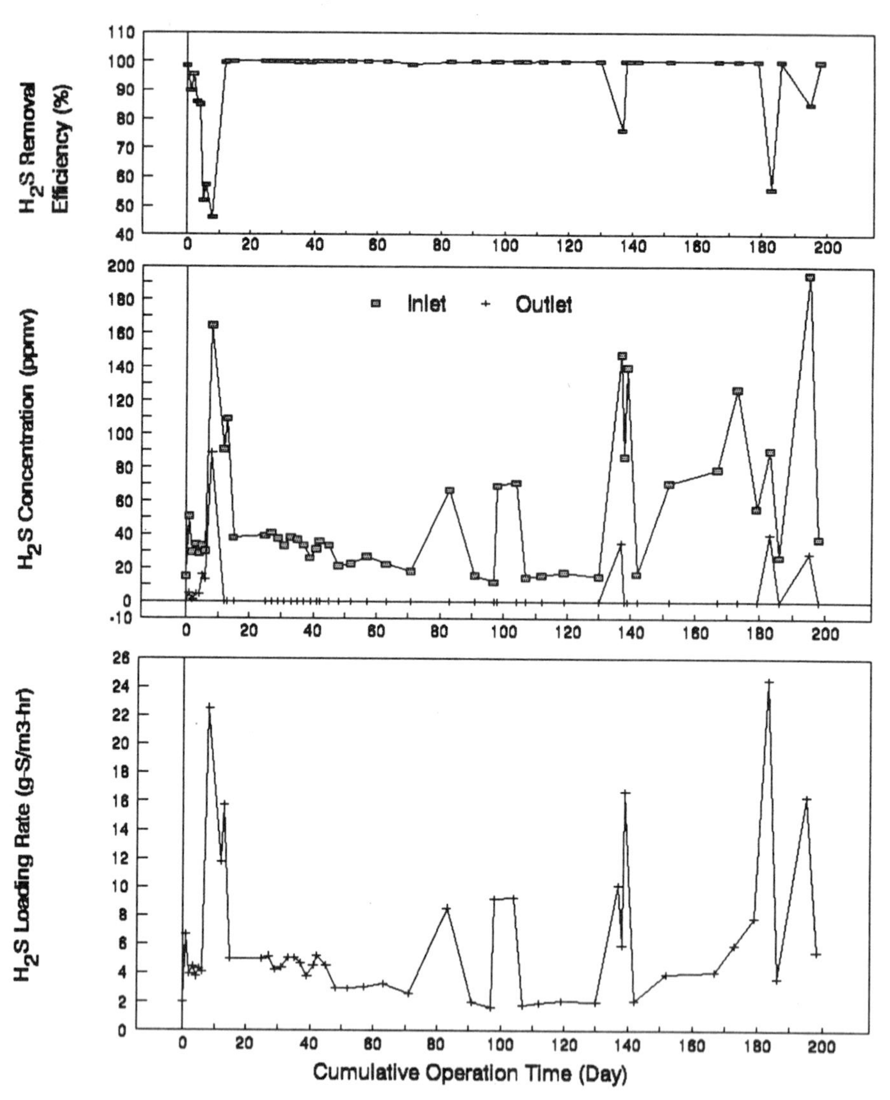


Figure 3. Biofilter Removal Efficiency During Long Term Laboratory Operation.

for the sulfur oxidizing bacteria to survive and propagate in order to establish the optimal population[37]. Thus, compost water contents in excess of 30% appeared to be necessary to both saturate the filter material and produce the desired liquid water film environment for the bacteria.

The compost was found to deteriorate in terms of accumulated acidity and sulfur content during operation. These phenomena were a result of the continuous conversion of $H_2S$, through several oxidation states, to sulfuric acid. If the bed is not treated then the 'active' portion of the biofilter becomes acidified and sulfated. Fortunately, this deterioration is not permanent and can be reversed by periodic washing of the bed with water. Although periodic washings can remove 35% of the accumulated sulfate and hydrogen ions, they appear to have only a small effect on pH, i.e. from 0.02 to 0.03 pH units. Nevertheless, periodic washings can maintain or restore the original removal efficiency, prevent bed overloading and contamination, remove accumulated fine particles and extend the operating life of the biofilter. Additional details of the laboratory studies briefly described here are reported elsewhere.[33,34]

Full-Scale Studies

Influent and effluent gas samples were collected and analyzed during the first 16 days of operation of the full-scale biofilter. During this start-up period $H_2S$ concentrations in the influent gas stream varied in the range from 156 to 229 ppm and off-gas concentrations were observed in the range from 0.05 to 0.40 ppm. The efficiency of $H_2S$ removal was greater than 99% for the sampling period because some of the compost in the mixture used for preparing the biofilter had been exposed previously to $H_2S$, thus, the initial physico-chemical reaction and conditioning stages were not observed. Subsequently, gas and compost samples were taken every quarter (3 months) to determine whether the filter bed efficiency and composition had deteriorated during operation. Physical and chemical properties of the 'aged' on ground biofilter are given in Table IV.

Table IV. Physical and Chemical Properties of the Aged Full Scale Biofilter.

| Date | Bulk Density (g/cc) | Water Content (%) | Organic Matte (%) | pH | Total N (mg/g) | Total C (mg/g) | Total S (mg/g) |
|---|---|---|---|---|---|---|---|
| 05/10/88 | 0.23 | 52.1 | 68.1 | 8.63 | 2.30 | 40.9 | 7.3 |
| 11/21/88 | 0.27 | 54.6 | 69.3 | 4.40 | -- | -- | -- |
| 05/16/90 | 0.30 | 45.6 | 66.5 | 2.62 | 1.89 | 34.3 | 44.5 |
| 12/20/90 | 0.27 | 54.6 | 66.5 | 1.60 | 1.58 | 36.6 | 71.0 |
| 02/05/91 | -- | 57.2 | 64.7 | 1.80 | -- | -- | 109 |

Additional tests were conducted to ensure that off-gas sampling at different locations were representative of the bed as a whole by comparing results obtained at four locations on top of the bed. Values obtained were consistently less than 0.2% of the inlet gas concentration, although individual values varied initially by as much as a factor of 3. Also, tests were carried out to ensure that the methods employed for collection and storage of gas samples did not influence the analytical results. These quality control procedures provided asssurance that sampling and analysis data completeness, accuracy and precision objectives were achieved and that representative gas samples were collected from the biofilter.

The influent gas humidifier worked effectively to maintain the filter bed moisture content in the range from 45 to 60%. No other water was introduced to the bed except that due to rain. As a result the top of the bed dried out significantly during long term operation. No marked changes in the bulk density and particle size distribution of the compost were observed during 27 months (11/88 to 02/91) operation of the biofilter. The organic matter content decreased by about 4.5%, presumably due to mineralization of the compost. Total nitrogen and carbon contents of the compost decreased at different rates resulting in an increase of the C/N ratio from 17.8 to 23.2. The most significant changes observed in the bed material during extended operation were the total sulfur (S) content and pH. The total sulfur content increased from 7.3 to 109 mg-S/g-compost on a dry basis. Also, after prolonged operation, parts of the biofilter bed showed a pronounced color change from dark brown to yellowish-white, which was accompanied by emission of an obnoxious odor. These observations were similar to those from laboratory studies, which were obtained during dry-out of the compost or by polllutant overloading.

The pH of the compost decreased significantly over several years from 8.6 to 2.6 as a result of continuous aerobic biodegradation of $H_2S$. Although this acidification of the compost did not appear to affect the overall $H_2S$ removal efficiency, considerable corrosion of the cement blocks in the retaining wall and plugging of the bed inlet-gas distribution vents in the SIEBO-stone base were observed. As a result of this obvious system deterioration, it was decided that the SIEBO-stone air distribution system should be replaced with components made from acid-resistant or anti-corrosive material.

## CONCLUSIONS

Both the laboratory and full-scale biofilter systems have been demonstrated to perform with suitable efficiency during long term operation provided that the systems are operated within specific ranges for important operating variables. Recommended ranges for operating variables when using compost biofilters for control of $H_2S$ emissions are presented in Table II. Maintenance procedures for compost beds in this application are minimal. The biofilters require washing with water from the top of the bed at weekly intervals to maintain an appropriate water content (>30% wt/wt) throughout the bed and to prevent acidity and soluble sulfur from accumulating to levels which are toxic to sulfur oxidizing bacteria. A sprinkler system capable of providing water at a rate of 40 liters/$m^2$-min. for 10 minutes every week should suffice.

Laboratory kinetic studies have shown that $H_2S$ removal is zero order in $H_2S$ at high concentrations (>400 ppm) and first order in $H_2S$ at low concentrations (<200 ppm). Presumably, the bioreaction occurs by diffusion of $H_2S$ into liquid water droplets, where the sulfur oxidizing bacteria are located.[37] At high concentrations of $H_2S$ in the gas phase the water droplets will be saturated with $H_2S$ and the rate of reaction will depend primarily on the microbial activity. However, at low $H_2S$ concentrations the availability of $H_2S$ for the microorganisims will depend on the rate of diffusion of $H_2S$ to the liquid surface and mass transfer across the gas-liquid interface, which depends on the $H_2S$ concentrations in the gas phase.

Although certain deficiencies in using compost-based biofilters have been noted, in general biofiltration has been demonstrated to be a simple, inexpensive, durable and effective method for $H_2S$ control at wastewater treatment plants. Similar systems could be designed, constructed and operated effectively in other industrial applications. Additional research is essential to provide detailed knowledge of the appropriate biofilter systems, operating conditions and maintenance procedures necessary to efficiently treat other dilute industrial waste gas streams containing organo-sulfur, volatile organic and air toxic compounds.

## REFERENCES


1. Kneer, F.X. "Device for the Reduction of Gaseous Organic Pollutants from Waste Gases", German Patent No.: DE 2445315 C2 BOID 53/54, West Germany, 1976.
2. Kneer, F.X. "Waste Gas Purification with a Biological System", Das Technische Umweltmagazin, West Germany, 20, October 1978.
3. Jaecklin, F.P. "The Deodorization of the Sewage Treatment Plant at Staz (St. Moritz) Using Soil Bed Filters", Stuttgarter Berichte zur Siedlungswasserwirtschaft, Vol 59, University of Stuttgart, West Germany, September 1976.
4. Gust, M., Sporenberg and E. Shippert. "Fundamentals of Biological Waste Gas Purification Part IV: Gas Cleaning by Microorganisms in Bio-Scrubbers", Staub-Reinhaltung der Luft, West Germany, pp. 308-314, September, 1979.
5. Gust, M., H. Grochowski and S. Sibirz. "Fundamentals of Biological Waste Gas Purification, Part V: Gas Cleaning by Microorganisms in Biofilters", Staub-Reinhaltung der Luft, West Germany, pp. 397-438, November 1979.
6. Ottengraf, S.P.P, A.H.C. Van Den Oever, and F.J.C.M. Kempenaars. Waste Gas Purification in a Biofilter Bed," in Innovations in Biotechnology, Houwink, E.H. and Van Dan Meer, R.R. eds., Elsevier Science Publishers, B.V., Amsterdam, The Netherlands (1984).
7. Koch, W., H.G. Liebe and B. Striefler. "Experiences with Biofilters for the Reduction of Odorous Air Emissions", Staub-Reinhaltung der Luft, West Germany, pp. 488-493, December 1982.

8. Ottengraf, S.P.P. and A.H.C. Van Den Oever. "Kinetics of Organic Compound Removal from Waste Gases with a Biological Filter", Biotechnol. Bioeng. 25, 3089-3102, (1983).
9. Bohnke, B. and D. Eitner. "Investigation and Comparsion of Different Kinds of Compost in a Mobile Biofilter", Gutachten i. A der AG Kompostabsatz NW, Aachen, West Germany, June 1983.
10. Eitner, D. "Investigations of the Use and Ability of Compost Filters for Biological Waste Gas Purification with Special Emphasis of Aspects", The Operation Time, GWA, Band 71, RWTH Aachen, West Germany, September, 1984.
11. Kneer, F.X. "Gas Cleaning using a Biofilter", CAV, West Germany, pp. 112, October 1985.
12. Gethke, H.G. "Capabilities of a Compost Filter System for Waste Gas Purification Using, as an Example, Waste Water Treatment Plants with Industrial Waste Water Loading", Paper presented at Colloquim on Odorants, Baden-Baden, West Germany, October, 1985.
13. Dragt, A.J. and S.P.P. Ottengraf. "Biofiltration: A New Technology in Air Pollution Control Experiences in the Netherlands", Paper No. 4-4508-23-10, Eindhoven University of Technology, Einhoven, The Netherlands. May 1985.
14. Don. J.A. and L. Feenstra "Odour Abatement Through Biofiltration", Paper presented at Symposium in Louvain-La-Neuve, Belgium, April 1984.
15. Don, J.A. "The Rapid Development of Biofiltration for the Purification of Diversified Waste Gas Streams", VDI Berichte 561, VDI Verlag Dusseldorf, West Germany, p. 63, 1985.
16. V.D.I. "Biological Waste Air Purification-Biofilters", VDI-3477, VDI Handbuch, Reinhaltung der Luft, Band 6, Dusseldorf, West Germany, December, 1984.
17. V.D.I. "Biological Waste Air Purification - Bioscrubbers", VDI 3478 VDI Handbuch, Reinhaltung der Luft, Band 6, Germany, July, 1985.
18. Zeisig, H.D. "Biofilter for Agricultural and Industrial Applications, Especially the Tobacco Industry", Paper presented at Colloquium on Odorants, Baden-Baden, West Germany, October 1985.
19. Klee, W.K., Lutzke and H. Vollmer, "Reduction of Odorous Emissions by Biofiltration Demonstrated Using as Examples, a Fiber Plate Production Facility, a Vulcanization Facility and a Foundry", Paper presented at Colloquium on Odorants, Baden-Baden, West Germany, October 1985.
20. Ottengraf, S.P.P. "Exhaust Gas Purification", in Biotechnology, Rehm, H.J. and Reed, G. Eds. Vol. 8; VCH Verlagsgesellschaft, Weinheim, FRG (1986).
21. Eitner, D. and H.G. Gethke. "Design, Construction and Operation of Biofilters for Odor Control in Sewage Treatment Plants", Paper No. 87-95A.6. Presented at 80th Annual Meeting of Air Pollution Control Association, New York, NY. June 21-26, (1987).
22. Carlson, D.A. and C.P. Leiser. "Soil Beds for the Control of Sewage Odors", J. Water Pollut Control. Fed., 38, 829 (1966).
23. Rands, M.B., D.E. Cooper, C.P. Woo, G.C. Fletcher, and K.A. Rolfe. "Compost Filters for $H_2S$ Removal from Anaenrobic Digestion and Rendering Exhausts" J. Water Pollut. Control Fed., 53, 185 (1981).
24. Prokop, W.H. and H.L. Bohn. "Soil Bed System for Control of Rendering Plant Odors", J. Air Pollut. Control Assoc., 35: 1332 (1985).
25. Kampbell, D.H., J.T. Wilson, H.W. Read, T. Thomas and T.T. Stocksdale. "Removal of Volatile Aliphatic Hydrocarbons in a Soil Bioreactor", J. Air Pollut. Control Assoc., 37: 1236 (1987).
26. Hartenstein, H.U. "Assessment and Redesign of an Existing Biofiltration System" M.S. Thesis, Univeristy of Florida, Gainesville, Fl., 1987.

27. Hartenstein, H.U. and E.R. Allen. "Biofiltration an Odor Control Technology for a Wastewater Treatment Plant", Project Report, Environmental Engineering Sciences Department, University of Florida, Gainesville, FL., October 1986.
28. Allen, E.R., H.U. Hartenstein and Y. Yang. "Review and Assessment of the Design and Operation of a Compost Biofilter System for Odor Control", Project Report, Environmental Engineering Sciences Department, University of Florida, Gainesville, FL., 1987.
29. Allen, E.R., H.U. Hartenstein and Y. Yang. "Identification and Control of Industrial Odorous Emissions at a Municipal Wastewater Treatment Facility", Paper No. 87-95A.4, Presented at 80th Annual Meeting of Air Pollution Control Association, New York, NY, June 21-26, 1987.
30. Ottengraf, S.P.P. "Biological Systems for Waste Gas Elimination," TIBTECH, 5, 132, 1987.
31. Leson, G. and A.M. Winer. "Biofiltration: An Innovative Air Pollution Control Technology for VOC Emissions", J. Air Waste Manage. Assoc., 41: 1045, 1991.
32. Yang, Y. and E.R. Allen. "Biofiltration Control of Odor Emissions in Wastewater Treatment Plants", Paper presented at Symposium on Biotechnology for Wastewater Treatment - II, 201st National Meeting ACS, Atlanta, GA., April 14-19, 1991.
33. Allen, E.R. and Y. Yang. "Biofiltration Control of Hydrogen Sulfide Emissions", Paper No. 91-103.10, Presented at the 84th Annual Meeting of Air and Waste Management Association, Vancouver, B.C. Canada, June 16-21, 1991.
34. Yang, Y. "Biofiltration for Control of Hydrogen Sulfide", Ph.D. Dissertation, University of Florida, Gainesville, FL., 1992.
35. David, M.B., M.J. Mitchell and J.P. Nakas. "Organic and Inorganic Constituents of a Forest Soil and Their Relationship to Microbial Activity", Soil Sci. Soc. Amer. J., 46, 847, 1982.
36. Starkey R.L. "Oxidation and Reduction of Sulfur Compounds in Soils", Soil Sci., 101, 297, 1966.
37. Ottengraf, S.P.P. "Theoretical Model for a Submerged Biological Filter", Biotechnol. Bioeng, 19, 1411, 1977.

# INORGANIC ION EXCHANGE MATERIALS FOR NUCLEAR WASTE EFFLUENT TREATMENT

Abraham Clearfield

Department of Chemistry
Texas A&M University
College Station, Texas 77843

## STATEMENT OF PROBLEM AND SCOPE

Over the past 48 years, nuclear defense activities have produced large quantities of nuclear waste that now require safe and permanent disposal. The major high-level radioactive waste accumulations are stored at Hanford in the southeastern desert of Washington State and at Savannah River in South Carolina. The magnitude of the problem is enormous. There are 177 storage tanks of a million-gallon plus size nearly filled with high level waste at Hanford alone. Much of the waste was placed in the tanks with no records kept of their content. Over a period of time new additions were made to the tanks as needed so that the inventory of content is at best approximate. In order to prevent corrosion of the tanks the pH was raised to above 13 which resulted in the precipitation of large quantities of hydroxides. Thus, Hanford must not only deal with huge amounts of liquid waste, but must also develop methods of treating the solid waste.

In all, nine plutonium production reactors were built at Hanford, starting in 1943 and lasting until 1963. The plutonium fuel from these reactors was reprocessed in on-site processing facilities using a solvent extraction process designated PUREX. In general, the nuclear fuel was dissolved in nitric acid and a separation of uranium and plutonium effected by organic phosphate complexing agents dissolved in an organic phase. The extraction of 1 kg of plutonium by the PUREX process produced over 340 gallons of liquid high level radioactive wastes, more than 55,000 gallons of low to intermediate-level radioactive wastes and required over 2.5 million gallons of cooling water [1]. Since 1948, the U.S. has produced about $10^5$ kg of plutonium for military use. At Hanford the high level radioactive wastes, containing most of the fission products, complexing agents, organic solvents and water were poured into the waste storage tanks and the pH raised to above 13. As a consequence the solution is 7-8M in $NaNO_3$. These tanks are of the single shell type and after a number of years of storage, several of them were found to leak. Subsequent to this discovery 28 newer tanks that were built were of the double shell type. The low to intermediate level radioactive wastes were put into cribs, which are like septic-tank drainage fields, and the cooling water was pumped into surface ponds[2].

The single shell tanks have leaked more than 3/4 of a million gallons of their waste into the surrounding soil [3]. These tanks continue to leak and pose a pressing immediate problem. There are about 250 million curies of radioactive material in all the holding tanks. This is five times the amount of radioactivity released into the environment by the Chernobyl accident. The waste stored in the cribs has overflowed and is moving towards the aquifer underlying the Hanford site. In addition the movement of radioactivity in the

*Industrial Environmental Chemistry*, Edited by D.T. Sawyer
and A.E. Martell, Plenum Press, New York, 1992

soil is greater than expected and is flowing towards the Columbia river. This river located to the north and east of Hanford, and the Yakima river, in the south, are the sources of drinking and irrigation water for the entire southeastern Washington area.

Another problem that must be faced is the danger of explosions in the tank. The presence of all the organic materials and water, coupled with high radioactivity, results in radiolytic decomposition to generate significant volumes of $H_2$. Smaller amounts of nitrogen oxides are also formed. Normally these gases are ventilated out of the tanks. However, in some tanks the gas is trapped between the sludge at the bottom and salt cake (largely sodium nitrate) which formed over the sludge. These gases increase in pressure raising the salt cake and then being violently released as a huge bubble. Another potential explosion hazard is present in about 20 tanks to which ferrocyanide had been added to precipitate $^{137}Cs$. The ferrocyanide is a reducing agent whereas the nitrates and nitrites are oxidizing agents. The combination could produce enough heat to cause a violent explosion of the $H_2$.

## ACTION PLAN FOR REMEDIATION

Battelle Pacific Northwest Laboratory (PNL) has been given the responsibility for long range research on environmental remediation and waste management at Hanford. The most urgent items are to be dealt with on a short time scale. The greatest danger to human beings is the threat of a violent explosion in the tanks. Such an eventuality could release large amounts of radioactive species and heavy metals to the surrounding area. A second threat stems from the migrations of contaminants to the aquifer or to the surrounding rivers. PNL, in conjunction with DOE, has developed several strategies to deal with these problems. Our focus in this paper is to deal with problems connected with the disposal of the present contents of the holding tanks and will not be concerned with soil remediation. The general strategy outlined by Battelle is shown in Figure 1. High level waste will be contained in a borosilicate glass whereas the low level waste will be enclosed in cement. The salt cake consists largely of sodium nitrate and nitrite and the sludge of insoluble hydroxides, but may also contain hydroxynitrates. Since the salt cake may contain trapped radioactive species it is to be dissolved and combined with the liquid waste. The major high level radioactive species in this combined solution is $^{137}Cs$ and its removal would allow the remaining solution to be treated as low level waste (LLW) and formed into cementitious grout. Removal of the $Cs^+$ is a formidable task because the solutions are about 8M in $Na^+$ and contain considerable amounts of Al, Cr, $K^+$ and other elements as shown in Table 1. Thus a highly selective ion exchanger or complexing agent is required.

The sludge and separated $^{137}Cs$ is to be vitrified directly into a borosilicate glass. Reference to Table 1 shows that the sludge contains large amounts of boron and silicon[4]. It is estimated that it would be required to prepare about 30,000 canisters of borosilicate glass and to store them deep underground. The projected cost of this phase of the operation is \$30 billion as compared to \$1.3-2 billion for the low level grout. Since the sludge contains most of the radionuclides, an enormous cost savings would result if the radionuclides could be removed from the sludge and treated separately. The remaining LLW would go to much less expensive grout. Given the complex nature of the sludge, a highly innovative chemical separations technology is required.

The sludge and separated $^{137}Cs$ is to be vitrified directly into a borosilicate glass. Reference to Table 1 shows that the sludge contains large amounts of boron and silicon[4]. It is estimated that it would be required to prepare about 30,000 canisters of borosilicate glass and to store them deep underground. The projected cost of this phase of the operation is \$30 billion as compared to \$1.3-2 billion for the low level grout. Since the sludge contains most of the radionuclides, an enormous cost savings would result if the radionuclides could be removed from the sludge and treated separately. The remaining LLW would go to much less expensive grout. Given the complex nature of the sludge, a highly innovative chemical separations technology is required.

Table 2 provides a summary of elements which would be targeted for removal from the sludge along with Pu and U. If the noble metals could be recovered efficiently, without creation of additional waste, then a valuable resource would be gained. Thus, even thoughplans have been formulated and are proceeding to dispose of defense waste, there still remains a strong incentive to find a technologically better, more efficient solution to

Table 1. Quantities of Chemicals in Single-Shell Tanks.

| Element | Liquid + Salt Cake g Mole | Sludge g Mole | Total g Mole | Total g Mole |
|---|---|---|---|---|
| Ag | 269 | 393,000 | 393,000 | |
| Al | 1,890,000 | 266,000,000 | 267,890,000 | 83,000,000 |
| Am | 0.01 | 300 | 300 | |
| As | 0 | 75,200 | 75,200 | |
| B | 2,480 | 62,300,000 | 62,300,000 | |
| Ba | 103 | 4,020,000 | 4,020,000 | |
| Bi | 1,560 | 1,040,000 | 1,040,000 | 1,250,000 |
| Ca | 5,910 | 35,600,000 | 35,600,000 | 3,200,000 |
| Cd | 652 | 660,000 | 660,000 | 36,000 |
| Cl | 0 | 1,090,000 | 1,090,000 | 1,130,000 |
| CN | 0 | 351,000 | 351,000 | 8,800,000 |
| Ce | 2 | 5,650 | 5,650 | |
| Co | 3 | 42,800 | 42,800 | |
| CO | 20,300,000 | 27,700,000 | 48,000,000 | 27,000,000 |
| Cr | 400,000 | 7,540,000 | 7,940,000 | 1,800,000 |
| Cs | 4,470 | 350 | 4,820 | |
| Cu | 77 | 550,000 | 550,000 | |
| F | 0 | 1,640,000 | 1,640,000 | 42,400,000 |
| Fe | 477 | 96,500,000 | 96,500,000 | 11,200,000 |
| Hg | 0 | 247,000 | 247,000 | 4,500 |
| I | 313 | 25 | 338 | |
| K | 8,560,000 | 0 | 8,560,000 | |
| Mg | 98 | 29,900,000 | 29,900,000 | |
| Mn | 10,300 | 6,860,000 | 6,870,300 | 2,200,000 |
| Na | 808,000,000 | 192,000,000 | 1,000,000,000 | 2,230,000,000 |
| Ni | 2,760 | 3,940,000 | 3,942,000 | |
| $NO_2$ | 74,100,000 | 13,000,000 | 87,100,000 | 102,000,000 |
| $NO_3$ | 100,000,000 | 45,100,000 | 145,100,000 | 1,540,000,000 |
| P | 73,200 | 32,300,000 | 330,320,000 | 91,000,000 |
| Pb | 1,180 | 1,200,000 | 1,200,000 | |
| Pd | 0 | 60,800 | 62,800 | |
| Pu | 9 | 72,100 | 72,100 | |
| Rh | 0 | 21,000 | 21,000 | |
| Ru | 0 | 98,000 | 98,000 | |
| Se | 964 | 18,300 | 19,300 | |
| Si | 18,800 | 298,000,000 | 298,000,000 | 15,300,000 |
| $SO_4$ | 0 | 8,190,000 | 8,190,000 | 5,200,000 |
| Sr | 1 | 30,100 | 30,100 | |
| Sr | 6 | 177,000 | 177,000 | 17,200,000 |
| Tc | 23,725 | 36,125 | 59,400 | |
| U | 870 | 7,260,000 | 7,260,000 | |
| Zn | 529 | 292,000 | 292,000 | |
| Zr | 1,610 | 28,000,000 | 28,000,000 | 2,700,000 |
| | $m^3$ | $m^3$ | $m^3$ | |
| Volume | 91,540 | 47,950 | 139,500 | |
| | g/cc | g/cc | | |
| Density | 1.37 | 1.52 | | |

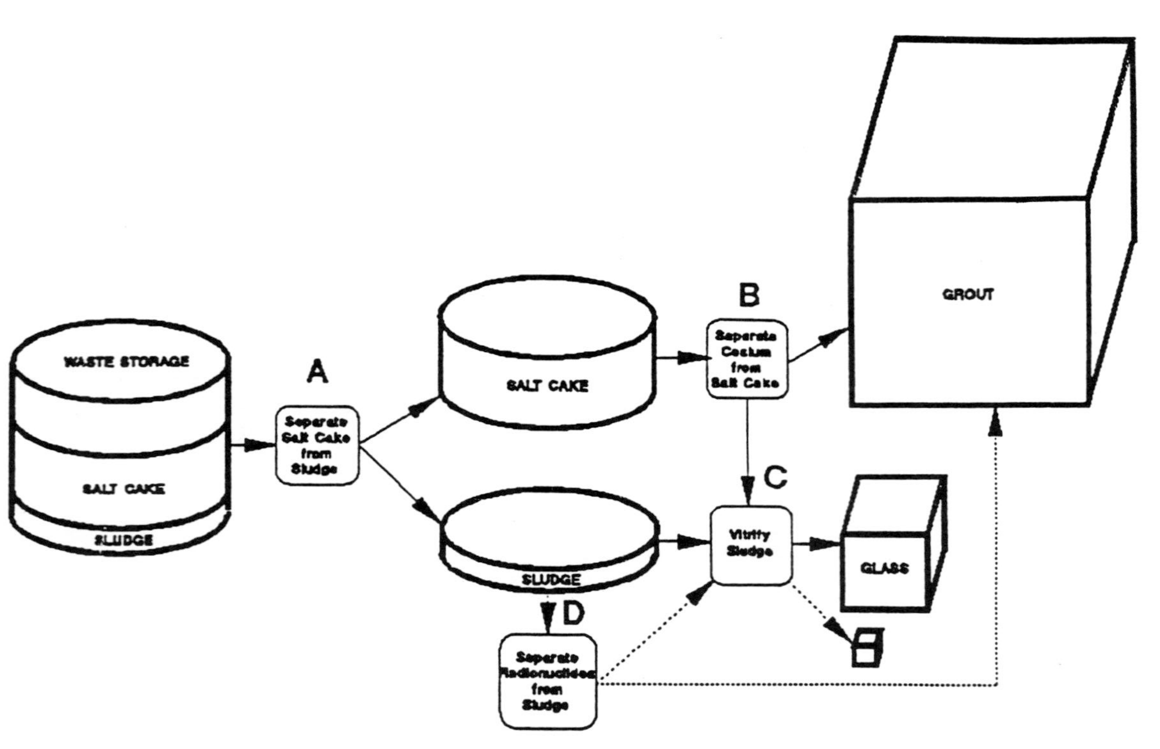


Figure 1. Schematic diagram of a disposal option for buried radioactive tank wastes. (Ref. 4).

Table 2. Possible Targeted elements from Old Hanford Waste

- Sr-90, Cs-137: power or irradiation sources(Separation treated in Other Sessions)
- Noble Metals (Pd, Ru, Rh): Strategic metals, catalysts
- Tc-99: Hazardous waste, catalyst
- Other Specialty Isotopes:

  Mo-98 as target for Mo-99 to in turn generate Tc-99m
  Zr-90 low neutron cross section isotope
  Y-90 from Sr-90 for medical use
  Others? (See handout for summary of isotopes)

the problem. It is estimated that the application of advanced chemical separations could save \$6-10 billion just by reducing the high level waste volumes from the sludge alone [4].

## NEW SORBANTS AND ION EXCHANGERS FOR USE IN NUCLEAR WASTE TREATMENT

### Specific Exchangers for $Cs^+$ Removal.

In the past, zeolites such as chabazite and mordenite were used to remove $^{137}Cs^+$ from waste streams even though the uptake of the exchanger for cesium was small in the presence of high levels of sodium ion[5-7]. We note in Figure 2 that the distribution coefficients for $Cs^+$ on these zeolites decreases rapidly as the concentration of $Na^+$ or $K^+$ increases[8]. At the level of 8M $Na^+$ only about 1% of the sites accept $Cs^+$. The cesium is also difficult to remove from the exchanger. We have prepared a series of monphenydiphosphonate phosphates of zirconium of general formula $Zr(O_3PC_6H_4PO_3)_x(HPO_4)_{2-2x}$. A schematic drawing of their structure is shown in Figure 3. These compounds were first prepared by Dines et al[9]. In Table 3 we have collected the distribution coefficients for a series of the phosphonate phosphates in which X varies from

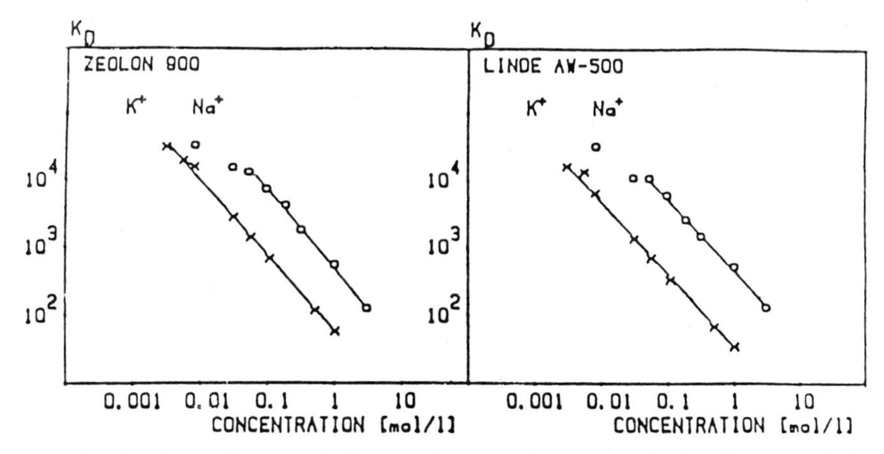


Figure 2. Distribution coefficient ($K_d$) of trace cesium on synthetic zeolites Zeolon 900 and Linde AW-500 from a nuclear waste solution ($[Na^+]_{initial}$ = 0.5 mmol/L) as a function of $Na^+$ and $K^+$ concentrations. Equilibration time 20 h. The ratio of solution volume to solid weight 100 mL/g. Zeolite grain size < 0.32 mm. (From Ref. 8 with permission).

Table 3. Kd values in $10^{-3}$ M solutions of the ions for a series of zirconium phenylphosphonate phosphates, $Zr(O_3PC_6H_4PO_3)_x(HPO_4)_{2-2x}$ from x=0.65(IP-1) to 0.15(IP-5).

| Sample No. | Ion | Li | Na | K | Rb | Cs | Mg | Ca | Sr |
|---|---|---|---|---|---|---|---|---|---|
| | | | | | | $K_d$(ml/g) | | | |
| IP-1 | | <1 | <1 | <1 | 38.0 | 65.1 | 26.8 | 21.8 | 208.0 |
| IP-2 | | <1 | 2.4 | 10.9 | 140.4 | 276.6 | 44.0 | 27.3 | 139.5 |
| IP-3 | | 5.8 | 5.0 | 96.9 | 288.0 | 2091 | 188.0 | 456.7 | 594.9 |
| 1P-4 | | 1.9 | 2.4 | 122.8 | 552.5 | 3347 | 319.8 | 586.8 | 613.0 |
| 1P-5 | | <1 | <1 | 133.0 | 220.0 | 4491 | 187.2 | 361.0 | 460.8 |

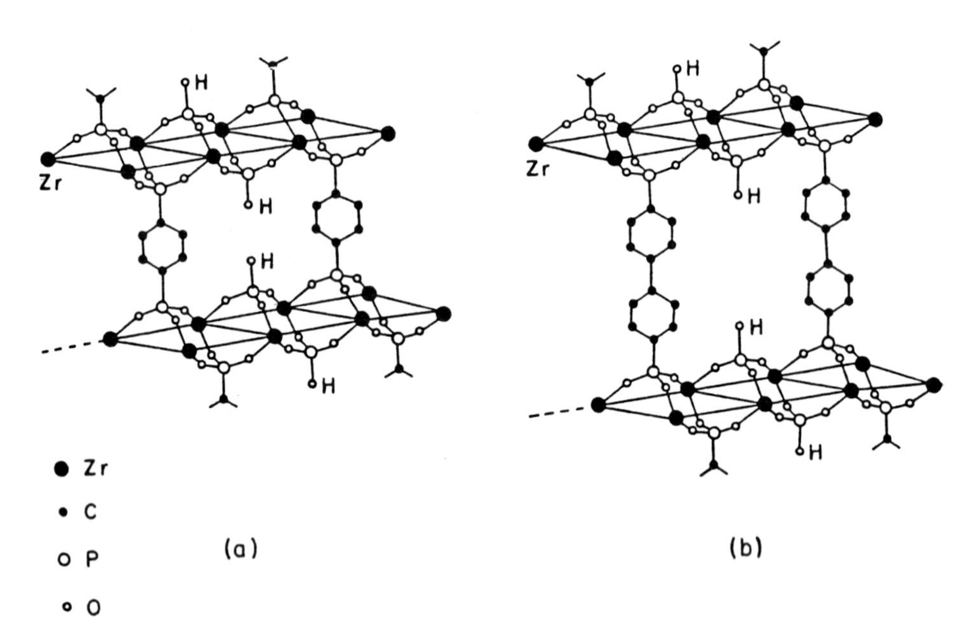


Figure 3. Schematic representation of cross-linked zirconium arylphosphonate phosphates: (a) $Zr(O_3PC_6H_4PO_3)_{0.5}(HPO_4)$; (b) $Zr(O_3PC_6H_4\text{-}C_6H_4PO_3)_{0.5}(HPO_4)$.

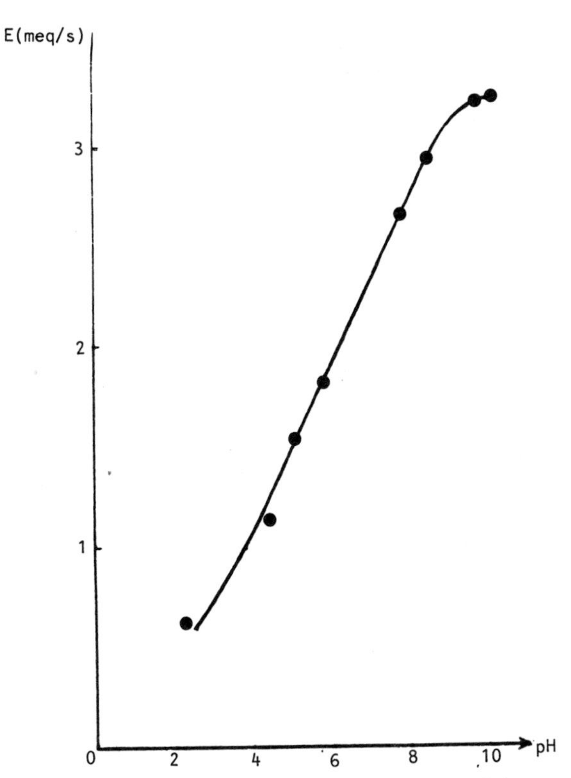


Figure 4. Potentiometric titration curve for $Zr(O_3PC_6H_4PO_3)_{0.45}(HPO_4)_{1.1}$ as uptake versus pH. Titrant: 0.1M(KCl + KOH).

0.65 to 0.15. We note that the uptake for $Na^+$ and $Li^+$ ion is negligible but that $K_d$ for $Cs^+$ is quite high. In the case of sample 1P-5 $Na^+$ was almost completely rejected and the affinity for other alkali and alkaline earth ions, except $Cs^+$, is moderate. The ion exchange capacity (IEC) for 1P-5 is 3.3 meq/g as seen by the $K^+$ titration curve in Figure 4. These phosphonate compounds are extremely stable to acid since they can be sulfonated in fuming sulfuric acid and are thermally stable to at least 350 °C in air. At high pH values, phosphate groups are removed by hydrolysis so these exchangers are best utilized at near neutral pH values.

The interlayer distance in the monophenyl derivatives is 9.6 Å while the layer thickness is 6.3 Å[10]. Thus, the free space is 3.3 Å which is just sufficient to allow an unhydrated $Cs^+$ to diffuse into the lattice ($r_{Cs+}$ = 1.7 Å). Since the hydrated radius of $Na^+$ is larger than 1.7 Å, there must be insufficient energy available at room temperature to dehydrate the $Na^+$ ions. The distance in the lateral direction between pillars is 5.3 Å for near neighbors, but if spaced by phosphate groups this value is 10.6 Å. The ring van der Waals radius is about 3.4 Å[11] so the lateral free space is ~7.2 Å.

A series of compounds with diphenyl groups has also been prepared. For these compounds the interlayer distance is 13.6 Å, large enough to allow hydrated ions to diffuse into the interlayer space. The $K_d$ values for $M^+$ and $M^{2+}$ ions on these compounds are not remarkable, but in general increase as the size of the ion increases. However, on sulfonation these compounds show a remarkable specificity for $Ba^{2+}$ (Table 4). This specificity must be chemical in nature because the values for the other alkaline earth metals were all roughly similar and relatively low. The high $Ba^{2+}$ selectivity may result from a combination of free space and strong binding of $Ba^{2+}$ ion to $SO_3^-$. This is illustrated in Figure 5B where sulfonate groups from adjacent pillars are in a position to complex a $Ba^{2+}$ ion between them.

Table 4. Kd values in $10^{-3}$M solutions of the ions for a series of sulfophosphonates, $Zr(O_3PC_6H_4SO_3HC_6H_4SO_3HPO_3)_x(HPO_4)_{2-2x}$, for which x decreases from 0.8 to 0.5 down the column.

| Sample No. | Ion | Li | Na | K | Rb | Cs | Mg | Ca | Sr |
|---|---|---|---|---|---|---|---|---|---|
| | | $K_d$(ml/g) | | | | | | | |
| 2PS-1 | | 6.7 | 2.3 | 54.7 | 96.0 | 154.5 | 70.1 | 111.3 | 135.2 |
| 2PS-2 | | 22.2 | 31.8 | 217.0 | 415.2 | 586.5 | 268.4 | 221.0 | 380.0 |
| 2PS-3 | | 19.3 | 34.1 | 401.0 | 831.5 | 1915.9 | 220.0 | 303.3 | 289.0 |
| 2PS-4 | | 27.5 | 52.1 | 521.0 | 1052.5 | 2853.5 | 341.5 | 430.5 | 453.8 |
| 2PS-5 | | 11.6 | 19.6 | 354.1 | 1052.5 | 2440 | 91.2 | 108.7 | 117.2 |

Table 5. Distribution coefficients for alkali and alkaline earth metal ions on exchangers $Zr(O_3PC_6H_4SO_3H)_x(HPO_4)_{2-x}$ (2 samples), amorphous zirconium phosphate, and AG 50W-X8 at 25 °C.

| Ion | $K_d$(mL/g) | | | |
|---|---|---|---|---|
| | MY-IV-95[a] x = 0.767 | MY-VI-2[a] x = 0.43 | Amorphous ZrP[a,b] | AG 50W-X8[c] |
| $Li^+$ | 110 | | 7 | 33 |
| $Na^+$ | 205 | | 11 | 54 |
| $K^+$ | 1500 | 650 | 120 | 99 |
| $Cs^+$ | 6500 | | 1600 | 148 |
| $Mg^{2+}$ | 21000 | 9800 | | 790 |
| $Ca^{2+}$ | 89000 | 37000 | | 1450 |
| $Ba^{2+}$ | 400000 | 190000 | | 5000 |

[a] $K_d$ at pH = 2.00 and a metal loading of 0.1 meq/g.

[b] Calculated from selectivity coefficients given in Reference [14].

[c] $K_d$ in 0.1 M $HNO_3$. Data from Reference [15].

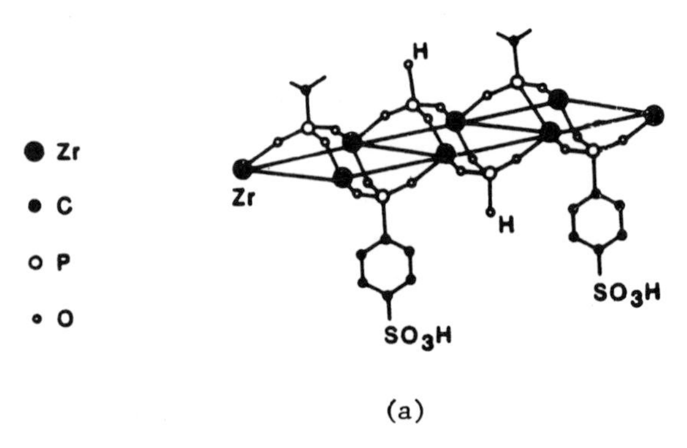


Figure 5. Schematic representation of sulfonated zirconium phenylphosphonates: (a) $Zr(O_3PC_6H_4SO_3H)(HPO_4)$

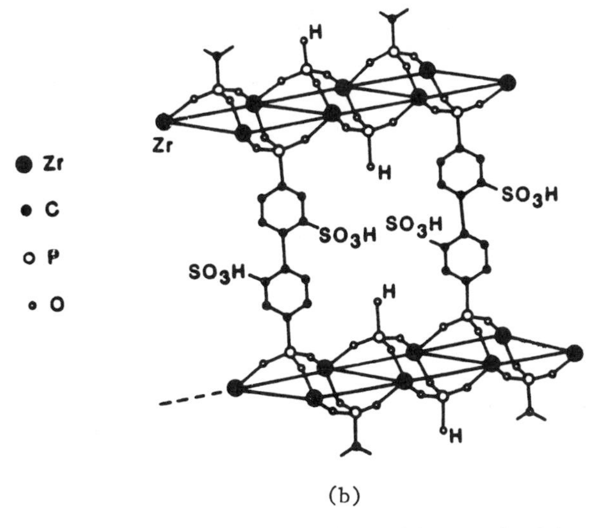


Figure 5. (b) $Zr[O_3PC_6H_3(SO_3H)_2C_6H_3SO_3H_2]_{0.5}(HPO_4)$.

We have also prepared zirconium sulfophenylphosphonate phosphates of the type shown in Figure 5a. These compounds swell to exfoliation in water but reprecipitate in the presence of polyvalent cations trapping them between the layers. Some ion exchange data for these type compounds is shown in Table 5.[12] It is seen that, while these compounds have a high affinity for $C_s^+$, they might be ideal for $Sr^{2+}$ sequestration as described in the next section.

## Strontium Specific Ion Exchangers

In the data shown in Table 5 it is clear that $K_d$ increases rapidly from $Mg^{2+}$ to $Ba^{2+}$. Since logK versus the ionic radii of the alkaline earth cations are straight lines for the MY samples, we estimate values of $K_d$ for MY-IV-95 and MY-VI-2 to be 158,500 and 75,800, respectively for $Sr^{2+}$. Compounds such as these would be broad spectrum sequestrants. That is, other polyvalent species would also be sequestered, but if their $K_d$ values are much lower than those for $Sr^{2+}$ complete removal of $Sr^{2+}$ is possible.

Another compound exhibiting very high selectivity for $Sr^{2+}$ is of the type shown in Figure 5b but with 3 phenyl rings partially sulfonated. Some preliminary $K_d$ values are collected in Table 6. it is seen that sample 3PS-2 has an extremely high $K_d$ value for $Sr^{2+}$ but only moderate values for $Ca^{2+}$ and $Ba^2$. This exchanger, being cross-linked, is totally insoluble in strong acid solutions and may also be stable in alkaline solution. Thus, it might be very suitable as a specific $Sr^{2+}$exchanger. This supposition is strengthened by the fact that these terphenyl cross-linked materials are rather indifferent in their affinity for some first row transition elements (Table 6).

## SEQUESTRATION OF COMPLEXES

The aqueous solutions present in the tanks contain large quantities of complexing agents. Thus, many of the metals in solution may not be present in the hydrated state but rather as metal complexes. The sulfonates shown in Figure 5B remove a broad spectrum of such positively charged species from solution. For example, we have precipitated

Table 6. $K_d$ values in $10^{-3}$M solutions for the listed ions for zirconium sulfoterphenyl diphosphonate phosphate with increasing phosphate content down the column.

| Sample No. | Ion | Li | Na | K | Rb | Cs | Mg | Ca | Sr |
|---|---|---|---|---|---|---|---|---|---|
| | | $K_d$(mL/g) | | | | | | | |
| 3PS-1 | | 17.9 | 18.3 | 169.0 | 270.3 | 4366.9 | -- | 1783.8 | 12680 |
| 3PS-2 | | 32.3 | 31.7 | 201.8 | 571.6 | 8900 | -- | 1880 | 63516 |
| 3PS-3 | | 33.9 | 47.7 | 381.7 | 851.4 | 9592.3 | -- | 1499.9 | 5523.5 |
| 3PS-4 | | 21.8 | 47.1 | 308.8 | 6446 | 10400 | -- | 1623.5 | 3732 |

species of the type $[M(NH_3)_6]^{n+}$from solution with the sulphophosphonates. A second type of compound that sequesters anionic species has also been prepared. They are aminophosphonates of general formula $Zr(O_3PCH_2(NHCH_2CH_2)_nNH_2)_2$. We have synthesized compounds with n=1-4. On protonation these compounds form colloidal dispersions which reprecipitate in the presence of polyvalent anionic species. Using these polyimines $[PtCl_4]^{2-}$, $Fe(CN)_6^{3-}$, $Fe(CN)_6^{2-}$, $[PMo_{12}O_{40}]^{3-}$, $WO_4^{2-}$, polyimines have been quantitatively removed from solution. Many similar species are present in the tank solutions as well as anionic technecium species and these could be recovered by application of the polyimine broad spectrum precipitating agents.

## CONCLUSIONS

We have demonstrated the synthesis of new materials with high selectivities for $Cs^{2+}$ and $Sr^{2+}$ and low affinity for $Na^+$. In addition sequestrants for both anions and cations present in nuclear waste solutions have been prepared. It is now necessary to demonstrate the utility of these materials in simulated nuclear waste solutions.

**Acknowledgment:** The author wishes to thank the State of Texas for a grant through its "Advanced Technology Program" in support of this research. Grateful acknowledgment is made to the authors scientific collaborators C.-Y. Yang, C. Yolanda Ortiz-Avila, Chhaya Bhardwaj and especially Guangzhi Peng and the U. S. China Program of NSF through Grant No. INT 8910902.

## REFERENCES

1. Office of Technology Assessment, U. S. Congress, "Long-lived Legacy: Managing High-Level and Transuranic Waste at the DOE Nuclear Weapons Complex," OTA-BP-)-83, U. S. Govt. Printing Office, Washington, D.C. (May 1991).
2. B. G. Levi, *Physics Today*, March 1992, 17.
3. Govt. Accounting Office "Nuclear Waste: Hanford Single-Shell Tank Leaks Greater than Estimated," GAO/RCED-91-77, Washington, D.C. (1991).
4. J. R. Morrey and J. L. Swanson, "A Primer on Hanford Defense Tank Wastes and Prospects for Advanced Chemical Separations," Battelle Pacific Northwest Laboratories, April 1991.
5. D. O. Campbell, E. D. Collins, L. J. King, J. B. Knaur and R. M. Wallace, ORNL/TM-7448 (1980).

6. E. D. Collins, D. O. Campbell, L. J. King, J. B. Knauer and R. M. Wallace, AIChE. Symp. Ser. 213, 78, 9 (1982).
7. L. L. Ames and J. L. Nelson, Report HW-74609 (1962) Battelle PNL, Richland, Wash.
8. R. Harjula and J. Lehto, Nucl. Chem. Waste Mgt. 6, 133 (1986).
9. M. B. Dines, P. M. DiGiacomo, K. D. Callahan, P. C. Griffith, R. H. Lane and R. E. Cooksey in Chemically Modified Surfaces in Catalysis and Electrocatalysis, J. S. Miller, Ed., ACS Symp. Ser. 192, ACS, Wash., D.C. 1982, p 223.
10. A. Clearfield and G. D. Smith, *Inorg. Chem.* 8, 431 (1969).
11. L. Pauling, "The Nature of the Chemical Bond," 3rd ed., Cornell University Press, Ithaca, N.Y. 1960.
12. L. H. Kullberg and A. Clearfield, *Solvent Extraction and Ion Ex.* 7, 527 (1989).
13. G.-Z. Peng and A. Clearfield, work in progress.
14 L. Kullberg and A. Clearfield, *J. Phys. Chem.* 85, 1578 (1981).
15. F. W. E. Strelow, R. Rothmeyer and C. J. C. Bothma, *Anal. Chem.* 37, 106 (1965).

## APPENDIX I

Abstract of a lecture not submitted in chapter form:

## ADVANCES IN BIOREMEDIATION OF ENVIRONMENTAL CONTAMINANTS

K. W. Brown and K. C. Donnelly

Soil and Crop Sciences Department
and Department of Vet. Public Health
College Station, Texas 77843

### ABSTRACT

Many organic chemicals readily undergo biodegradation in the environment. While higher organisms can degrade some organic chemicals, microorganisms play the major role in degrading naturally occurring and contaminant organics in surface waters, soils, and groundwaters. The necessity of developing economical, environmentally safe methods of decontaminating sites of spills associated with industrial activities and post waste disposal activities, as well as the need to develop better methods for the disposal of wastes we are presently producing has resulted in a great increase in the interest and in the use of biodegradation.

Many of the efforts have involved the use of procedures to alter the environment to enhance the rate of biodegradation. The date, less success has been achieved by the introduction of selected microorganisms, although some organisms have been developed which are capable of degrading xenobiotic chemicals. The enzymatic mediated pathways of biodegradation of many environmental contaminants are now understood. The environment contains large number of diverse microbial species, many of which may be involved simultaneously or sequentially in the degradation of a given compound. The biodegradation rate constants are now known for an increasing number of organic chemicals in a variety of media including surface waters, soils, anaerobic sediments and groundwater. Information is being developed on the influence of environmental parameters including temperature, nutrient supply, substrata concentration, the presence of associated substrates, redox potential, salinity, etc. which can be used to predict rates of biodegradation in the field and the response of the rates to environmental manipulation.

Readers wanting more information on the subjects described in this abstract are invited to communicate directly with Dr. Brown.

APPENDIX II

# POSTERS

S. N. Ahmadi, Y. C. Huang, S. S. Koseoglu and B. Batchelor, Food Protein Research, Texas A&M Univeristy

*Micellar Enhanced Ultrafiltration of Heavy Metals Using Lecithin*

Dr. Ramesh C. Bhardwaj, Department of Chemistry, Texas A&M University

*Electrochemical Oxidation of Waste*

Dian Chen, Ramunas J. Motekaitis, Derek McManus and Arthur E. Martell, Department of Chemistry, Texas A&M University, ARI Technologies, Palatine, Illinois

*LoCAT Process for the Removal of $H_2S$ from Natural Gas with Fe(III)-NTA as Catalyst*

Tanya Lewis and David H. Russell, Department of Chemistry, Texas A&M University

*Characterization of Water Soluble Organics: High Resolution, Mass Spectrometry*

J. B. Shapiro and Emile A. Schweikert, Department of Chemistry, Texas A&M University

*Methodoly for Direct Field Analysis of Organic Contaminants*

Wilfredo Delgado-Morales, Mysore S. Mohan, J. Drew Ilger and Ralph A. Zingaro

*Identification of Analysis of Arsenic Compounds in Natural Gas*

Eric J. Munson, Ali A. Kheir, Greg Oliver and James F. Haw, Department of Chemistry, Texas A&M University

*High-Temperatures In Situ Solid-State NMR Studies of Catalytic Systems*

Sharon Taylor-Myers, David B. Ferguson and James F. Haw, Deparment of Chemistry, Texas A&M University

*Motion of Small Molecules in Polymers by $^{15}N$ NMR*

Jerry L. White, Larry W. Beck and James F. Haw, Department of Chemistry, Texas A&M University

*Hydrogen-Bonding and Exchange in Zeolites by Mas $^{1}H$ NMR*

INDEX